Advances in Comparative and Environmental Physiology 20

Volumes already published

Vol. 8: From the Contents: Dormancy in Aquatic Invertebrates-Root Effect – Red Blood Cell Functions in Fish Diving Mammals (1991)

Vol. 9: Volume and Osmolality Control in Animal Cells. Edited by *R. Gilles, E.K. Hoffmann,* and *L. Bolis* (1991)

Vol. 10: Comparative Aspects of Mechanoreceptor Systems Edited by *F. Ito* (1992)

Vol. 11: Mechanics of Animal Locomotion Edited by *R. McN. Alexander* (1992)

Vol. 12: Muscle Contraction and Cell Motility: Molecular and Cellular Aspects. Edited by *H. Sugi* (1992)

Vol. 13: Blood and Tissue Oxygen Carriers Edited by *Ch.P. Mangum* (1993)

Vol. 14: Interaction of Cell Volume and Cell Function Edited by *F. Lang and D. Häussinger* (1993)

Vol. 15: From the Contents: Salivary Gland Secretion – Nematocyst Discharge – Adaptations in Decapodan Crustaceans – Steroid-Sensitive Areas Mediating Reproductive Behaviors (1993)

Vol. 16: Ion Transport in Vertebrate Colon Edited by *W. Clauss* (1993)

Vol. 17: Effects of High Pressure on Biological Systems Edited by A.G. *Macdonald* (1993)

Vol. 18: Biomechanics of Feeding in Vertebrates Edited by *V.L. Bels, M. Chardon* and *P. Vandewalle* (1994)

Vol. 19: Electrogenic Cl^- Transporters in Biological Membranes Edited by *G.A. Gerencser* (1994)

Vol. 20: From the Contents: Motile Activities of Fish Chromatophores – Epithelial Transport of Heavy Metals – Heavy Metal Cytotoxicity in Marine Organisms – Comparative Pulmonary Morphology and Morphometry – Molecular Adaptations in Resistance to Penicillins (1994)

Volumes in preparation

Mechanisms of Systematic Regulation. Vol. I: Respiration and Circulation. Vol. II: Acid–Base Regulation, Ion Transfer and Metabolism. Edited by *N. Heisler*

Comparative Immunology: Invertebrate Immune Responses.
Vol. I: Cells and Molecular Products. Vol. II: Cell Activities and the Environment. Edited by *E.L. Cooper*

Advances in Comparative and Environmental Physiology 20

With Contributions by

J.L. Arpigny · J. Coyette · S. Davail · G. Feller · E. Fonzé
E.C. Foulkes · J.-M. Frère · R. Fujii · S. Génicot · Ch. Gerday
B. Joris · J. Lamotte-Brasseur · J.N. Maina · E. Narinx
M. Nguyen-Disteche · N. Oshima · A. Viarengo · Z. Zekhnini

With 56 Figures, Some in Colour

Springer-Verlag
Berlin Heidelberg New York
London Paris Tokyo
Hong Kong Barcelona
Budapest

ISSN 0938-2673
ISBN-13:978-3-642-78600-6 e-ISBN-13:978-3-642-78598-6
DOI: 10.1007/978-3-642-78598-6

Softcover reprint of the hardcover 1st edition 1994

Typesetting: Macmillan India Ltd., Bangalore 25
SPIN: 10124783 31/3130/SPS – 5 4 3 2 1 0 – Printed on acid-free paper

Foreword to the Series

The aim of the series is to provide comprehensive, integrated reviews giving sound, critical and provocative summaries of our present knowledge in environmental and comparative physiology, from the molecular to the organismic level.

Living organisms have evolved a widespread range of basic solutions to cope with the different problems, both organismal and environmental, with which they are faced. A clear understanding of these solutions is of course of fundamental interest for all biologists, zoologically or medically oriented. They can be best comprehended in the framework of the environmental and/or comparative approaches. These approaches demand either wide surveys of animal forms or a knowledge of the specific adaptive features of the species considered. This diversity of requirements, both at the conceptual and technological level, together with the fact that physiology and biochemistry have long been mainly devoted to the service of medicine, can account for the fact that these approaches emerged only slowly amongst the other new, more rapidly growing disciplines of the biological sciences.

The field has now gained the international status it deserves and the organization of a series devoted to it appeared timely to me in view of its actual rapid development and of the interest it arouses for a growing number of biologists, physiologists, and biochemists, independently of their basic, major orientation.

Liège, Belgium, Spring 1988 *Raymond Gilles*

List of Editors

Contents

Chapter 4

Comparative Pulmonary Morphology and Morphometry:
The Functional Design of Respiratory Systems
J.N. Maina

Chapter 5

Molecular Adaptations in Resistance to Penicillins
and Other β-Lactam Antibiotics
J. Coyette, M. Nguyen-Distèche, J. Lamotte-Brasseur,
B. Joris, E. Fonzé, and *J.-M. Frère*

Chapter 6

Molecular Adaptations of Enzymes from Thermophilic and Psychrophilic Organisms
J.L. Arpigny, G. Feller, S. Davail, S. Génicot, E. Narinx, Z. Zekhnini, and *Ch. Gerday*

Chapter 1

Factors Influencing Motile Activities of Fish Chromatophores

R. Fujii and *N. Oshima*[1]

Contents

[1] Department of Biomolecular Science, Faculty of Science, Toho University, Funabashi, Chiba 274, Japan

Advances in Comparative and Environmental Physiology, Vol. 20

1 Introduction

The beautiful coloration, conspicuous patterns and their spectacular changes displayed by many poikilotherms have always attracted attention. The animals themselves utilize these features as "aposematic" or "advertisement" coloration. Inconspicuous colors and patterns are conversely exploited as "cryptic" or "concealing" coloration. Many people simply appreciate these phenomena, while zoologists and physiologists can also enjoy studying them. For the animals themselves, however, such colors, patterns and reactions represent strategies of the utmost importance for the survival of individuals or of species. For example, "protective" coloration, which constitutes part of the cryptic coloration mentioned above, is useful for avoiding attacks by predators, while conspicuous displays function to frighten predators. On many occasions, delicate and subtle changes in hues and patterns are used for communication with conspecifics (cf. Cott 1940; Bagnara and Hadley 1973; Needham 1974; Fujii 1993a). These phenomena are especially highly evolved in fish among the vertebrates, and in particular among members of the class Osteichthyes. The chromatic systems of Osteichthyes have developed extraordinarily sophisticated properties during the evolution of this class of vertebrates over the course of more than 400 million years.

Colorations of animals are due to the presence of pigmentary substances and/or the light-reflecting microstructures within the integument (cf. Needham 1974; Fox 1979; Fujii 1993a, b). That is, the colors of the skin are generated by the absorption, reflection and scattering of light of certain wavelengths by the pigments and microstructures. The compounds responsible for the generation of color are normally packed within specialized organelles, the "chromatosomes", which are produced and stored in pigment cells, called "chromatophores" in the skin. Usually, chromatosomes are found in the cytoplasm of the chromatophores. However, they are sometimes detectable in the epidermal cells (keratinocytes), whence they have been transferred from the epidermal chromatophores.

Changes in hue or in pattern of coloration are crucial for certain animals, as mentioned above. Slower changes can be attributed to the increases or decreases in the actual number of chromatophores, and/or in the net amount of pigmentary material within them. The low rate of change is mostly due to the fact that the phenomena are associated with the synthesis or degradation of pigmentary materials. Such changes are generally referred to as the "morphological color changes", and they are outside the scope of the present review. The more rapid changes are known as the "physiological color changes". These changes are caused by the motile activities of chromatophores. In the present article, we shall be concerned with the effects of various factors, both intrinsic and external, on the chromatophores. Fish, in particular those of smaller size, serve as excellent materials for experimental studies of these rapid changes because chromatic systems are well developed as a result of ethological requirements that are essential for survival. In addition, the pigmented cells are easily and clearly observable in the skin. The results presented in this article are mostly obtained from experiments on such small fish.

In earlier monographs or review articles pertinent to the present subject (e.g., Parker 1948; Waring 1963; Fujii 1969; Bagnara and Hadley 1973), a rather wide variety of "fish" has been dealt with, including lower classes of Pisces, of which Chondrichthyes is a representative. In recent years, however, not much progress has been made on the pigment cells of these groups. In this article, therefore, the descriptions are limited for the most part to chromatophores of bony fish. Thus, those who are interested in the pigment cells of lower fish are advised to consult earlier publications, such as those mentioned above. From these articles, they may possibly glean some information relevant to their future research. The available information about the very complicated systems associated with pigment cells in "higher" fishes, as discussed in this article, may also, we hope, be useful for such studies.

2 Chromatophores

2.1 Varieties of Chromatophores

In poikilothermic vertebrates, chromatophores can appropriately be classified into five groups. Among them, those that assume their characteristic colors by absorbing light include the melanophores, the erythrophores and the xanthophores, while those that reflect light are known as leucophores and iridophores (Fujii 1969, 1993b). Before discussing their particulars, we should, however, touch briefly upon the more general characteristics of these cells.

The term "chromatocyte" is not used, and chromatophore is the prevailing current term for pigment cells. The chromatophore is definitely a specific type of cell, and each cell contains most of the cell organelles commonly found in animal cells. Details of the cytological features of chromatophores can be found in a number of papers, references to which are cited in various review articles (Bagnara and Hadley 1973; Schliwa 1984; Obika 1986; Fujii 1993b). There is evidence that all the pigment cells of vertebrates, including those of fish, originate from the neural crest. Sometimes categorized as "paraneurons", these terminally differentiated cells still retain some characteristics of neurons, and the dendritic shape assumed by most chromatophores may be related to the origin of these cells.

In the integument of fish, chromatophores are present mostly in the dermis. Among the various chromatophores, only melanophores are frequently found in the epidermis, especially in larger fish. One exceptional example, however, is provided by the epidermal xanthophores that were recently reported in the Antarctic blenny, *Trematomus bernacchii* (Obika and Meyer-Rochow 1990). Since capillaries do not extend far enough, epidermal cells are rather remote from hormonal influences. By contrast, the cells in the dermis are easily accessible to controlling nerve fibers and by blood-borne hormones. Chromatic systems in lower vertebrates may, thus, have evolved with their effector cells predominantly in the dermis because the ability to change color rapidly and efficiently must be very important in smaller fish in particular.

2.1.1 Melanophores

Melanophores are melanin-containing, dark chromatophores of the dendritic type. As is the case in other lower classes of vertebrates, melanophores are the most common type of chromatophore in fish. The chromatosomes within them are called "melanosomes". The melanosomes are round or slightly ellipsoidal, with a diameter of about 0.5 μm.

As compared with other types of chromatophore, the melanophores usually exhibit higher motile activities. The rapid aggregation or dispersion of melanosomes takes place in response to nervous and hormonal cues. Thus, the melanophores play a leading role in the rapid changes in color exhibited by many species of fish. The ease of experimental handling of such cells in vivo and in vitro has resulted in their use in most studies of the influence of various factors on integumental coloration and on chromatophores.

2.1.2 Erythrophores

Erythrophores are reddish dendritic chromatophores found in the dermis of several species of fish. They are dominant in the skin of reddish fish and they are also densely distributed in the red patches that are frequently seen on the skin of tropical freshwater fish and coral-reef dwellers. Their overall configuration and morphological features are fundamentally the same as those of melanophores (Fujii 1993b).

The compounds that absorb light to generate reddish tint are carotenoids and pteridines. Organelles containing these pigmentary substances are normally called carotenoid vesicles (carotenoid droplets) and pterinosomes, respectively. Many erythrophores contain both carotenoid vesicles and pterinosomes. However, cases of one and the same organelle containing both kinds of pigment have not been reported. Prior to complete chemical identification of the included pigment, it may be safest to call the reddish organelles "erythrosomes". Usually, the dimensions of pterinosomes are similar to those of melanosomes. Resembling lipid droplets, by contrast, carotenoid vesicles are quite variable in size.

In many cases, erythrophores have been reported to participate in the active changes in color of skin. In particular, the pterinosomes and the carotenoid vesicles have been shown to move into the cell body or towards the periphery of the cell in response to various signals.

2.1.3 Xanthophores

Xanthophores are dendritic chromatophores with a yellowish tint. Usually, they are smaller than melanophores and erythrophores.

Since the distinction between cells of this category and the erythrophores is not completely clear, several researchers have actually included them in the category of xanthophores. In fact, it is sometimes hard to name orange-colored chromatophores appropriately. A distinction may, however, be justified because not only the hues, but also the distribution and physiological characteristics are rather different between erythrophores and xanthophores. Xanthophores are usually distributed with melanophores. That is, the xanthophores function to add a yellowish tone to black, giving rise to the brownish hue of the integument that is very commonly seen in fish. Frequently, they appear to form a yellow network by connecting with one another at the tips of their dendritic processes. Such a feature may be related to some kind of functional syncytium.

As the erythrophores, the pigmentary compounds responsible for the generation of yellowish hues are carotenoids and/or pteridines, and pterinosomes and carotenoid vesicles are the carrier organelles. Other cytological features of xanthophores are described in a recent review article (Fujii 1993b).

2.1.4 Leucophores

Like the chromatophores mentioned above, leucophores are dendritic chromatophores, but their optical properties are entirely different from those of the chromatophores described above: leucophores are light-scattering chromatophores. In various papers and review articles, quite a few researchers have called cells in this category "iridophores", including them with the iridophores proper, which are described below (Sect. 2.1.5). These two types of cell should, however, be treated separately because not only their morphology, but also their light-reflecting properties are essentially different (cf. Fujii 1993b).

The distribution of these cells seems to be fairly limited. The cells in this category seem to exist exclusively in teleosts, possibly even only in the dermis of a very few cyprinodontiform species since, to date, reliable descriptions of their presence can only be found in the case of the killifish *Fundulus*, the medaka *Oryzias* and the guppy, *Poecilia.*

The organelles responsible for the cellular reflectivity are called "leucosomes". They are similar in size to pigmentary inclusions in other light-absorbing chromatophores. Enclosed by a single limiting membrane, each leucosome is a globular organelle and scatters incident light in all directions.

Although the numbers of dendritic processes are usually smaller than those of other light-absorbing chromatophores, the overall dimensions of a leucophore are similar to or frequently even larger than those of the latter cells. Their architectural features are fundamentally similar to those of other chromatophores (Obika 1988; cf. Fujii 1993b). It should be noted here that a leucophore is very frequently found to be located concentrically just beneath a melanophore. The same nerve fibers control both cells (Iwata et al. 1981; Obika 1988).

Like the chromatosomes in light-absorbing chromatophores, leucosomes move centripetally or centrifugally within the leucophores in response to various signals which may be either intrinsic or experimentally applied cues. In general, the direction of their movement is opposite to that seen in light-absorbing chromatophores. That is, the leucophores and the other light-absorbing chromatophores behave in quite the opposite manner. In view of the contribution of the light-reflecting and absorbing chromatophores to the darkness of the skin, these reciprocal movements can be easily understood in terms of bringing about effective dark-to-light changes in the color of the skin.

2.1.5 Iridophores

Iridophores are unusual cells and they contain a number of light-reflecting flat crystals throughout their cytoplasm. The crystals are monoclinic, consisting mostly of guanine. They are very frequently found in the skin of fish and are especially common on whitish and silvery belly surfaces. Usually, they are devoid of dendritic processes, but they fulfill an indispensable role in generating the very high reflectivity of the skin. The large crystalline platelets tend to form parallel stack(s) in the cytoplasm. In each stack, the spacing between adjacent platelets is strikingly uniform. Having refractive indices of no less than 1.83, the platelets are highly refractive. The light reflectance at the surfaces of each platelet is, therefore, high. The characteristic hues generated by interference of rays of light at such alternating layers are generally called "interference colors", and they represent the "physical" or "structural" coloration. Considering that many such biological systems can be treated as multilayered thin-film interference phenomena of the ideal type, Land (1972) presented an analysis of the reflection of light from the stack of platelets. Indeed, the highly reflective whitish or metallic sides and belly skins of many teleosts can be explained by such a mathematical treatment. The iridophores in such skins are static cells and do not exhibit motile responses to stimulation of any kind (Fujii 1993b).

Recently, "motile" iridophores have been described in several teleostean species (cf. Fujii 1993b). As in the static iridophores, the platelets are piled up to form stacks, but the spacing between the platelets can vary in response to controlling signals. Thus, the multilayer interference system in the motile cells should be treated differently from systems of the ideal type.

In many of the motile iridophores examined thus far, platelets have always been very thin, being no thicker than the plasma membrane. In all cases examined, the reflectivity was very low. The most characteristic feature was the higher purity of the color generated. Actually, many species of fish that display beautiful fluorescent-like hues possess such iridophores, which contain a number of very thin platelets that are piled up in a regular fashion and are responsible for non-ideal interference phenomena (Fujii 1993a). Representative examples of such cells include those responsible for the characteristic cobalt-blue to greenish hues of the coral-reef damsel-fishes (Kasukawa et al. 1987; Fujii et al. 1989) and for the blue-green hue of the stripe of the neon tetra (Nagaishi et al. 1990) and other related tetras. Actual measurements of the reflected light revealed sharp spectral peak. However, these iridophores are often motile and spectral peaks shift in response to certain stimuli. Observations in vivo further indicated that these cells play a predominant role in changes in color as a result of their motile activities. Detailed analyses of the optical events in such iridophores have been presented in a separate article (Fujii 1993b).

When the distance between platelets in a pile increases, such as after nervous stimulation, the motile iridophores reflect light of longer wavelengths. When the spacing between the platelets decreases, conversely, the spectral peak shifts towards shorter wavelengths. The former response has been designated the "LR response", being an abbreviation of the "Longer-wavelength light-Reflecting response", while the latter response has been designated the "SR response", being an abbreviated version of "Shorter-wavelength light-Reflecting response" (Oshima et al. 1989; Fujii 1993b). In later sections, these terms will repeatedly be employed to describe the reaction of the iridophores to certain signaling cues.

Dendritic motile iridophores have recently been described in some gobiid fishes, including the dark sleeper *Odontobutis obscura obscura* (Iga and Matsuno 1986). Like the iridophores of many amphibians, reflecting platelets aggregate into the perikarya or disperse to dendritic processes in response to neural or hormonal stimuli. When the platelets are aggregated in the perikarya, the cells appear bluish in color. However, the same cells look yellowish when the platelets are dispersed. The bluish tone is considered to be due to the presence of organized piles of platelets, which form gradually during the process of aggregation of platelets. A bluish color of high purity can be explained by multilayered thin-film interference from the piles of platelets (Fujii et al. 1991a).

2.2 Motile Activities of Chromatophores

2.2.1 Motile Mechanisms

Aggregation of pigment in the perikaryon and dispersion throughout the cytoplasm are the characteristics of the cellular motility of an ordinary dendritic chromatophore, as mentioned above. To date, a number of investigations have been performed to examine the mechanism of chromatophore motility and they

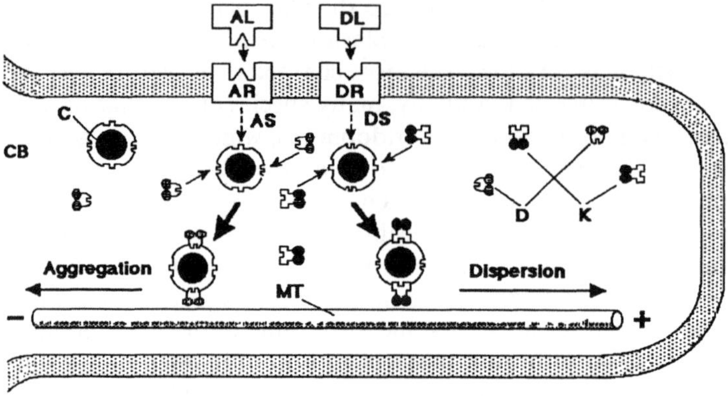

Fig. 1. Diagrammatic representation of the possible mechanism of translocation of pigment in a dendritic chromatophore. *AL* Extracellular chromatosome-aggregating ligand; *AR* aggregation-specific receptor; *AS* intracellular aggregation signal; *C* chromatosome; *CB* cell body; *D* cytoplasmic dynein; *DL* extracellular chromatosome-dispersing ligand; *DR* dispersion-specific receptor; *DS* intracellular dispersion signal; *K* kinesin; *MT* microtubules. (Fujii 1993b)

have been reviewed by Schliwa (1984) and Obika (1986). Historical theories of interest have also been reviewed in an earlier article by Fujii (1969). Very recently, one of the present authors has offered some general considerations of the current theories related to the mechanisms of motility of both dendritic and non-dendritic chromatophores (Fujii 1993b).

Agreement has recently been reached about the motility of dendritic chromatophores. In brief, it is apparent that the microtubules that radiate from the cell body towards the periphery of the dendritic processes play an active role. A sliding force between the surface of chromatosomes and the microtubules causes the movement of chromatosomes in a centripetal or centrifugal direction (Fig. 1). Presumably, a set of motor proteins, similar or identical to cytoplasmic dynein and kinesin, are involved in the centripetal (retrograde) and centrifugal (orthograde) translocation of pigment, respectively (cf. Fujii 1993b).

Studies on round iridophore (Sect. 2.1.5) indicate that their characteristic motile activities are also dependent primarily on microtubular systems inside the cell (Oshima and Fujii 1987; Oshima and Nagaishi 1992; cf. also Fujii 1993b), although the precise mechanisms underlying the characteristic motile responses remain still to be clarified. Very recently, Nagaishi and Oshima (1992) extended their analysis of the motility of iridophores in neon tetras, and they concluded that both tubulin-dynein and actin-myosin systems may be functioning cooperatively.

2.2.2 Intracellular Substances That Affect Cellular Motility

As one of the major subjects of the present article, the effects of first messengers, namely, hormones, neurotransmitters and their modulators, on the motile activities of chromatophores will be discussed in a later section of this review (Sect. 5). However, we would like to introduce here the current concepts that relate to the way in which extracellular information brought by extracellular signaling ligands is transmitted to the motor proteins that are directly responsible for the motility of chromatophores. Such a mechanism of so-called "signal transduction" is being studied intensively in a variety of effector systems, but the results for chromatophores are still meager.

It is now generally accepted that signal transduction in effector cells is executed via changes in the cytosolic concentrations of second messengers. Cyclic adenosine 3′,5′-monophosphate (cAMP), Ca^{2+} ions, and inositol 1,4,5-trisphosphate (IP_3) have been shown to function as second messengers in teleostean chromatophores. A decrease in the concentration of cAMP leads to the aggregation of pigment in chromatophores (Novales and Fujii 1970; Fujii and Miyashita 1976a), whereas an increase in the concentration of Ca^{2+} ions (Luby-Phelps and Porter 1982; Negishi and Obika 1985a; Oshima et al. 1988) or of IP_3 (Fujii et al. 1991b) results in the same response. The reverse changes in concentrations are associated with the dispersion of chromatosomes.

The dynamics of these intracellular messengers are shown schematically in Fig. 2. Adrenoceptors of the alpha$_2$ type mediate a decrease in levels of cAMP via inactivation of adenylate cyclase, with resultant aggregation of pigment. The same response to stimulation of adrenoceptors of the alpha$_1$ type, by contrast, is

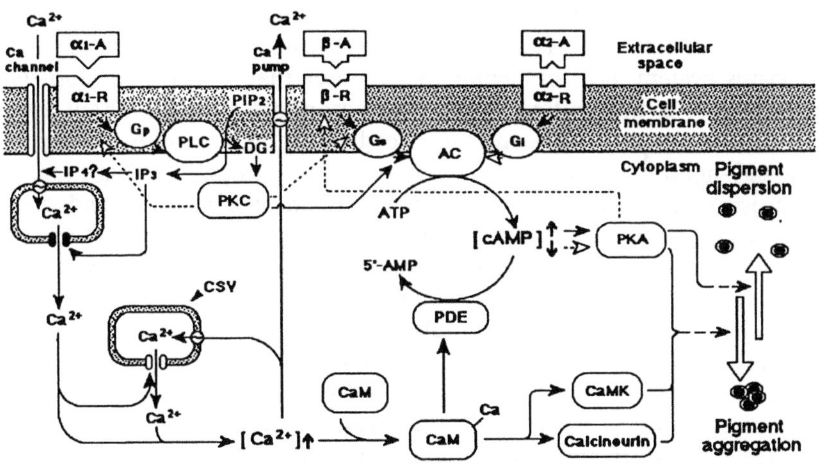

Fig. 2. Diagrammatic representation of the system of signal transduction responsible for the adrenergic control of the movement of chromatosomes. *α-A* α-Adrenergic agonist; *α-R* α-adrenoceptor; *AC* adenylate cyclase; *β-A* β-adrenergic agonist; *CaM* calmodulin; *CaMK* calmodulin-dependent protein kinase; *CSV* Ca^{2+}-storing vesicle; *DG* diacylglycerol; *Gi* inhibitory G protein; *Gp* phospholipase-activating G protein; *Gs* stimulatory G protein; *IP_3* inositol 1,4,5-trisphosphate; *IP_4* inositol 1,3,4,5-tetrakisphosphate; *PDE* 3′,5′-cyclic-nucleotide phosphodiesterase; *PKA* protein kinase A; *PKC* protein kinase C. (Fujii 1993b)

caused by the activation of phospholipase C (PLC), which catalyzes the breakdown of phosphatidyl inositol 4,5-bisphosphate (PIP_2) in the membrane. The IP_3 produced acts to release Ca^{2+} ions from the intracellular Ca^{2+}-storage compartment (vesicular elements of the smooth endoplasmic reticulum). In a few chromatophores, such as the erythrophores of the squirrelfish *Holocentrus*, Ca^{2+} ions may also be supplied from outside the cell through Ca^{2+} channels (Luby-Phelps and Porter 1982). G-proteins (cf. Fujii 1993b), calmodulin (Oshima et al. 1990a) and calcineurin (Thaler and Haimo 1990, 1992) are also involved in the motile responses of chromatophores.

Investigations of such issues have frequently been made with model cells whose cell membranes had previously been permeabilized by treatment with a surfactant, such as digitonin or Brij 58 (cf. Schliwa 1987). Ionophores, such as A23187, have also been employed to manipulate intracellular levels of Ca^{2+} ions (Luby-Phelps and Porter 1982; Negishi and Obika 1985a; Oshima et al. 1988). In some studies, specific compounds are microinjected directly into the cells. For example, Fujii and Miyashita (1976a) injected cAMP iontophoretically into the melanophores and observed the dispersion of melanosomes. In permeabilized cells, exogenously applied second-messenger molecules invade the cells to trigger the cellular motility. Thus, they may appropriately be considered as substances that influence the state of chromatophores.

2.3 Control of Chromatophore Motility

In order to adapt to environmental conditions, many fish exploit chromatic strategies (Sect. 7; cf. also Fujii 1993a). Having been processed in the central nervous system, environmental information that is perceived mostly by the lateral eyes is transmitted to the chromatophores, with resultant appropriate chromatic reactions. By contrast, circadian changes in color are dependent on the day-night alternation of environmental brightness. The light sensor in this process is now believed to be the pineal organ. Being also an endocrine gland, the extraoptic photoreceptor or pineal body secretes melatonin during the dark phase of the day to induce the aggregation of pigment in chromatophores.

Several factors may influence chromatophores in the skin either directly on the cells or indirectly by the stimulation of some part(s) of the body. As a factor of major importance, light is treated separately in Sect. 3. Other factors that influence the state of chromatophores are discussed in Sect. 4. Severance of the nerves en route to the effector cells, namely the chromatophores, is also discussed because the results of such treatment provide important clues to the general mechanisms of the effector systems.

The motile activities of chromatophores of teleosts are regulated effectively by endocrine as well as nervous systems. As the main focus of this article, recent results on the effects of hormonal substances and neurotransmitters are reviewed in Sect. 5. The relative weight given to hormonal and neural regulation varies among groups or species of fish. Endocrine control is predominant in

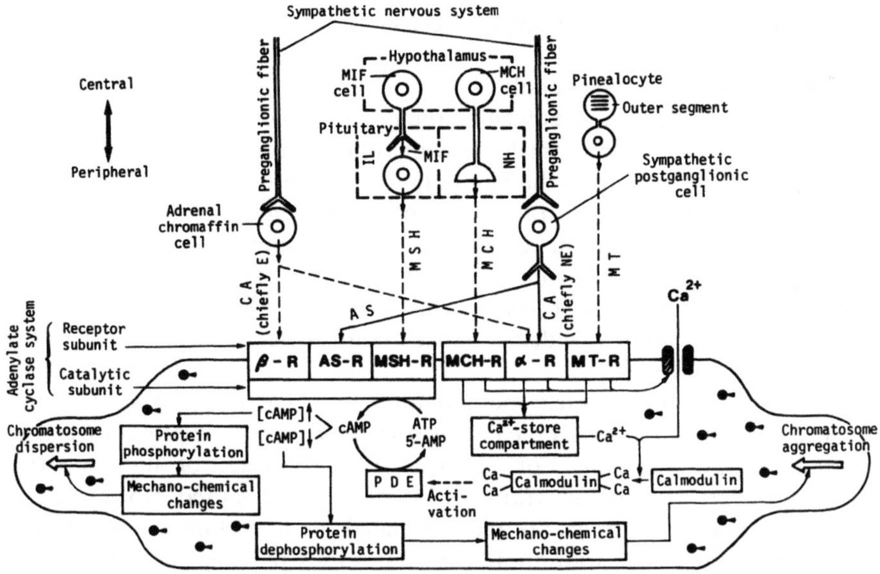

Fig. 3. Diagrammatic representation of the regulatory system for motile activities of melanophores and other light-absorbing chromatophores in teleosts. *AS* Adenosine; *AS-R* adenosine receptor; *CA* catecholamines; *E* epinephrine; *IL* intermediate lobe of the pituitary; *MCH* melanin-concentrating hormone; *MCH-R* MCH receptor; *MIF* MSH-release inhibiting factor; *MSH* α-melanophore-stimulating hormone; *MSH-R* MSH receptor; *MT* melatonin; *MT-R* melatonin receptor; *NE* norepinephrine; *NH* neurohypophysis; *PDE* phosphodiesterase; *α-R* alpha-adrenoceptor; *β-R* beta-adrenoceptor. (Fujii and Oshima 1986)

some fish, while in others regulation is effected solely by the nervous system. Between these two extremes, there are more common cases in which both hormonal and nervous mechanisms are operative. In addition, other biogenic substances may act as local hormones or autacoids to influence the state of chromatophores. Data about these substances are still scarce. Since these are important problems to be explored in the near future, however, a brief discussion about the possible involvement of such substances is included in Sect. 5.

In order to facilitate an understanding of the entire system for regulation of chromatophores, we present a diagram of the systems that control the motile activities of chromatophores in teleosts (Fig. 3). This diagram was drawn primarily to show the systems that are involved in the hormonal and neural control of dendritic chromatophores of the light-absorbing type. Because of the different optical properties of the effector cells, the regulatory systems for light-reflecting chromatophores are naturally organized somewhat differently from those for the light-absorbing cells. Although the fundamental details are analogous, the above diagram cannot, therefore, be applied as it stands to the control of leucophores or motile iridophores.

2.4 Measurement of Chromatophore Responses

If we are to understand the details of the mechanisms involved in the motility of chromatophores (Sect. 2.2), as well as the complicated systems for controlling this motility (Sect. 2.3), accurate descriptions of the states of the chromatophores and the reliable recording of changes in these states are essential. Photomicrography, using either monochromatic or color film, if necessary, can provide a great deal of information about the state of chromatophores. Recently, videotape recording has become a popular method for following cellular responses. For visualization of low-contrast cells, for example xanthophores or amelanotic melanophores, video-enhanced contrast techniques appear to hold considerable promise.

A convenient photometric method for measuring transmittance through a restricted area of a microscopic field is now widely used for physiological and pharmacological analyses. For studying the responses of light-absorbing chromatophores, an ordinary transmission microscope is used. However, dark-field, epi-illumination optics are recommended for recording the responses of leucophores and motile iridophores. A continuous planimetric method, using a video-image analysis system, also appears to be useful, especially for examinations of cells in culture. Activities of motile iridophores, such as those of neon tetras and blue damselfish (Sect. 2.1.5), can be monitored precisely by recording changes in spectral reflectance from the cells. Current optoelectronic devices for rapid measurements of spectra, adapted to an epi-illumination microscope, allow such activities to be followed.

Classical but still useful methods, as well as more modern ones, have been reviewed previously by one of the present authors (Fujii 1969, 1993b). For more detailed descriptions of photoelectric methods, for example, the reader can refer to some original papers, such as those by Oshima and Fujii (1984) and Oshima et al. (1984).

3 Effects of Light on Chromatophores

3.1 Effect of Lightness and Albedo on Chromatophores in Vivo

Color-change responses of teleost fish can be divided into two principal types: "primary responses", in which chromatophores respond directly to incident light, and "secondary responses", which are controlled by the nervous system and/or hormones. The latter responses are the usual chromatic adaptation mediated by the eyes. In the light, fish are either pale or dark, with their coloring depending on their surroundings: they are pale on a white background, dark on a black one. That is, when a fish is on a white background and the background is irradiated from above, the eyes of the fish receive in part light directly from the overhead source and in part the reflected light from the white background.

When the fish is on a black background and illuminated from above, only the direct light from above enters its eyes because of the low level of light reflection from the black background. Thus, the ratio of the intensities of the light from above and below ("albedo") is probably the major factor that determines the color of the integument.

In complete darkness, a fish becomes pale or takes on intermediate tints. In many teleost species such as *Chasmichthys gulosus* and *Synodontis nigriventris*, the melanosome-aggregating action of melatonin, which is secreted from the pineal body that is believed to be very accessible to environmental light, has been confirmed (Fujii 1961; Nagaishi et al. 1989). Hanyu and Niwa (1970) actually detected spontaneous neural discharges from the pineal vesicles of some fish species, and these discharges were inhibited solely by light stimulation. In the neon tetra, *Paracheirodon innesi*, the red and brown colors produced by erythrophores and melanophores were found to be markedly reduced at night, suggesting the regulation of coloration by an endogenous circadian rhythm (Lythgoe and Shand 1983). In practice, a demonstration in vitro by Hayashi et al. (1993) indicated that the *Paracheirodon* erythrophores are mainly controlled by melatonin. Thus, in the dark, the body color of fish seems to be under humoral regulation for the most part.

A peculiar color change of the goby, *Odontobutis obscura obscura*, was reported by Naitoh and Iwata (1976). When a fish was exposed to intense light (10,000–30,000 lx), its body color darkened, irrespective of the background color. Since the removal of the eyes caused failure of the reaction and since the melanophores did not respond directly to illumination, this phenomenon can be considered to be a modification of a secondary response.

A prolonged period of background adaptation induces a morphological color change, namely, an increase in the number of melanophores and/or in the net content of melanin, which is followed by the darkening of fish body (Odiorne 1957). From studies with the rainbow trout, Rodrigues and Sumpter (1984) pointed out the possible involvement of alpha-MSH in morphological color changes. Furthermore, in gobies adapted to a white background, both the number of leucophores and their guanine content increased, while the reverse changes occurred in fish adapted to a dark background (Fries 1958).

3.2 Effect of Environmental Hues on Chromatophores in Vivo

It is well known that the dermal chromatophore unit contributes to the characteristic green color of frogs and lizards that are fully adapted to a green background such as the green leaves of plants. We may ask, then, how teleost chromatophores in vivo behave in response to a change in the color of the surroundings? This problem remains to be analyzed in detail. However, it can easily be understood that a change in the peak wavelength of light received by their eyes should affect the chromatic states of teleosts, just as a change in the intensity of incident light influences this state.

In an earlier study of *Fundulus*, Fries (1931) found that, on a blue background, the melanophores were mainly dispersed and the xanthophores aggregated, while on yellow the xanthophores were dispersed and the melanophores aggregated. On an illuminated white background, both melanophores and xanthophores were aggregated, and both were dispersed on a black one. Very recently, the skin color of the dark sleeper, *Odontobutis obscura obscura*, adapted to a blue, yellow or ochre background was examined by Fujii et al. (1991a). On the blue background, the fish became dark, being of the same color as fish on a black one. Microscopic observations of the skin revealed that the melanophores dispersed their pigment and the "motile" iridophores aggregated their light-reflecting organelles to give a blue tint. On the yellow or ochre background, fish were light yellowish-brown in color, being the same color as fish exposed to a white background. That is, pigment granules were aggregated in the melanophores, while the reflecting platelets were dispersed in the iridophores, producing a yellowish tint. For the mechanism of generation of color by the iridophores, the reader is referred to Sect. 2.1.5 or to the article by Fujii et al. (1991a). Thus, in a colorful environment, the behavior of each chromatophore species can be expected to be quite complicated.

Some teleost species are known to possess a large variety of chromatic possibilities. Parker (1948) described representative examples in his writings: *Fundulus* usually assumes a simple series of tints between steel grey and pearly white, but it may also exhibit subordinate tones of yellow, red, green and blue. The flatfishes also assume blue, green, yellow, orange, pink or brown tints, in addition to the extremes of ashen grey and black. Of course, the fishes change not only their tints, but also their chromatic patterns. Therefore, such teleost species appear to exhibit very subtle hues in response to changes in the environmental colors around them (cf. Fujii and Oshima 1986).

The varied color changes of the integument may be based on the differences among chromatophore species with respect to sensitivity to neurotransmitters, hormones or light, the direction of pigment movement and the time required for a specific response. For instance, a demonstration in vitro showed that the dispersion of reflecting organelles in *Odontobutis* iridophores was caused by sympathetic stimulation, and the application of alpha-MSH gave rise to the aggregation of the organelles (Iga and Matsuno 1986). Both the dispersion and aggregation were very slow: more than 30 min were required for the maximum responses. The melanophores and xanthophores of this teleost species responded in quite the opposite way to the iridophores, and the responses of these chromatophores were much more rapid than those of the iridophores. Very recently, working on the tilapia, Oshima et al. found for the first time that the pituitary hormone prolactin causes melanosome aggregation, while the xanthophores respond to the hormone by the pigment dispersion, resulting in yellowing of the fish body (unpubl. data). Of course, the xanthophores respond to neurotransmitters, alpha-MSH, melanin-concentrating hormone (MCH) and melatonin in the same manner as the melanophores do.

As mentioned above, little research has been directly concerned with the analysis of the effects of environmental hues on fish chromatophores in vivo. If we make good use of various modern techniques, our understanding of this subject should increase steadily in the future.

3.3 Direct Effects of Light on Chromatophores

The pioneering study of Spaeth (1913) on the action of light on teleost chromatophores indicated that visible light had no detectable effect on the melanophores in isolated scales of *Fundulus heteroclitus.* Using the strips of skin from *Salmo gairdnerii* of 1 to 2 years of age, Robertson (1951) obtained similar results. Therefore, until comparatively recently, the primary color response, the direct reaction of the chromatophores to illumination, has been considered to be observed in blinded fish and larval forms (Duspiva 1931; Tomita 1936; Parker 1948). In 1978, Wakamatsu reported that some of the cultured melanophores from embryos and from two month-old fish of the black platyfish, *Xiphophorus maculatus*, responded directly to light by the aggregation of melanosomes. No melanophore was light-sensitive initially, but about 20% of the melanophores that had been cultured for 5 to 10 days exhibited light-sensitivity. The cells showed maximum sensitivity to light at wavelengths of about 410 nm and were almost insensitive to wavelengths longer than 500 nm, results that suggest a possible involvement of pteridines in the light-induced response of the cultured cells (Wakamatsu et al. 1980). Unlike the platyfish cells in culture, the melanophores from young fish (larval stage) of *Rhodeus ocellatus ocellatus* responded to light by pigment dispersion, and melanosomes were aggregated under dark conditions (Ohta 1983). The wavelength of light effective for melanosome dispersion was about 420 nm (Ohta and Muramatsu 1988).

Iga and Takabatake (1983) examined the responses of integumentary melanophores of "adult" fish to illumination, and they concluded that the melanophores of *Zacco temmincki* were light-sensitive, although the sensitivity differed among individuals. Both innervated and denervated melanophores in scales isolated from *Zacco* attained the fully aggregated state in darkness and responded to light (about 600 lx) by the dispersion of melanosomes. In cultured melanophores from adult *Zacco*, the dark-induced aggregation response was conspicuous 2 days after initiation of culture (Naora and Iga 1989). In addition, cultured melanophores of "adult" *Oryzias latipes* were found to respond directly to light in the same manner as *Zacco* melanophores (Negishi 1985). The responses to illumination appeared after 1 day in culture, and the degree of melanosome translocation depended on the intensity of the incident light. The most effective wavelength for inducing the dispersion of melanosomes was close to 415 nm in the medaka, whereas *Zacco* melanophores showed the maximum sensitivity to light with a peak wavelength of 525 nm, suggesting that a porphyropsin was active in the light response (Naora et al. 1988).

The responses of chromatophores other than melanophores to changes in illumination have been studied by some investigators. For example, Ohta and Sugimoto (1980) described the light-induced dispersion of leucosomes within *Oryzias* leucophores, which suggested a possible protective role for the cells against harmful radiation. Recently, Kawai (1989a) reported the sensitivity to light of denervated xanthophores of the medaka, whereas neither melanophores nor leucophores responded to light (3900 lx). When cells were irradiated, carotinoid droplets within the xanthophores migrated centripetally. In darkness, they were dispersed. The most effective wavelengths were near 400 nm, which is in accord with the results obtained with cultured melanophores of the platyfish and medaka. Using isolated scales of an Antarctic teleost, *Trematomus bernachii*, Obika and Meyer-Rochow (1990) observed the pigment aggregation in the majority of dermal xanthophores in total darkness, while the melanophores remained dispersed under both light and dark conditions.

The light-sensitive iridophores in the integument of the lateral stripes of the cardinal tetra *Cheirodon axelrodi* (Rohrlich 1974) and the neon tetra *Paracheirodon innesi* (Lythgoe and Shand 1982) are worthy of discussion here. The stripe shows the diel color changes in vivo, from green in the daytime to violet at night, in response to the change in the luminous intensity of the surroundings. Irradiation causes a shift in the spectal reflectance to longer wavelengths in isolated skin of the neon tetra, as is also the case in intact live fish (Nagaishi et al. 1990). The neon tetra iridophores contain two stacks of light-reflecting guanine platelets (each platelet being about 5 nm thick). A change in illumination may be followed by a change in the distance between adjacent platelets in each stack, leading to the shift in the spectral peak of light reflected from the iridophores. The "motile" iridophores are also under the control of sympathetic postganglionic fibers (Nagaishi and Oshima 1989). In addition to the lateral stripes of the tetras, an early description indicates that the blue spots of the male killifish *Fundulus heteroclitus*, in which iridophores are distributed, change from blue to red in response to the direct action of light (Foster 1933).

Light may act as a trigger to induce changes in intracellular levels of cAMP in cultured melanophores (Wakamatsu et al. 1980; Negishi 1985), although, in the xanthophores of the medaka, Ca^{2+} released from intracellular stores may be involved in the light-induced aggregation of pigment (Kawai 1989b). Note, in this context, that modification of the activity of phosphodiesterases by light has been demonstrated in the outer segment of rod cells of amphibia and mammals (Bitensky et al. 1975). While working on cultured melanophores of *Oryzias latipes*, Negishi (1988) observed that the peripheral regions of cells were retracted in the dark but extended in the light. Furthermore, the pigment dispersion in response to light irradiation was prevented by antimitotic reagents, while the dark-induced aggregation was not. Therefore, it seems likely that cycles of disassembly and assembly of microtubules in the cytoplasm may be involved in the light-dependent movement of melanosomes in cultured melanophores.

With respect to the phenomenon of photoreception, Imaki-Lamer and Chavin (1975) described unique structures in the melanophores of the coelacanth *Latimeria chalumnae*, a living Mesozoic "fossil". Groups of parallel membranes observed within the cytoplasm resembled, in terms of their gross morphology, the lamellae of the outer segment of photoreceptor cells in the pineal organ and in the retina. The membranes, therefore, were considered to represent photoreceptor structures responsible for physical color change, although the motility of the cells was not examined. In case of *Zacco* melanophores, it seems likely that photoreceptive sites are distributed over the entire surface of cells on the basis of the results of experiments in which the responses of a single cell to local light stimulation in various parts of the cell were tested (Iga and Takabatake 1986). In contrast to this possibility, an interesting phenomenon was reported by Kawai (1989b). In case of the xanthophores of the medaka, when the locally irradiated area exceeded about 30% of the total surface area of the cell, the centripetal movements of xanthosomes were observed not only in the stimulated area but also in unirradiated areas. Furthermore, this pigment-aggregating response occurred in other xanthophores which had not been irradiated, probably as a result of a transmission of Ca^{2+} to adjoining xanthophores from the target cell. Immunocytochemical studies on the light-sensitive iridophores of the neon tetra suggest that a rhodopsin or a closely related molecule is present in these cells and acts as a light receptor (Lythgoe et al. 1984). After dark adaptation, blue light was most effective in inducing color change, while light at wavelengths longer than 570 nm was relatively ineffective. Lythgoe and his colleagues have suggested that the activation of visual pigment by photons of light might cause the opening of sodium channels in the cell membrane, with a resultant increase in the thickness of the cytoplasm layers as a consequence of the inflow of water. A contradictory conclusion was proposed by Nagaishi et al. (1990), namely, that photoreception by the iridophores of the neon tetra may be followed by a simultaneous change in the angles of inclination of all the light-reflecting platelets, which, in turn, causes a shift in the spectral peak of reflected light (cf. Sect. 2.1.5).

At present, we know little about the direct responses of fish chromatophores to light. In particular, the mechanism by which cellular motility arises after photoreception remains to be clarified. The functional significance of the primary color response is not well understood either. However, it seems possible that the pigmentary layer may be useful for protecting deeper tissue components from harmful radiation from the sun, at least in case of dispersal of chromatosomes under illumination.

3.4 Effect of Ultraviolet Rays on Chromatophores

Using isolated scales of *Fundulus heteroclitus*, Spaeth (1913) showed that exposure of the melanophores to ultraviolet (UV) light within a range of 185–290 nm induced pigment aggregation. Such reaction was reversed under a condition of visible light. Working on the goby, *Chasmichthys gulosus*, Fujii et al. (1973) again analyzed the effects of UV rays on the melanophores in split-fin preparations, which were irradiated from the dermal side. Until the total dosage reached 1.3×10^5 erg mm^{-2}, the magnitude of the melanin aggregation response of melanophores increased. At higher doses, the effect became irreversible. On the other hand, the responsiveness of the melanophores to norepinephrine, a neurotransmitter released from nerve endings, decreased upon increasing the exposure to UV light, and melanosome movements were not detectable in the melanophores irradiated with doses above 2×10^5 erg. This phenomenon might be due to the irreversible change in the colloidal state of the cytoplasm, namely, a sol-gel transformation. In addition, transmitter-releasing function of and conduction by nerve fibers were affected by UV radiation, although the UV sensitivity differed between these processes. When UV rays were applied from the epidermal side, a much higher dose was required to affect the activity of the melanophore system, implying that the overlying layers, namely, the epidermis and subepidermal collagenous layer, might protect the system from the harmful actions of short-wave photons.

4 Effects of Other Physical Factors on Chromatophores

4.1 Ionizing Radiation

In studies of adult xanthic goldfish, several authors reported that melanophores appeared in the skin and fins after X-ray irradiation and that the melanophores gradually increased in number and size over the course of a few weeks (Egami et al. 1962). From their results, Egami and his collaborators concluded that ACTH secretion from the pituitary was stimulated by exposing fish to X-rays, which led to a morphological color change. In newly formed melanophores, neural stimuli failed to induce pigment aggregation. Three weeks or more after irradiation, however, the melanophores became responsive to such stimuli, suggesting the completion of the innervation normally seen in black fish (Etoh and Egami 1963).

Etoh (1963) also examined the effects of gamma-irradiation on the melanin-aggregating nervous system in the crucian carp. The conductive part of nerves was much more sensitive to the radiation than was the presynaptic transmitter-releasing portion. The melanophores themselves were very resistant to gamma rays.

4.2 Hydrostatic Pressure

How does a change in hydrostatic pressure affect the state of chromatophores? Such changes occur when fish sink below their normal level or rise above it.

Marsland (1944) confirmed that the pigment-dispersed condition of the melanophores in *Fundulus* scales, immersed in NaCl solution, was maintained even at very high pressures. Although pulsation (cf. Sect. 5.4) of the melanophores could be caused at fairly high pressures, the amplitude of the ebb and flow of melanosomes gradually decreased as the pressure increased. Further analyses by Marsland and Meisner (1967) indicated that very high hydrostatic pressure caused the melanosome dispersion, inhibiting the action of pigment-aggregation stimuli. Moreover, such an effect of high pressure was not apparent in deuterated solutions. Since D_2O is an agent accelerating gelation of the protoplasm, it was concluded that the dispersion of pigment was primarily due to cytoplasmic solation.

By contrast, Robertson (1951) demonstrated the pigment-aggregating effect of high pressure in vivo, suggesting the possible role of K^+ liberated from the muscle tissue under the integument.

A decrease in hydrostatic pressure may also disturb the normal physiological conditions of animals, but chromatic states at lower pressures have not been described. With large-scale decreases in pressure, the gases dissolved in the water of an animal's body may appear as such in the tissues, as a result of which the tissues might become fragmented.

4.3 Temperature

In general, it has been accepted that high temperature causes the melanosome aggregation, low temperature the dispersion of pigment and an intermediate one allows these changes to be subject to illumination (cf. Parker 1948). Working with *Fundulus heteroclitus*, Smith (1928) made some very interesting observations: when two fish, one of which had been kept in cold water and the other in warm, were transferred to water at an intermediate temperature of 20 °C, the fish from the cold environment showed pallor and the fish from the warm water darkened. Later, Robertson (1951) and Ando (1962) reported similar results, from studies of trout melanophores and embryonic *Oryzias* melanophores, respectively. A contradictory observation was made by Pye (1964), who found that a local application of warm water induced the maximum dispersion of pigment in *Phoxinus* melanophores, while a jet of cool water brought about the aggregation of melanosomes. It has not yet been clarified whether such responses to temperature differences are produced via the nervous system or are direct reactions of the pigment cells themselves.

Pulsation of *Fundulus* melanophores was found also to be affected by temperature differences (Smith 1931). The upper limit of pulsation was between 30 °C and 34 °C, and such movements failed to appear below 10 °C. With a rise of temperature above 10 °C, the frequency increased, the maximum frequency being observed at 27 °C.

Recently, Obika and Meyer-Rochow (1990) studied the chromatophores of an Antarctic teleost that lives under the sea-ice north of the Ross Ice Shelf, where water temperature remains about $-1.9\,^{\circ}C$ throughout the year. In phosphate-buffered saline (about $10\,^{\circ}C$), dermal melanophores and xanthophores in isolated scales became dispersed, and KCl and epinephrine caused the aggregation of pigment within a few minutes. These pigment cells were shown to have well-developed arrays of microtubules in the cytoplasm. These findings may suggest that, when we examine the various factors affecting the state of chromatophores, special attention should be paid to the natural environmental conditions to which the fish in question have accommodated themselves.

It was pointed out by Spaeth (1913) that the melanin-aggregating action of K^+ is accelerated by an increase in temperature. At present, it is still generally accepted that the responsiveness of pigment cells to various chemical agents is enhanced with increases in temperature.

4.4 Osmolarity

Evidence for the effect of hypotonicity on melanophores was first obtained by Spaeth (1913) in studies of *Fundulus*. When isolated scales were immersed in distilled water, the melanosome aggregation was observed. Recently, the response of tilapia melanophores in hypotonic solution was reexamined (Fujii and Ikeda unpubl. data). In both denervated and cultured melanophores exposed to hypotonic media, the reversible aggregation of pigment was confirmed. From these observations, they supposed that the decrease in the tonicity of the bathing solution was accompanied by an influx of water into melanophores, which might induce the swelling of membranes of Ca^{2+}-storing organelles within cells, with a resultant leakage of Ca^{2+}. Alternatively, a decrease in the concentration of cytosolic cAMP caused by an inflow of water into pigment cells might be responsible for the melanosome aggregation.

In motile iridophores of the neon tetra and blue damselfish, the spectral curves shifted towards longer wavelengths as the tonicity of the medium was reduced (Lythgoe and Shand 1982; Kasukawa et al. 1987). Such reaction of the iridophores may be a simple passive process: an increase in the thickness of cytoplasmic layer sandwiched between light-reflecting platelets may result from an influx of water into the cells.

The hypertonicity of the bathing solution prevented the Brownian movements of melanosomes in isolated scale melanophores (Kamada and Kinosita 1944). Moreover, such melanophores did not respond to various melanin-aggregating or -dispersing agents.

5 Effects of Chemical Substances on Chromatophores

5.1 Hormonal Substances

Various substances, both biogenic and artificial, affect chromatophores. The effects of hormonal principles, in particular, have been the focus of considerable interest for many years, and there are many pertinent review articles, such as those by Pickford and Atz (1957), Waring (1963), Bagnara and Hadley (1973), Fujii and Oshima (1986) and Baker (1991). In the present article, therefore, we shall review relevant work only briefly, paying particular attention to recent developments. The effects of principles that modulate the responses of chromatophores to hormonal stimuli have seldom been reviewed. Together with the agents that modulate nervous cues, these factors will be discussed in a separate section (5.3).

5.1.1 Melanophore-Stimulating Hormone

In 1957, Pickford and Atz published an extensive review of the pituitary control of fish chromatophores. The more recent reviews cited above are also very useful. In the present review, our citations are mainly restricted to subsequent publications.

Resembling those in other classes of poikilothermic vertebrates, melanophores of fish are usually very responsive to melanophore-stimulating hormone (MSH), which is secreted from the intermediate lobe of the pituitary. Numerous workers have reported that MSH effectively disperses melanosomes in a number of species within both Chondrichthyes and Osteichthyes. Using a synthetic, purified sample of MSH, Fujii and Miyashita (1982) showed that both the dermal and the epidermal melanophores of a siluroid catfish, *Silurus asotus*, responded to MSH by dispersion of melanosomes.

The pigment-dispersing action of MSH is not restricted to the melanophores, and several examples of such effects can be found in the reviews cited above. Recent results include those of Matsumoto et al. (1978) who showed clearly that swordtail erythrophores in primary culture responded to MSH. Working on the medaka *Oryzias latipes*, Sugimoto et al. (1985) found that, even in amelanotic melanophores which lack mature melanin, dispersal of colorless premelanosomes occurred in response to MSH. Leucophores of medakas responded to MSH by the dispersion of leucosomes (Negishi and Obika 1980; Oshima and Fujii 1985). This result was rather unexpected, since leucophores normally respond to first-messenger signals in the opposite way to light-absorbing chromatophores.

In some cases, such as those cited by Pickford and Atz (1957), MSH fails to disperse pigment, as has recently been confirmed. In connection with this issue, we note that Baker et al. (1984) found, in both the eel (*Anguilla anguilla*) and the rainbow trout (*Salmo gairdneri*), that the plasma level of MSH changed rapidly in response to the changes in background coloration. However, the level did not change during background adaptation in the flounder (*Pleuronectes flesus*). Working on the rainbow trout, Rodrigues and Sumpter (1984) came to the conclusion that MSH is not involved in the physiological control of color changes but may be solely involved in morphological color changes. Thus, the role played by this hormone in fish seems to vary considerably depending on the species.

Recent analyses indicated that the effects of MSH on fish melanophores may be mediated by specific receptors on the effector membrane (Fujii and Miyashita 1982). Ca^{2+} ions are required for the action of this peptide hormone (Fujii and Miyashita 1980; Iga and Takabatake 1982). The dependence on Ca^{2+} ions of the action of MSH has also been demonstrated for bright-colored chromatophores, namely, the erythrophores of the platyfish, the xanthophores and leucophores of the medaka (Oshima and Fujii 1985). Thus, the requirement for Ca^{2+} ions of the action of MSH may be a common property in fish. Among the several types of signal to chromatophores, MSH is the only one to require Ca^{2+} ions for its effect. Presumably, Ca^{2+} ions are required for the binding of the ligand to the active site of the receptor.

Dendritic motile iridophores of the dark sleeper goby (*Odontobutis obscura obscura*) respond to MSH by aggregation of their platelets (Iga and Matsuno 1986; Fujii et al. 1991a). Round motile iridophore of the damselfish type (Oshima et al. 1989) or the neon tetra type (Nagaishi and Oshima 1989) reacted to the peptide by exhibiting the SR response (Sect. 2.1.5). In both cases, the direction of the response is the opposite of that to sympathetic cues, as is the case for light-absorbing chromatophores.

5.1.2 Melanin-Concentrating Hormone

It has long been postulated that the chromatophores of lower vertebrates are under the influence of two hormones with reciprocal functions (Pickford and Atz 1957; Warning 1963; Baker and Ball 1975; Baker 1991). The hormone that disperses pigment is, of course, MSH as discussed above. The antagonistic hormone was first demonstrated in the pituitary, as well as in the hypothalamus, of the siluroid catfish *Silurus asotus* by Enami (1955). He referred to this principle as "melanophore-concentrating hormone" or "MCH". Although the abbreviation is the same, the term "melanin-concentrating hormone" is currently employed to denote this principle because it is not the melanophores but the organelles within them that are concentrated.

Kawauchi et al. (1983) purified MCH from chum salmon pituitaries and showed it to be a cyclic heptadecapeptide. It induced discernible aggregation of melanosomes in tilapia at 1 nM (Wilkes et al. 1984), in damselfish at 100 pM (Oshima et al. 1985) and in freshwater eel at 10 pM (Abrao et al. 1991). Past studies have indicated that it is effective in melanophores of almost all tested species of teleost (Kawauchi et al. 1983; Wilkes et al. 1984; Oshima et al. 1985; Nagai et al. 1986).

Working on the blue damselfish *Chrysiptera cyanea*, Oshima et al. (1985) analyzed the pharmacologic mechanism of action of this peptide on melanophores, and they came to the conclusion that is acts directly on the cells via specific receptors on the cell membrane of effector cells. Abrao et al. (1991) showed more recently that the stimulation of receptors for MCH causes activation of phospholipase C in the membrane. The diacylglycerol that is produced activates protein kinase C, and subsequently the aggregation of melanosomes occurs.

Chromatophores other than melanophores have also been shown to be responsive to MCH (Oshima et al. 1986a). Erythrophores of the platyfish (*Xiphophorus maculatus*) and swordtail (*X. helleri*), as well as xanthophores and amelanotic melanophores of the medaka (*Oryzias latipes*), respond quite similarly to the melanophores of many fishes. The action of MCH on the motile iridophores of the damselfish type (Oshima et al. 1989) and of the goby type (Iga and Matsuno 1986) is quite the opposite of that of MSH, as is the case for light-absorbing chromatophores. In every case, specific receptors are considered to be responsible for transducing the signal.

As an exceptional case, the leucophores of the medaka were found to respond to MCH by dispersion of pigment (Oshima et al. 1986a). Since MSH has been found to induce pigment dispersion in this type of cell (Sect. 5.1.1; Negishi and Obika 1980; Oshima and Fujii 1985), this result was rather unexpected. In addition, this process required Ca^{2+} ions, as does the action of MSH. Presumably, MCH acts on receptors for MSH on the cell membrane, having the same effect as MSH. Thus, the dual-hormone hypothesis is not applicable here.

5.1.3 Melatonin

Lerner et al. (1958) isolated a potent principle from bovine pineal gland, that caused blanching of frog skin. They identified it as *N*-acetyl-5-methoxytryptamine and named it melatonin. It is synthesized from 5-hydroxytryptamine (5-HT), via reactions catalyzed by *N*-acetyltransferase (NAT) and hydroxyindole-*O*-methyltransferase (HIOMT). The activity of NAT becomes higher at night, while it is low in the daytime. Thus, a circadian change occurs in the rate of production of the hormone and in its release. Observations to date also indicate that the pineal gland functions as an extraoptic receptor for the entrainment of dark-to-pale color changes of animals to the light-dark cycle in the environment (cf. Fujii and Oshima 1986).

The pigment-aggregating action of melatonin on fish chromatophores was first demonstrated by Fujii (1961), who worked on the melanophores of the gluttonous goby, *Chasmichthys gulosus*. His observation that denervated melanophores were also responsive to the hormone led him to conclude that the hormone acts on the cells directly. Since then, various studies on melanophores, as well as on bright-colored chromatophores, have been carried out (cf. Fujii and Oshima 1986). Fujii and Miyashita (1978) showed that the amine acts on both dermal and epidermal melanophores of the catfish *Silurus* via specific receptors on the melanophore membrane. Comparing the effectiveness of substances metabolically related to melatonin, they further concluded that the 5-methoxy group on the indole ring may be essential for the action of melatonin.

Apparently, there are differences between species in terms of sensitivity to melatonin (cf. Fujii and Oshima 1986). Even within an individual, moreover, differential responsiveness is evident among areas of the skin or even among melanophores (Reed 1968; Fujii and Taguchi 1969). These observations suggest that melatonin may be involved in the formation of various colored patterns. Using the common pencil fish, *Nannostomus beckfordi*, Reed (1968) found that melatonin functions in the formation of a nighttime spotted pattern that is different from the well-recognized striped pattern seen in the daytime.

Comparing the day and the night patterns of the same species of pencil fish, Nishi and Fujii (1992) recognized that parts of the skin darkened under the influence of melatonin. From their pharmacological analysis, they concluded that the melanophores have receptors hat mediate the dispersal of melanosomes when stimulated by melatonin. They designated the receptor "beta-melatonin receptors". The conventional receptors for melatonin should, therefore, be designated "alpha-receptors", when both kinds of receptor are discussed at the same time.

Effects of melatonin on chromatophores other than melanophores have been reported. For example, Matsumoto et al. (1978) observed aggregation of pigment in erythrophores of the swordtail (*Xiphophorus helleri*) in primary culture. Satake (1980) detected the same effect in goldfish erythrophores after intracranial injection of the principle. Motile iridophores of neon tetras (Oshima et al. 1989) and of blue-green damselfish (Kasukawa et al. 1987) responded to the amine by the LR response. In all cases, the direction of the response was the same as that to sympathetic stimuli.

Leucosomes in the leucophores of the medaka aggregated in response to melatonin (Obika 1976, 1988). Being the opposite of that to sympathetic signals, the response of these chromatophores seemed to be rather peculiar because light-reflecting and -absorbing chromatophores usually behave in opposite ways. Sometimes, we encounter cases in which the responses to various first messengers differ among different types of chromatophore. These phenomena are probably a reflection of the fact that fish are endowed with abilities that allow them to realize various delicate, intermediate hues which could not be generated in the absence of such differential control mechanisms.

5.2 Neuronal Substances

Like those of hormones, the effects of neuronal substances on chromatophores have been studied for many years (Fujii 1961, 1969; Fujii and Novales 1972).

Reliable data have, however, only been obtained quite recently. Thus, early articles on this subject, either research reports or review articles, are not very useful. A recent review by Fujii and Oshima (1986) may be of use to those who wish to follow progress in this field prior to the mid-1980s. In the present review, we intend to pay particular attention only to recent advances in this field.

Chromatophores of lower fish are now regarded as being solely under the control of endocrine systems (Parker 1948; Waring 1963; Fujii 1969). In Osteichthyes, by contrast, chromatophores of many species have clearly been shown to be regulated by both nervous and endocrine signals (Fujii and Novales 1972; Fujii and Oshima 1986). The extent of the dependence on hormonal or nervous regulation differs among species of animals, kinds of chromatophore, and even among chromatophores at different sites on the integument of the same individual. For example, chromatophores of the eel seem to be solely under the control of the endocrine system (cf. Parker 1948), whereas the motile iridophores of the blue damselfish are regulated exclusively by sympathetic postganglionic fibers (Kasukawa et al. 1986a). Between these two extreme cases, there are common examples of both systems operating together to varying extents.

It is not at all difficult to determine whether or not particular chromatophores in the skin are regulated by the nervous system. Chromatosomes in light-absorbing dendritic chromatophores become dispersed throughout the cytoplasm when a piece of skin is equilibrated in physiological saline (cf. Sect. 5.4). If the electrical stimulation of nerve fibers to the skin gives rise to the aggregation of pigment within the chromatophores, we can safely conclude that the cells are under the control of the nervous system. In dendritic light-reflecting chromatophores (leucophores and the dendritic iridophores of the goby type; Sects. 4 and 2.1.5), such stimulation induces the reverse movement, namely, the dispersion of light-reflecting organelles, if a system for nervous control exists. In non-dendritic motile iridophores, nervous stimulation induces a change in the arrangement of reflecting platelets so that they reflect light of longer wavelengths (LR response; Sect. 2.1.5). In the dendritic iridophores of the goby type (Sect. 2.1.5), platelets disperse into the processes upon nervous stimulation. In all cases, the state assumed by chromatophores that have been equilibrated in physiological saline normally reflects the state when nervous signals are absent. If a detectable response is observed to nervous stimuli, the chromatophores can, therefore, be safely considered to be under the control of the nervous system.

Current investigations suggest that, in living fish, firing of nerves at relatively low frequencies (e.g., ca. 1 Hz) can maintain chromatophores in an intermediate state of pigment aggregation. Higher rates of firing result in a more aggregated state, while a lower frequency or the cessation of impulses causes dispersion of pigmentary organelles (Fujii and Miyashita 1975; cf. Fujii and Oshima 1986).

Recent studies indicate that the parasympathetic system is not involved in the motile responses of chromatophores, although several earlier researchers, such as Parker (1948), strongly advocated the involvement of this system in the dispersal of pigment in melanophores. We now know that the sympathetic division of the autonomic system is solely responsible for the regulation of chromatophores in fish (cf. Fujii and Oshima 1986). The peripheral neurotransmitter that signals chromatophores was naturally supposed to be adrenergic. Observing that an adrenergic antagonist, dibenamine, interfered with the pigment-aggregating action of sympathetic stimuli, Fujii (1961) was able to demonstrate the adrenergic nature of transmission to the melanophores of the goby *Chasmichthys*. Since dibenamine is known to block alpha-adrenoceptors, the transmission could be regarded as alpha-adrenergic in nature. Using a variety of alpha-blockers developed thereafter, many later workers have come to the same conclusion (e.g., Fujii and Miyashita 1975; cf. Fujii and Oshima 1986). Kumazawa and Fujii (1984) recently showed that a catecholamine transmitter is actually liberated from the fibers: they detected the release of radiolabeled norepinephrine in response to nervous stimulation.

Several researchers have attempted to determine the subtype of alpha-adrenoceptors on chromatophores. Some of them have reported that $alpha_2$-agonists are more effective than $alpha_1$-agonists, and that transmission is more easily blocked by $alpha_2$-blockers than by $alpha_1$-blockers (e.g., Andersson et al. 1984; Morishita 1987). These workers came to the conclusion that the pigment-aggregating adrenoceptors are of the $alpha_2$ type. It was also found, however, in many investigations, that a considerable aggregation of pigment occurred in response to $alpha_1$-agonistic stimuli. Similarly, $alpha_1$-type adrenolytics always had an inhibitory effect on the aggregation of pigment. Thus, the participation of $alpha_1$-adrenoceptors should be reexamined.

In some species the aggregation of pigment may be triggered by an increase in levels of Ca^{2+} ions in the cytosol (Sect. 3.2). In addition, Fujii et al. (1991b) demonstrated lately the involvement of inositol 1,4,5-trisphosphate (IP_3) in the aggregation of pigment in tilapia melanophores. In many types of cell, IP_3 has been shown to induce the release from intracellular storage compartments of Ca^{2+} ions into the cytosol. Moreover, we are now aware that $alpha_1$-adrenergic stimuli activate phospholipase C, which catalyzes the production of IP_3. These observations indicate that $alpha_1$-adrenoceptors are functional, at least in some cases.

Chromatophores other than melanophores have also been shown to be under the control of the sympathetic system. For instance, the erythrophores of the swordtail, *Xiphophorus helleri* (Matsumoto et al. 1978) and those of the squirrelfish *Holocentrus ascensionis* (Luby-Phelps and Porter 1982) have been shown to be under the strong influence of the nervous system. Comparing the physiological characteristics of xanthophores with those of melanophores and leucophores on the scales of the medaka *Oryzias latipes*, Iwata et al. (1981) showed that xanthophores responded in quite the same way as melanophores. Therefore, the cellular receptors for hormones and neurotransmitters of the xanthophores seem to be identical with those of melanophores.

The nervous mechanism that controls the leucophores of the medaka has been analyzed. Fujii and Miyashita (1979) and Iga (1983) showed clearly that dispersal of leucosomes takes place when the controlling nerves are stimulated. The receptors concerned are of the beta-adrenergic type (Obika 1976; Iga et al.

1977). Later pharmacological analyses by Yamada (1980) indicated that the subtype of the receptors is beta$_1$. Iga (1979) noticed that leucosomes in leucophores aggregated in response to catecholamines, when beta-adrenoceptors had previously been blocked, and concluded that the response was mediated by adrenoceptors of the alpha type. Morishita and Yamada (1989) characterized these receptors as being of the alpha$_2$ type. It remains to be determined whether receptors of this type actually function in vivo.

It was recently found that, in siluroid catfishes, the peripheral transmission to the melanophores is cholinergic, even though the postganglionic fibers to the cells are sympathetic in the usual manner (Fujii and Miyashita 1976b; Fujii et al. 1982). Replacing alpha-adrenoceptors, muscarinic cholinoceptors play an exclusive role in transducing nervous signals to the effector cells (Fujii and Miyashita 1976b; Kasukawa and Fujii 1984). Surveys have been made to examine the presence of cholinoceptors in catfish in the order Siluriformes. In families closely related to Siluridae, melanophores are often endowed with both adrenergic and cholinergic receptors, both of which mediate the aggregation of melanosomes (Kasukawa and Fujii 1985; Kasukawa et al. 1986b). Moreover, in these fishes, the neurally evoked aggregation of pigment is mediated by alpha-adrenoceptors, as it is in many common teleosts. The physiological role of the cholinoceptors in these fishes still remains to be investigated.

It was first suggested by Fujii and Miyashita (1976a) that adenosine or adenine nucleotides might take part in the control of pigment dispersal in fish chromatophores. They found that non-cyclic adenyl compounds, which they were using as control compounds, were even more effective than cyclic adenosine 3′,5′-monophosphate (cAMP) in dispersing pigment in melanophores of guppies. Using guppies and siluroid catfish, Miyashita et al. (1984) extended this pharmacological analysis and came to the conclusion that the pigment-dispersing action of these nucleotides was mediated by adenosine receptors since the effects could easily be antagonized by methylxanthines, specific blockers of adenosine receptors.

Working on the melanophores of tilapias, Kumazawa et al. (1984) detected the apparent liberation of ATP from chromatic nerves in response to electrical stimulation. They concluded that ATP is released as a co-transmitter from the postganglionic sympathetic fibers together with the true transmitter, norepinephrine. The concurrent release of the true transmitter and the co-transmitter from the fibers to chromatophores has been confirmed in experiments with radiolabeled compounds (Kumazawa and Fujii 1984, 1986).

The peripheral nervous mechanism, as characterized to date, is shown schematically in Fig. 4. The true transmitter, norepinephrine, acts to induce a rapid aggregation of melanosomes via mediation by alpha-adrenoceptors on the membrane. Then most of it is quickly removed by re-uptake into the nervous elements. The remainder is either removed via the general circulation or is inactivated by catechol *O*-methyltransferase (COMT) and monoamineoxidase (MAO). ATP molecules, after their concurrent release with norepinephrine are gradually dephosphorylated by the ATPase and 5′-nucleotidase in the synaptic

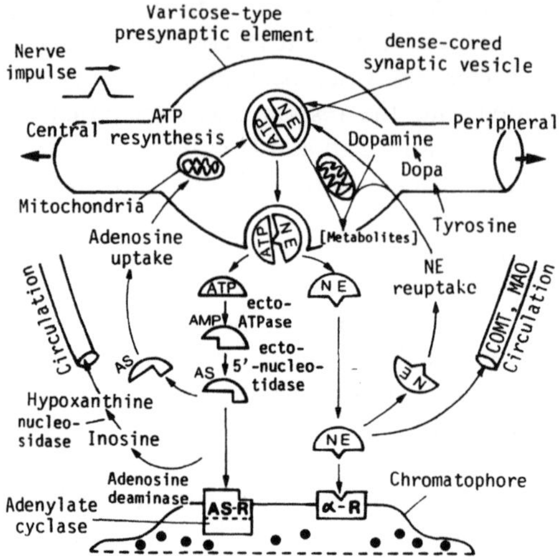

Fig. 4. Diagrammatic representation of the dynamics of events that involve both the true transmitter and the co-transmitter liberated from the sympathetic postganglionic fiber in teleosts. *AS* Adenosine; *AS-R* adenosine receptor; *COMT* catechol *O*-methyltransferase; *MAO* monoamine oxidase; *NE* norepinephrine; *α-R* α-adrenoceptor. (Fujii and Oshima 1986)

cleft. The resultant nucleoside, adenosine, survives for a longer period in the synaptic cleft and functions to reverse the influence of the true transmitter, namely, to cause the re-dispersion of pigment, via specific receptors for adenosine on the effector membrane. Most of the nucleoside is finally removed by re-uptake into presynaptic nervous elements, and the remainder is carried away by the circulation.

One important aspect of the proposed dual-transmitter theory is that there is a substantial difference, in terms of action, between the true transmitter and the co-transmitter. The effect of norepinephrine disappears very quickly, while that of the co-transmitter lasts much longer. After the cessation of nervous excitation, the latter can effectively reverse the action of the former. Rapid dispersion of pigment after nervous excitation is realized in this way. The rapid changes observable in living animals may also be controlled by the same mechanism. An identical explanation has been presented for the regulation of melanophores of the blue damselfish *Chrysiptera cyanea* (Kasukawa et al. 1986a) and the blue-green damselfish *Chromis viridis* (Oshima et al. 1989). The motile responses of amelanotic melanophores of medakas are also regulated in the same way (Sugimoto et al. 1985).

Recent studies on medakas indicate that leucophores respond to adenosine by dispersion of leucosomes (Oshima et al. 1986b). Specific adenosine receptors of the A_2 type mediate this response. However, the direction of the movement of leucosomes in response to the co-transmitter is the same as that elicited by the true transmitter. In fact, the recovery from the effect of norepinephrine occurs very slowly. The search for the role of the co-transmitter in controlling pigmentation of this type of cell should, therefore, include other functions.

The involvement of the dual-transmitter system in the control of motile iridophores may be analogous to that in melonophores. The motile iridophores of blue damselfish (Kasukawa et al. 1986a), blue-green damselfish (Oshima et al. 1989) and neon tetras (Nagaishi and Oshima 1989) respond to adenine derivatives of adenine with the SR response, which is the opposite of the response elicited by alpha-adrenergic stimuli (the LR response).

5.3 Substances That Modify Hormonal and Nervous Signals

In addition to the hormonal and neuronal substances that have definite physiological effects on chromatophores (Sects. 5.1.1 and 5.1.2), various biogenic substances have been shown in influence the state of chromatophores. Some may act directly on the effector cells, while others act indirectly by activating or inhibiting the liberation of transmitters or hormones. Naturally, such substances are considered not to have crucial roles in the regulation of the motile activities of chromatophores. Nevertheless, further investigations may reveal that at least some of them are indispensable for the control of chromatophores since it has become more and more apparent that chromatophores of fish are subject to incredibly delicate and subtle control (cf. Sect. 7). In the present section, we shall briefly describe the current status of studies on such substances. Detailed descriptions may be found in the original papers or in review articles relevant to the subject, of which only a few representatives are cited here (Fujii 1961, 1969, 1993b; Healey and Ross 1966; Bagnara and Hadley 1973; Fujii and Oshima 1986). Earlier papers are cited in the text or in the Reference sections of these reviews.

Adrenocorticotropic hormone (ACTH) has pigment-dispersing action, which can rather naturally be understood if we recall that ACTH includes the amino acid sequence of MSH. In addition, both hormones share a common active site. Melanophores and other bright-colored chromatophores responded to ACTH. The chromatosomes in the xanthophores of the goldfish (Obika et al. 1978) and the mudsucker goby *Gillichthys mirabilis* (Sage 1970), and cultured erythrophores of the swordtail *Xiphophorus helleri* disperse in response to this peptide (Matsumoto et al. 1978).

Prolactin (PRL), another peptide from the anterior lobe of the pituitary, was shown to disperse pigment in the xanthophores of *Gillichthys*, with resultant yellowing of the fish (Sage 1970). Using two highly purified molecular species of PRL from tilapia, Oshima et al. (1990c) confirmed their effects on the xanthophores of the medaka, *Oryzias latipes*. They further recognized that the peptide acted to aggregate melanin in the melanophores of the same species. Yellowing of the skin is, thus, realized very effectively in response to PRL.

Reports on the actions of other anterior pituitary hormones, including FSH, LH, TSH and GH, were still few in number, but the results obtained to date indicate that these principles may be without effect on melanophores, such as those of a freshwater characin, *Astyanax* (Rasquin 1958). By contrast, the pigment in xanthophores has been shown to be influenced by these peptides, for example, dispersion of xanthosomes was observed within the cells of a gobiid,

Bathygobius (Rasquin 1958). Together with the results for PRL described above, these observations suggest possible differential hormonal control of different types of chromatophore.

Robertson (1949) suggested that thyroid hormone may have a pigment-aggregating effect on the melanophores of live rainbow trout, and thyroxine was later reported to aggregate melanosomes in the melanophores of some teleosts, for example, the brown trout *Salmo trutta* (Woodhead 1966).

Catecholamines, other than the sympathetic neurotransmitter norepinephrine (Sect. 5.2), have been reported to affect the distribution of pigment in some chromatophores. Among them, epinephrine has been the most intensively studied. Recent reports on the pigment-aggregating action of this compound include those by Fujii (1961) and Fujii and Miyashita (1975). Mediated by alpha-adrenoceptors, the action of epinephrine is fundamentally identical with that of norepinephrine.

Injecting the synthetic beta-adrenergic agonist isoproterenol intraperitoneally, Reed and Finnin (1972) were able to observe a transient dispersion of pigment in melanophores of angelfish, and they suggested the presence of beta-adrenoceptors on the melanophores. Using isolated pieces of fin from the top minnow *Gambusia affinis*, Colley and Hunt (1974) also demonstrated the presence of beta-receptors on the melanophores. Miyashita and Fujii (1975) found that a number of sympathomimetic amines, including epinephrine, could cause the reversible dispersion of pigment in the melanophores of split pieces of tail fin from the guppy *Poecilia reticulata*. The effect was especially remarkable when alpha-adrenoceptors were blocked. They further noted that these substances at higher doses elicited aggregation of pigment, with the effect being mediated by alpha-adrenoceptors. The dispersion responses could easily be antagonized by beta-adrenolytic drugs. It was concluded, therefore, that beta-adrenoceptors are definitely involved in the darkening responses of fish. The same conclusion was subsequently reached by Obika (1976), Iga (1983) and several other researchers for a variety of teleosts. Morishita et al. (1985) classified the beta-receptor that mediates dispersion of melanosomes in *Oryzias* as being of the beta$_2$ type. Matsumoto et al. (1978) showed that erythrophores of the swordtail have beta-adrenoceptors that mediate dispersion of the pigmentary organelles within the cells. Erythrophores and xanthophores of many species may possess beta-receptors hat are involved in dispersal of their pigmented organelles. The motile iridophores of fish such as the blue-green damselfish are also endowed with receptors of this type, which are involved in induction of the SR response (Oshima et al. 1989). Using an autoradiographic technique, Komatsu and Yamada (1982) visualized the distribution of beta-adrenoceptors on the melanophores of *Oryzias*: they were distributed almost uniformly over the entire cell.

The endogenous catecholamine responsible for the darkening response of live fish must be epinephrine, which may possibly be secreted from the chromaffin cells of adrenal tissues. The "excitement darkening", which will be touched upon later in this review, may be representative of the beta-receptor-mediated reactions (Sect. 7). A sudden outflow of the amine from the chromaffin cells is considered to induce this remarkable change in coloration.

Enami (1955) recorded a peculiar case of a teleost, namely the siluroid catfish *Silurus asotus*, in which epinephrine applied either in vivo or in vitro always induced darkening of the skin. Recent pharmacological analyses of the same species indicate that the adrenoceptors on both dermal and epidemal melanophores are solely of the beta type (Fujii et al. 1985). Melanophores of another siluroid, the translucent glass catfish, *Kryptopterus bicirrhis*, have also been shown to have adrenoceptors of this type (Fujii et al. 1982). As mentioned above, the nervous supply to the melanophores of these siluroids is cholinergic (Sect. 5.2).

Biogenic monoamines other than norepinephrine and epinephrine have been shown to influence the chromatophores of teleosts. Among them, dopamine has several times been shown to aggregate the pigment in several types of melanophore (Scott 1965; Healey and Ross 1966; Iga 1968). In most cases, alpha-adrenoceptors appear to mediate its actions (Fujii and Miyashita 1975). By contrast, melanophores of a siluroid catfish, *Silurus asotus*, which are unusual in that they receive cholinergic innervation (Sect. 5.2), respond to dopamine by dispersion of melanosomes (Miyashita and Fujii 1977). Specific receptors for dopamine may be involved, since haloperidol and fluphenazine effectively block the action of this amine.

Tyramine, another biogenic trace amine, was shown to act on the presynaptic portions of adrenergic nerves to cause release of adrenergic transmitter, which in turn caused aggregation of pigment in melanophores of the cuckoo wrasse (Scheline 1963). Robertson (1951) reported that histamine, an autacoid imidazole amine, had a slight pigment-aggregating effect on the melanophores of rainbow trout. Healey and Ross (1966) stated, however, that it had no effects on the melanophores of *Phoxinus* minnows. Miyashita and Fujii (1975) also found that it had no pigment-dispersing effect on the melanophores of guppies.

5-hydroxytryptamine (5-HT, serotonin), a bioactive indoleamine, has been examined for its action on chromatophores. It had a moderate pigment-aggregating effect on guppy melanophores (Miyashita and Fujii 1973). By contrast, it had no effect on melanophores of *Silurus* catfish (Fujii and Miyashita 1978). No pigment-dispersing action could be detected in guppy melanophores (Miyashita and Fujii 1975). The potent melanosome-aggregating action of the pineal hormone, melatonin, has been dealt with above (Sect. 5.1). Precursors of melatonin, 5-HT and *N*-acetyl-5-hydroxytryptamine, had no effects on *Silurus* melanophores, but two of its metabolites, 5-methoxytryptamine and 6-hydroxymelatonin had moderate pigment-aggregating effects (Fujii and Miyashita 1978).

Although expected to play some role, amino acids have only rarely been examined for their action on chromatophore systems. Recently, Miyashita and Fujii (1975) examined the abilities of various amino acids to aggregate or to disperse pigment in guppy melanophores, but none of the tested amino acids influenced the state of melanophores. However, further surveys on many varieties of fishes are needed if we are to identify amino acids that might function to modulate the control of chromatophore motility.

Robertson (1951) noted that adrenocortical steroids had practically no direct effects on melanophores. By contrast, Rasquin (1958) reported that cortical steroids had discernible effects on melanophores and xanthophores in some teleosts. Although the results are still confusing, the positive effects reported to date may be attributable to feedback inhibition of the release of ACTH or MSH from the pituitary (cf. Fujii and Oshima 1986).

The effects of sex steroids on the motility of chromatophores have also been examined. Olivereau and Olivereau (1985) injected 17-alpha-methyltestosterone into freshwater male eels (*Anguilla anguilla*). They detected darkening of the skin, and confirmed the dispersion of pigment in the melanophore by light microscopy.

In recent years, receptors for opioid peptides have been shown to exist in the brain, as well as in the peripheral tissues, of vertebrates. These peptides have also been shown to inhibit liberation of transmitters from nerve terminals. Therefore, a role for such neuropeptides in modulating the primary effects of endocrine or nervous signals on chromatophores might be expected. A possible relationship between endogenous opioid peptides and the secretion of MSH has been examined. From the first experiments on the chromatic system in fish, Satake (1980) reported that an intracranial injection of naloxone, a specific inhibitor of opiate receptors, induced aggregation of pigment in xanthophores of goldfish. The effect was antagonized by met-enkephalin (met-E). Levina and Gordon (1983) reported that melanophores and xanthophores of the zebrafish (*Brachydanio rerio*) responded to MSH and to met-E by dispersion of pigment, and that the effect of met-E developed later and faded more slowly. Naloxone inhibited the action of met-E. Apparently, a central mechanism is involved in the met-E-induced darkening of the skin. Recently, Carter and Baker (1980) reported that either the pars distalis or the neurointermediate lobe of the pituitary actually contains substantial opiate activity. Working on the melanophores of the guppy, Fujii (unpubl. data) detected a transient aggregation of melanosomes in response to enkephalins. At present, only little is known about the role of opioid peptides in the chromatic physiology of fish. However, given the proven importance of these peptides in the ethology of animals, it may not be unreasonable to assume that these peptides play some role in the control of integumentary color changes.

In higher vertebrates, various neuropeptides and local hormones have been identified and shown to affect neuronal activities. However, few experiments to detect the effects of these substances on chromatophores of fish have been reported. Very recently, endothelin was found to have a discernible pigment-aggregating effect on zebrafish melanophores (Fujii et al. unpubl.).

The possible role of prostaglandins (PGs), a group of physiologically active derivatives of fatty acids which act as local hormones that affect neurally controlled effector cells, is also of interest. In 1974, Abramowitz and Chavin noted that PGs produced dispersion of pigment in melanophores of black goldfish in vitro. Current progress in studies of the physiology of PGs includes the demonstration of their roles as local hormones. Some species of PGs have been shown to induce increases in levels of cAMP in effector cells. It is highly probable, therefore, that these derivatives of fatty acids take part in modulating the mechanisms that control the motility of chromatophores.

In a number of autonomic effector systems, presynaptic mechanisms for regulation of the liberation of neurotransmitters have been demonstrated. By contrast, only few studies have been reported on the nervous system controlling the chromatophores. Using the tilapia *Oreochromis niloticus*, Oshima (1989) recently succeeded in showing that adenine derivatives inhibit the release of adrenergic transmitter, possibly by decreasing the rate of entry of Ca^{2+} ions into the presynaptic portions of the fibers. To date, however, neither inhibition via alpha$_2$-adrenoceptors nor acceleration (presumably via beta-adrenoceptors) of the outflow of the transmitter has been proven.

As summerized above, a number of chemical substances have been suggested to modulate the actions of endocrine as well as nervous systems. However, a description of the exact roles and physiological significance of many of these compounds awaits future research. Possibly forming part of complicated but effectively organized networks of signal circuitry, many molecular species may participate in the exquisitely fine-tuned control of the motility of various chromatophores and in the production of the various hues of the integument.

5.4 Other Chemical Factors Influencing Chromatophores

Spaeth (1913) was the first to describe how a solution of KCl or other neutral salts elicited the aggregation of pigment in the melanophores of the killifish, *Fundulus heteroclitus*. Since then, several researchers have examined the effects of various inorganic ions on the chromatophores of fish (cf. Fujii 1959a, b, 1969). In accord with Spaeth's results, most later workers have reported that cations of alkaline and alkaline-earth metals have pigment-aggregating effects on melanophores of various species of teleosts. The pigment-aggregating action of these ions on chromatophores was considered to be direct. Primarily from observations that denervated melanophores were unresponsive to pigment-aggregating cations, and that the liberation of the sympathetic neurotransmitter was involved in the action on innervated cells, Fujii (1959a) concluded that these cations do not act directly on the melanophores but act on the presynaptic elements of the pigment-aggregating nerves. The liberated transmitter then brings about the aggregation of pigment in the melanophore. Fujii (1961) also found that the effect of K^+ ions could be blocked by treatment with the adrenergic blocking agent dibenamine. Furthermore, Fujii and Novales (1968) showed that the response to K^+ ions could still be observed after treatment with tetrodotoxin, which effectively blocked the conduction of pigment-aggregating nerves. The results also indicated that active electrical events due to the entry of Na^+ ions are not involved in the motile activity of the cells.

Fujii and Taguchi (1969) found that various pigment-aggregating or -dispersing drugs retained their effects even in K^+-rich saline, when the melanophores had previously been denervated, or when the release of the transmitter was inhibited. Their observations showed that the motile responses of chromatophores are quite independent of the electrical potential across the membrane or of electrical activities there, since the membrane is almost completely depolarized in the K^+-rich medium. The conclusion supports the current view that the stimulation of membrane receptors is transduced via G-proteins with resultant changes in the intracellular levels of second messengers (Sect. 2.2). Ionic channels in the cell membrane, which are directly associated with the changes in potential across the cell membrane, as well as ionic channels that are voltage-dependent, seem not to be involved in the process of signal transduction of pigment cells.

The mode of action of most alkaline metal and alkaline-earth ions other than K^+ has been shown to be identical to that of K^+ ions (Fujii 1959a). However, the mechanism underlying the pigment-aggregating action of Li^+ ions may be different because responses are observed only after a significant delay (Fujii 1959a). Namoto and Yamada (1983) observed similar effects in *Oryzias* melanophores and concluded that the sequence of events was due to the inhibition of adenylate cyclase by Li^+ ions.

The physiological-saline bathing solution represents the medium that surrounds cells in vivo and it acts as the environment around chromatophores. If we restrict ourselves to those for fishes, we have several recipes but all bathing solutions used are composed mainly of NaCl and water. The effects of Na^+ ions are, therefore, rather difficult to evaluate. When a piece of skin loaded with chromatophores is equilibrated in physiological saline, the chromatosomes usually become dispersed throughout the cytoplasm. By analogy with the relaxation of muscle cells in saline, the dispersion has been assumed to represent the resting state of chromatophores. If the cell is not exposed to any signaling cues, the chromatosomes may distribute themselves rather evenly within the cytosol as a result of the Brownian movement, a passive process. Nowadays, however, it has become rather commonly accepted that the dispersion of pigment in Na^+-rich medium is an active process that reverses aggregation (Fujii 1959b; cf. also Sects. 5.2, 5.4). In other words, the dispersion of pigment in saline may be maintained by an active mechanism. The MSH molecules remaining in the tissue space might be responsible for the dispersion. However, such hormonal substances can easily be washed out into the bathing solution. It is well known, incidentally, that the rapid blanching in saline of an excised piece of amphibian or reptilian skin is due to aggregation of pigment in melanophores caused by the efflux of MSH molecules from the tissue space. The pigment-aggregated state of the melanophores is, therefore, considered to be the resting state in amphibians and reptiles. Without residual MSH molecules, how can the melanosomes remain in a dispersed state in the melanophores of fish? The answer can be found in the innervation of these cells, to which the chromatophores of amphibians and reptiles are not subject.

Spontaneous quantal leakage of neurotransmitters from the presynaptic element is a rather common event, which can be confirmed electrophysiologically in excised tissues. The miniature endplate potentials recorded at motor endplates may be a good example of results of the spontaneous and random efflux of transmitter from motor terminals. Such events can reasonably be assumed to occur at the junctions between sympathetic neurons and chromatophores. Both true transmitter and co-transmitter are released at the same time. The leakage may last as long as the nerves remain functional, usually for no less than several hours. The true transmitter, norepinephrine (acetylcholine, in the case of siluroid catfish; cf. Sect. 5.2), is either quickly taken up by the nerves or is inactivated by metabolizing exoenzymes, namely, catechol-*O*-methyltransferase (COMT) in the case of norepinephrine, and acetylcholinesterase in the case of ACh. By contrast, the co-transmitter, ATP, even though it is gradually dephosphorylated to adenosine, survives much longer in the tissue space and functions actively to disperse pigment. The Na^+-rich bathing solution may provide a favorable milieu for the leakage, survival and the action of the co-transmitter. In addition, the re-uptake of the true transmitter might be accelerated in the presence of Na^+ ions around the chromatophores.

An increase in the level of intracellular Ca^{2+} ions is known to be critically involved in the motile activities of many types of cell, and chromatophores may be no exception (Sect. 2.2). For example, the process of aggregation of pigment in the erythrophores of the squirrelfish *Holocentrus ascensionis* (Luby-Phelps and Porter 1982) and in the melanophores of the medaka *Oryzias latipes* (Negishi and Obika 1985a) requires extracellular Ca^{2+} ions. In chromatophores of some other species, elevation of cytosolic concentrations of Ca^{2+} ions seems to be brought about by these liberation of the ions from the endoplasmic reticulum where they are stored, since the aggregation of pigment takes place in the absence of extracellular Ca^{2+} ions (Fujii and Taguchi 1969; Fujii et al. 1991b).

The liberation of neurotransmitters from pigment-aggregating nerves, by contrast, has a strict requirement for extracellular Ca^{2+} ions (Fujii and Novales 1972; Kasukawa and Fujii 1984). Mg^{2+} ions can partially replace Ca^{2+} ions. Ions for other alkaline earths, including Sr^{2+} and Ba^{2+}, can also restore the transmission. Nervous impulses depolarize presynaptic membranes. The depolarization results in the opening of Ca^{2+}-channels through which extracellular Ca^{2+} ions enter the presynaptic portion of the nerve. Presumably via a process that involves calmodulin, transmitters are released that subsequently induce the aggregation of pigment (Sect. 5.2).

Treatment of a piece of skin with Ba^{2+}-rich medium causes alternating cycles of aggregation and dispersion of pigment in chromatophores. Referred to as "pulsation", this interesting phenomenon was first analyzed by Spaeth (1913). It is now known that treatment with Ba^{2+} ions is not the necessary condition, and it is the lack of Ca^{2+} ions that is primarily responsible for the rhythmic activity. The phenomenon might, therefore, be regarded as corresponding to the spontaneous, repetitive contraction of muscle fibers seen in Ca^{2+}-free medium.

However, using the crucian carp *Carassius*, Watanabe (1961) showed that the pulsation of melanophores is due to the spontaneous, rhythmic liberation of pigment-aggregating transmitter from sympathetic fibers. Observing the synchronous pulsations of chromatophores on a scale of *Oryzias*, Iwata et al. (1981) concluded that the same fiber controls many of melanophores, xanthophores and leucophores which together constitute a motor unit.

Cationic choline, which has frequently been employed as a substitute for Na^+ ions was recently shown to have an actively pigment-dispersing effect (Oshima et al. 1992), and it seems to act inside the melanophores.

Inorganic and small organic anions have only infrequently been studied, probably because they appear not to be directly involved in cellular motility (Spaeth 1913; cf. Fujii 1969). However, the viability of the cells in vitro must be influenced to varying extents by various anions.

In addition to their characteristic agonistic or antagonistic actions, autonomic drugs often have significant side effects. For example, dibeamine and phenoxybenzamine, both alpha-adrenolytic haloalkylamines, have potent pigment-dispersing activity (Fujii and Taguchi 1969). Among other alpha-adrenolytics, ergotamine (Fujii 1961) and tolazoline (Miyashita and Fujii 1975), by contrast, each have melanosome-aggregating effects. Making good use of this property of the latter drug, Miyashita and Fujii (1975) were able to detect the beta-adrenoceptors that mediate dispersal of melanosomes in guppy melanophores.

A known parasympatholytic agent, atropine, has very effective melanosome-dispersing activity (Fujii 1960; Watanabe 1960). Scopolamine, a closely related compound, has the same effect. Nagaishi et al. (1992) showed that atropine dispersed leucosomes in *Oryzias* leucophores and induced the LR response in the motile iridophores of neon tetras. They concluded that the site of action of scopolamine is inside the cells. In the cholinergically controlled melanophores found in catfish, these belladonna alkaloids act as parasympatholytics in standard fashion (Sect. 5.2; Fujii et al. 1982; Kasukawa and Fujii 1985). Recently, experiments were performed in attempts to characterize the muscarinic receptors on the melanophores of catfish (Hayashi and Fujii 1991). Several synthetic muscarinic antagonists were employed and the receptors were characterized as being of the M_3 type.

For the analysis of mechanisms of chromatophore motility, various other drugs have also been exploited. Antibodies against actin were effective in blocking both the aggregation and the dispersion of pigment in swordtail erythrophores (Akiyama and Matsumoto 1983). Cytochalasin B (CB), an inhibitor of actin, has been used several times in studies of the possible involvement of actin filaments in the movement of chromatosomes, and marked inhibitory effects of this drug on melanophore motility has been reported by several workers, such as Obika (1976). By contrast, motile responses of *Oryzias* leucophores (Obika 1988), of *Xiphophorus* erythrophores (Matsumoto 1979) and of *Oreochromis* melanophores in culture (Oshima et al. 1990b) were not influenced by CB. CB partially inhibited the motility of iridophores of neon tetras (Oshima and Nagaishi 1992), while the iridophores of blue damselfish were insensitive to this drug (Oshima and Fujii 1987).

With regard to the action of alkaloids, a number of studies have been performed, and only a few works are cited here. In these studies, antimitotic drugs have frequently been employed. For example, colchicine markedly retarded the aggregation, but accelerated the dispersion of melanosomes in *Fundulus* melanophores (Wikswo and Novales 1969). Schliwa and Bereiter-Hahn (1973) confirmed the disturbance of pigment migration following the treatment of colchicine and vinblastine. Indirect immunofluorescence staining showed that the radial arrangement of microtubules within cultured melanophores of the tilapia disappeared after the cells had been exposed to colchicine or vinblastine (Oshima et al. 1990b). Both drugs reversibly inhibited the movements of motile iridophores of blue damselfish (Oshima and Fujii 1987) and of neon tetras (Oshima and Nagaishi 1992). Vinblastine at higher concentrations rendered melanophores punctate and immobile (Obika et al. 1978). In the dendritic processes of such immobile cells, large vinblastine crystals were frequently observed being formed. Both lumicolchicine and colchicine inhibited the aggregation of pigment granules in melanophores and xanthophores of *Fundulus* and *Oryzias*, and Obika et al. (1978) assumed that they blocked the translocation of pigment at the membrane level. These alkaloids are now known to interact with a component of plasma membrane.

Erythro-9-[3-(2-hydroxynonyl)]-adenine (EHNA), an inhibitor dynein ATPase, blocked the aggregation of pigment in melanophores of the medaka *Oryzias latipes* (Negishi et al. 1985) and in those of the tilapia *Oreochromis niloticus* (Oshima et al. 1990b). EHNA also inhibited both the aggregation and dispersion of leucosomes in *Oryzias* leucophores (Obika 1988). In addition, the extent of the LR responses of the motile iridophores of blue damselfish (Oshima and Fujii 1987) and of neon tetras (Oshima and Nagaishi 1992) was reduced by the inhibitor.

Nocodazole, a potent microtubule-disrupting agent, inhibited the Ca^{2+}-induced aggregation of melanophores inclusions in *Oryzias* melanophores (Negishi and Obika 1985a). Hexylene glycol had the same effect. Since numerous cytoskeletal elements were seen in random orientation after the treatment with nocodazole, its effect may be due to the modification of microtubular organization in the melanophores.

Methylxanthines, namely, caffeine and theophylline, inhibit the activities of cyclic-nucleotide phosphodiesterase which catalyzes the conversion of adenosine 3′:5′-cyclic monophosphate (cAMP) to inactive 5′-AMP. In many cases, the dispersion of chromatosomes due to methylxanthines has been explained as being a result of an increase in cytosolic levels of cAMP. However, methylxanthines are also known to block receptors for adenosine. Making use of this latter property, Oshima et al. (1986c), working with motile iridophores of the blue damselfish, showed that exogenously applied cyclic nucleotides act on receptors for adenosine on the cell membrane. Thus, the effects of cyclic nucleotides, as well as those of methylxanthines, on chromatophores have to be analyzed with care.

Forskolin, a stimulator of adenylate cyclase, induced the dose-dependent dispersion of melanosomes in *Oryzias* melanophores (Namoto and Yamada 1987).

Luby-Phelps and Porter (1982) utilized a Ca^{2+}-ionophore, A 23187, in their studies of the mechanism of erythrophore motility in squirrelfish, and they concluded that an experimental elevation of cytosolic levels in Ca^{2+} ions induced aggregation of erythrosomes. Negishi and Obika (1985a) confirmed this effect using the melanophores of *Oryzias*. Oshima et al. (1988) applied A 23187 to erythrophores and melanophores of the platyfish *Xiphophorus helleri* in culture, and obtained similar results. In these same chromatophores, W-7, an inhibitor of calmodulin, inhibited the aggregation of chromatosomes (Oshima et al. 1988). Furthermore, both the aggregation of pigment in melanophores and the LR response of iridophores of blue damselfish were effectively inhibited by the drug, while H-7, an inhibitor of protein kinase C (PKC), did not affect the responses of these chromatophores. This result suggests that PKC is not involved in the motile activity of the chromatophores (Oshima et al. 1988). These data indicate that the cellular motility is regulated by enzymes that are under the control of the Ca^{2+}-calmodulin system.

In the killifish *Fundulus heteroclitus*, Fujii and Novales (1968) showed that tetrodotoxin effectively blocked the conduction of melanin-aggregating fibers, but not the responsiveness of melanophores per se. Their results also indicated that Na^{+}-dependent action potentials are not involved in the responsiveness of the effector cells. Local anesthetics, such as procaine, dispersed melanophore inclusions in *Oryzias* (Watanabe 1961). Like tetradotoxin, procaine blocked the conduction of chromatic fibers (Fujii and Novales 1968).

Surfactants are known to have major effects on cells, by eluting lipid components from the cell membrane. In dilute solution, however, they have sometimes been suitable for the preparation of permeabilized chromatophores that are useful as model systems for the analysis of intracellular mechanisms (cf. Schliwa 1987). For example, Brij 58, a synthetic surfactant, has frequently been used for this purpose. Using Brij-permeabilized cells, Clark and Rosenbaum (1982), Oshima et al. (1990b) and other workers have examined the mechanochemical system in melanophores. Fujii et al. (1991b) used this procedure to introduce IP_3 into melanophores to confirm its action on Ca^{2+}-storage sites (Sect. 2.3).

Pandey et al. (1981) reported that an organophosphorus insecticide, malathion, caused the aggregation of pigment and a possible decrease in the melanin content of the skin of a tilapia *Oreochromis mossambicus*. However, the effects on chromatophores of such synthetic organic compounds, as well as the anticipated oligodynamic actions of heavy metal ions, have not been studied extensively (cf. Chavin 1972). Present in the drainage from dwellings or industrial plants, these synthetic compounds frequently pollute coastal waters as well as groundwater. Since they are located near the surface of fish, pigmentary systems are susceptible to the contaminating compounds in the environment. Pandey et al. (1981) suggested the possible usefulness of these pigmentary systems for assessing the extent of pollution. Of course, water pollution is a serious problem that affects the taste of aquatic animals for human consumption. Furthermore, marked declines in fish populations have been reported on occasion. In land waters, a number of invaluable species have been reported to have become extinct as a result of the pollution caused by human activities. Investigations into the effects of pollutants on pigmentary systems of fish may now be anticipated.

6 Effects of Lesions in the Nervous System

6.1 Lesions in the Central Nervous System

In both neural and hormonal regulation, external cues from the environment, as well as signals from various internal organs, are processed in the central nervous system to generate efferent signals directed at chromatophores. Although still only poorly understood, such central mechanisms have been touched upon in earlier articles to which the reader can refer (Iwata and Fukuda 1973; Fujii and Oshima 1986; Baker 1991).

In experiments related to fish vision, enucleation of lateral eyes has frequently been performed. In fish treated in this way, chromatic responses to background colors and patterns are impaired, but changes in coloration in response to environmental lightness can still be achieved (e.g., von Frisch 1911). Blindfolded fish behave similarly. It has been shown that, upon an increase in environmental lightness, the rate of secretion of melatonin from the pineal gland decreases, whereas this rate increases in the dark. As described in Sect. 5.1, circadian changes in color of live fish are induced in this way. Naturally, the influence of light on the chromatic responses of many fish are almost completely lost after pinealectomy or extirpation of the pineal region of the brain.

In order to follow the pigment-aggregating motor pathway, von Frisch (1911) made incisions in various parts of the central nervous system, as well as in peripheral regions en route to melanophores, in the minnow *Phoxinus*. As the criterion for identification of the motor connection from the center, he adopted the darkening of the skin that extended peripherally from the cut. He concluded that, from the center in the medulla, the tract passes posteriorly through the spinal cord. Then the fibers emerge at the level of the fifteenth vertebra through rami into bilateral sympathetic chains. Running anteriorly as well as posteriorly, they finally reach the melanophores. Fundamentally identical results have been obtained in some other teleosts (Gray 1956; Scott 1965).

With the intention of clarifying the central nervous mechanisms regulating chromatophore motility, Iwata and Fukuda (1973) performed experiments in which they stimulated parts of the central nervous system electrically. In other experiments, they transected the spinal cord and observed the complete darkening of the body. Their conclusion from extensive experiments was that, although the nervous circuits are rather complex in the central parts, mononeuronic sympathetic control of solely responsible for the nervous regulation of chromatophores.

6.2 Cutting of Peripheral Nerves

As touched upon above, severing of the peripheral nerves to the skin induces rapid darkening of the areas innervated by such nerves. This phenomenon is, of course, due to the dispersion of the dark pigmentary organelles, the melanosomes, within the melanophores in these areas. From their observations of such denervated dark bands in some teleostean and elasmobranch fish and other results, Parker and his colleagues came to the conclusion that the response was caused by the repetitive firing of putative parasympathetic melanin-dispersing fibers at the cut ends of neurons. For a number of references relevant to this "double innervation theory", readers can refer to a monograph by Parker (1948). Several later researchers expressed support for Parker's now discarded hypothesis (Fujii 1993b).

It has now been established that the nervous control of chromatophores in teleosts is exerted solely by the sympathetic component of the autonomic nervous system, and that the excitation of fibers induces the aggregation of pigment in light-absorbing chromatophores, the dispersion of leucophore inclusions, and the LR response of motile iridophores, with resultant blanching of skin (Sect. 5.2). Severing of fibers to the chromatophores naturally results in the interruption of central tonic influences on the effector cells. By analogy to the phenomenon of muscular relaxation due to damage to motor fibers, the consequent darkening of the skin can easily be understood to be the result of the dispersion of chromatosomes in light-absorbing chromatophores.

The idea that the generation of the dark band is a totally passive process may not be totally correct, as we can see when we consider the sequence of events in more detail. Melanophore-stimulating hormone (MSH) may act to disperse pigment in melanophores even within areas in which efferent fibers had been disconnected from the center, because an anastomosing supply of capillaries remains intact around the chromatophores. Until the time at which the disconnected peripheral nerves lose their functions, furthermore, nervous transmitters leak out through the presynaptic membranes (cf. Sect. 5.2). The true transmitter, norepinephrine, soon disappears, while the adenine-derived co-transmitter, which persists for a longer time around the chromatophores, may actively cause dispersion of the dark pigment. This phenomenon lasts for about a day under normal conditions. After the functional degeneration of the axonal cylinder distal to the cut, chromatophores are mainly under the influence of MSH. Upon prolonged adaptation of the fish to a white background, the dark band gradually becomes pale like the rest of the skin, as Parker (1948) originally reported. An increased titer of melanin-concentrating hormone (MCH) may be involved in this process.

Working on the gluttonous goby *Chasmichthys* and the Nile tilapia *Oreochromis*, Fujii et al. (1992) recognized that denervated dark band becomes pale during the night. Microscopic observations confirmed the aggregation of pigment in the melanophores in the band region. Even in the daytime, the same reversal of the coloration of the band was inducible by an intraperitoneal injection of melatonin. Fujii et al. concluded that the blanching of the band during the night was due to the elevated titer of circulating melatonin (Sect. 5.1).

With an incision on the skin or on part of a fin, denervated chromatophores can be obtained rather easily and are distributed posteriorly from the site of the cut. Chromatophores on a scale can also be denervated by pulling the scale out and reinserting it into the same pocket. Some researchers have been using chromatophores on scales that have been pulled out and kept immersed in physiological saline for more than a day as denervated chromatophores (Iwata et al. 1959).

Observations of the denervated band on the fin allow the mechanisms that underlie the generation and lightening of the dark area to be analyzed in great detail because the fibers that innervate the chromatophores run almost parallel to the fin-rays. Sympathectomy of chromatophores by injection of 6-hydroxydopamine is also successful in some cases (Iga and Takabatake 1982).

Denervation of chromatophores has frequently been reported to cause hypersensitization to pigment-motor substances (Fujii 1960, 1961; Fujii et al. 1992). For example, Fujii (1961) reported hypersensitivity of melanophores of the gluttonous goby to norepinephrine, epinephrine and melatonin. As many experimenters have recognized, the spontaneous aggregation of melanosomes is also very frequently seen when melanophores are deprived of a nervous supply (Fujii unpubl. observ.). Such observations might be related to the loss of spontaneously released adenylic co-transmitter from sympathetic fibers, which actively disperses melanosomes in innervated melanophores (Sect. 5.2). Hypersensitization may also be explained by the same process.

7 Ecological and Ethological Factors Influencing Chromatophores

The ethological significance of integumentary coloration and the changes in such coloration in animals, including fish, have been discussed in earlier publications such as those by Cott (1940), Baerends and Baerends-van Roon (1950) and Needham (1974). Very recently, one of the present authors also presented a concise description of these issues, to which readers can refer (Fujii 1993a).

During the phylogeny of fish, various ecological factors in their habitats must have had an adaptive influence on the radiation of species, and, naturally, such influences must have affected the variations in colorations as well as the colored patterns of their integument. Even when we restrict our discussion to one species of fish, differences in environmental conditions continue to result in rather wide variations in phenotypic characteristic, which can be considered to be regional variations in pigmentation and can be categorized as a kind of morphological color change. Variations in patterns of coloration are frequent among fish from different habitats. Sometimes, however, modifications of the systems for physiological color changes may have developed over a long period of time, for example, alterations in the relative weight of the hormonal and nervous control of chromatophores. The conversion from adrenergic to cholinergic of the peripheral transmission to chromatophores, as touched upon in Sect. 5.2, may also be a phenomenon to be included in this category. In any case, in their natural environment, rapidly moving fish are exposed to almost instantaneous and dramatic changes in photic conditions, which include changes in brightness, albedo, background patterns and so on. By contrast, the conditions around other fish, such as bottom dwellers or oceanic migratory fish, do not fluctuate as much and do not require that the fish have the ability to change their hues rapidly and widely. In this way, both differences in behavioral patterns among fish and variations in environmental factors should affect the nature of the systems that control the physiological color changes.

Chromatic responses in vivo during adaptation to various backgrounds have been dealt with in relation to the photic environment (Sects. 3.1 and 3.2). There, environmental photic information is perceived mainly by the lateral eyes and partly via extraoptical pathways. In the latter process, the light is perceived by the pineal gland (Sect. 5.1) or by the chromatophores themselves (Sect. 3.3). Except in the latter case, the input information is sent to the central nervous system, where it is integrated and processed. Efferent signals are then transmitted hormonally (Sect. 5.1) and/or via sympathetic pathways (Sect. 5.2) to the effector cells, the chromatophores. The photic environmental conditions have a critical influence on the chromatic reactions of the fish. The physiological responses include background adaptation, circadian changes in color and the changes required for coping with ethological encounters.

Being probably the best-known chromatic strategy for survival of an animal is probably "cryptic" or "concealing" coloration (cf. Fujii 1993a). With such coloration, predators can approach their prey unnoticed. For prey animals, such coloration allows them to escape from their predators. In the latter case, the coloration is more popularly known as "protective coloration". These coloration phenomena include the body hues and the pigmentary patterns. Flatfishes, gobies and many other benthic fish adapt so well to their background that they are hardly visible, even in small aquaria. Incidentally, their cryptic colors and colored patterns are often static and do not exhibit detectable changes. In such cases, these phenomena have often been considered to be examples of "concealing mimicry", or "mimesis". The terms mimicry and mimesis usually apply to the resemblance of an animal to some other object, either animate or inanimate, not only in terms of colors and color patterns, but also in terms of form, posture, movement, and even smell. It is, therefore, more appropriate to use the expression "chromatic mimesis". A good example among many reported phenomena is that of the leaf fish (e.g., *Monocirrhus polyacanthus*; Nandidae) which mimic withered leaves.

Dynamic physiological changes for adaptation to background are also very common. On darker backgrounds, fish become darker, while on lighter backgrounds they become paler. These color changes proceed not only homogeneously on the skin but often also heterogeneously. When adapting to a mottled background, flatfish become similarly mottled within few seconds. Although usually to a lesser extent, flatheads and gobies, for example, can adapt to the substratum in the same way. The darkening of a part of skin is due to the dispersal of chromatosomes within the light-absorbing chromatophores, whereas the blanching is due to the aggregation of the pigment. Sometimes the light-reflecting chromatophores, namely the leucophores and motile iridophores, function to supplement the effects of the light-absorbing chromatophores via movements of organelles in the opposite direction (Sects. 2.1.4, 2.1.5). As detailed above (Sect. 5.2), rapid aggregation of chromatosomes in light-absorbing chromatophores and dispersion of light-reflecting organelles are due to the action of the true transmitter of the sympathetic postganglionic fibers, norepinephrine, whereas the effects of the true transmitter are reversed by the action of adenosine, the co-transmitter from the same fibers. More gradual color changes during adaptation to background are mainly due to effects of the endocrine systems. Melanophore-stimulating hormone is responsible for darkening, while melanin-concentrating hormone is now considered to be involved in blanching (Sect. 5.1; cf. also Fujii and Oshima 1986 ; Fujii 1993b).

Being rather independent of the background chromatic adaptation, a circadian change in hues or patterns can be detected in many species of fish: during the night, the pineal gland secretes melatonin at a much higher rate than in the daytime. Thus, the chromatosomes in light-absorbing chromatophores become aggregated during the night, and the body skin blanches (Sect. 5.1). Contrary to the general principle that the light-absorbing chromatophores and the reflecting ones behave in opposite ways, the light-scattering organelles in the leucophores of a small freshwater fish, the medaka *Oryzias latipes*, actually aggregate in response to melatonin (Obika 1976, 1988). That is, the pigmentary organelles of all three species of chromatophore, namely, the melanophores, the xanthophores and the leucophores, in the skin of the medaka aggregate during the night. Such a situation results in the skin becoming transparent. Presumably, such nocturnal transparency functions to lower the possibility of the fish being located by predators.

Regional differences in the sensitivity of chromatophores to melatonin have frequently been reported (Sect. 5.1; cf. Fujii and Oshima 1986). Reed (1968) showed that the characteristic "night" pattern of the pencilfish *Nannostomus beckfordi* could be produced by immersing the fish in a melatonin-containing solution. These observations indicate that melatonin is involved in the formation of color patterns in many fish. Within the region where the skin blanches in response to melatonin, chromatophores may be richly endowed with receptors for melatonin, while in the areas where blanching does not take place, receptors for melatonin may be sparse on the cell membranes of chromatophores.

One kind of "aposematic" or "advertisement" coloration is the "recognition" coloration that is used for mutual identification among individuals of the same species. Coloration of this type functions as a releaser or a key stimulus for intraspecific social interactions (cf. Fujii 1993a). Among such visual signals, there are many examples of static colors and color patterns. In many highly visual, social species of teleosts, by contrast, there are also instances where rapid changes in colors and patterns are part of a strategy for coping with various ethological encounters, including territorial fights against other males, attraction of mates, and communication with young. Cichlid species, in particular, exploit such strategies (e.g., Baerends and Baerends-van Roon 1950). For example, Muske and Fernald (1987) showed that the facial dark stripe, the "eyebar", which signals territorial ownership and aggressive intent in territorial males of the mouth-brooder tilapia, *Haplochromis burtoni*, appears and disappears very rapidly, and that these changes are due to the sympathetic activity. In fishes belonging to groups other than Cichlidae, similar examples have also been reported. Kohda and Watanabe (1982) recently showed that, during agonistic behaviors, the freshwater serranid *Siniperca kawamebari* displays a number of patterns depending on circumstances. Another example was reported by Naitoh et al. (1985), who studied the common freshwater goby, *Rhinogobius brunneus*. While examining the adaptation to black-and-white checkerboard backgrounds, they found that numerous fibers regulate dermal melanophores differentially and systematically. Thus, the rapid changes in color exploited for intraspecific communication are mostly achieved via nervous control of melanophores.

The brilliant coloration of a coral-reef dweller, the blue damselfish *Chrysiptera cyanea*, changes under certain ethological conditions. Usually, these fish are cobalt-blue in color, as indicated by their common name. Sometimes, however, they become very dark with shades of violet, while at other times they are greenish yellow. Studies on the light reflectance from the skin indicate that the peak of the spectral reflectance shifts to the near ultraviolet (UV) region, and to a longer-wavelength region, i.e., green to yellow (Kasukawa et al. 1987). These changes are due to the motile responses of iridophores just under the epidermis (Sects. 1.2.5, 1.3), which are regulated by the sympathetic nervous system (Sect. 5.2). Recent studies on the blue-green stripe of the neon tetra *Paracheirodon innesi* also revealed that similar changes occur in vivo as well as in vitro (Nagaishi et al. 1990). In the neon tetra, however, the changes are primarily due to the effect of light on the light-sensitive motile iridophores (Sect. 3.3), although the sympathetic system acts to modify the cellular activities. In neither case has it been clarified whether the fish perceive reflected UV light from the skin of mates as visual signals.

Another kind of recognition coloration is the nuptial coloration that is seen mostly in males of various species of fish during the breeding season and is usually restricted to part of the body surface. Categorized as a morphological color change, nuptial coloration has been shown to be induced by the action of androgenic steroid hormones (Bagnara and Hadley 1973).

It is known that, when excited, some animals undergo blanching while others darken. These conditions are called "excitement pallor" and "excitement darkening", respectively, but the physiological significance of these responses remains to be clarified. The direction of the reaction seems to depend on the species. Physiological analyses utilizing modern techniques have not been carried out. It may be safe to assume, however, that both phenomena are results of a sudden outflow of catecholamines (mainly epinephrine) from the chromaffin cells of adrenal tissues (cf. Fujii and Oshima 1986). If the adrenoceptors on integumentary melanophores of a certain species are of the alpha type, the response would be the aggregation of melanosomes, and blanching would be observed. Conversely, if the receptors are mostly of the beta type, the response would be observed as darkening. Further correlated ethological and physiological analyses are needed to elucidate the mechanisms responsible for these interesting phenomena.

8 Conclusions and Perspectives

The reader may have been surprised to learn that so many factors, both intrinsic and extrinsic, can influence the distribution of pigment in the chromatophores of fish. Individual chromatophores are endowed with various receptors and other devices for receiving numerous cues. If we consider only the biogenic mediators that affect the chromatophores, we can enumerate many that correspond to specific receptors on these cells (Sects. 5.1, 5.2, 5.3). In addition, several, types of chromatophore are sensitive to light and other physical factors. They must also possess systems that allow them to react to such cues (Sects. 3, 4).

Why have chromatophores developed to perceive such a large number of signals? The answer to this question must be sought in the roles of the chromatophores in the adaptive behaviors of the fish (Sect. 7). That is, variously defined types of coloration, for example, cryptic and aposematic, as well as the changeable patterns employed for visual communication, must be of the utmost importance to the survival of fish in a broad sense. Fish apparently need exquisitely fine-tuned mechanisms for controlling their integumental colored units, in particular when complicated patterns of coloration are regulated. Such requirements may also explain why melanophores, among the various chromatophores, usually possess more of species of receptors than other cells: the melanophores commonly play the most important part in generating the remarkable and yet subtle changes in hues or shades, as well as in color patterns.

Chromatophore systems have many advantages as systems from which to glean a more profound understanding of various biological processes. The rapid responses of chromatophores allow us to observe the influence of a hormonal or a neuronal substance with ease under the microscope. Moreover, single cells possess receptors of several types, as mentioned above. The integrating mechanisms inside a cell for processing of multiple inputs can be studied by investigating paraneurons of this type. Exploiting such features, we may be able to resolve many difficult problems in various areas of biology. We can anticipate that the knowledge obtained from the chromatophore-controlling systems will facilitate the elucidation of many, more general physiological processes.

Aquarists have been making efforts to obtain more beautiful, fancy strains of aquarium fish, mostly by mating suitable pairs or by locating individuals with spontaneous mutations. Currently, the techniques of genetic engineering are allowing remarkable progress to be made in the alteration of phenotypic characteristics via introduction of foreign genes. In the pigmentary system, the detection of phenotypes is rather easy because they can be confirmed visually. Microinjection or electroporation techniques may prove useful for the transfer of genes for expression of desirable colors or patterns from other species, not only within the same genus, but even from species belonging to other taxa. Recently, such experiments have actually been performed on some species, including the medaka *Oryzias latipes*: the gene for tyrosinase, namely the key enzyme for the synthesis of melanin, cloned from the mouse, was injected into unfertilized eggs of a pale variety of this species, and apparent melanization was detected in amelanotic melanophores (Matsumoto et al. 1991). That is, transgenic dark individuals with active mammalian tyrosinase were obtained. Hereafter, other genetic factors that affect coloration will surely be utilized in such experiments. Furthermore, techniques advanced by experimentation on the pigmentary systems should be applicable to other genetic characteristics that are more difficult to manipulate.

The information in the present article may be of use to those engaged in ethological, ecological, or even environmental studies, in particular in such studies of aquatic animals. To date, the progress made in physiological studies of chromatophores at the cellular level has not been recognized or taken into account by many researchers in the fields of behavioral and organismal biology. Therefore, we hope that the present review will provide useful clues towards an understanding at the physiological or cellular level, of phenomena that are presently being examined at the organismal level.

Acknowledgments. We thank Dr. M. Sugimoto for his help during the preparation of the manuscript. This work was supported in part by grants from the Ministry of Education, Science and Culture of Japan, to N.O. (nos. 03640617 and 04640674) who wrote Sections 3 and 4 and to R.F. (nos. 02640570 and 04640675) who wrote the other parts of this article.

References

Abramowitz J, Chavin W (1974) In vitro effects of prostaglandins upon melanosome dispersion in the skin of black goldfish *Carassius auratus* L. Prostaglandins 4: 805–818

Abrao MS, Castrucci AML, Hadley ME, Hruby VJ (1991) Protein-kinase C mediates MCH signal transduction in teleost, *Synbranchus marmoratus* melanocytes. Pigm Cell Res 4: 66–70

Akiyama T, Matsumoto J (1983) The blockade of pigment displacement in swordtail erythrophores by microinjection of antiactin antibody. J. Exp Zool 227: 405–411
Andersson RGG, Karlsson JO, Grundström N (1984) Adrenergic nerves and alpha$_2$-adrenoceptor system regulating melanosome aggregation within fish melanophores. Acta Physiol Scand 121: 173–179
Ando S (1962) responses of embryonic melanophores of the wild medaka (*Oryzias latipes*) to various stimuli. Embryologia 7: 169–178
Baerends G, Baerends-van Roon JM (1950) An introduction to the study of the ethology of cichlid fishes. Brill. Leiden
Bagnara JT, Hadley ME (1973) Chromatophores and color change. Prentice-Hall, Englewood Cliffs
Baker BI (1991) Melanin-concentrating hormone: a general vertebrate neuropeptide. Int Rev Cytol 126: 1–47
Baker BI, Ball JN (1975) Evidence for a dual pituitary control of teleost melanophores. Gen Comp Endocrinol 25: 147–152
Baker BI, Wilson JF, Bowley TJ (1984) Changes in pituitary and plasma level of MSH in teleosts during physiological colour change. Gen Comp Endocrinol 55: 142–149
Bitensky MW, Miki N, Keirns JJ, Keirns M, Baraban JM, Freeman J, Wheeler MA, Lacy J, Marcus FR (1975) Activation of photoreceptor disk membrane phosphodiesterase by light and ATP. In: Drummond GI, Greengard P, Robison GA (eds) Advances in cyclic nucleotide research, vol 5. Raven Press, New York, pp 213–240
Carter DA, Baker BI (1980) The relaionship between opiate concentration and cellular activity in part distalis and neurointermediate lobe of the eel (*Anguilla anguilla*) pituitary. Gen Comp Endocrinol 41: 225–232
Chavin W (ed) (1972) Responses of fish to environmental changes. Thomas, Springfield
Clark TG, Rosenbaum JL (1982) Pigment particle translocation in detergent-permeabilized melanophores of *Fundulus heteroclitus*. Proc Natl Acad Sci USA 79: 4655–4659
Colley AM, Hunt EL (1974) Cyclic AMP and adrenergic receptors in *Gambusia affinis* melanophore response in vitro. Gen Comp Endocrinol 23: 154–163
Cott HB (1940) Adaptive coloration in animals. Oxford University Press, New York
Duspiva F (1931) Beiträge zur Physiologie der Melanophoren von Fischembryonen. Sitzungsber Akad Wiss Wien Math-Naturwiss Kl I 140: 553–596
Egami N, Etoh H, Tachi C, Aoki K, Arai R (1962) Role of the pituitary gland in melanization in the skin of the goldfish, *Carassius auratus*, induced by X-ray irradiation. Proc Jpn Acad 38: 345–347
Enami M (1955) Melanophore-concentrating hormone (MCH) of possible hypothalamic origin in the catfish, *Parasilurus asotus*. Science 121: 36–37
Etoh H (1963) The effects of γ-irradiation on the physiological response of the nerve-melanophore system in *Carassius carassius*. Zool Mag 72: 277–282 (in Japanese)
Etoh H, Egami N (1963) Responses to Na^+, K^+, atropine and adrenaline of melanophores induced by X-irradiation in the fin of the goldfish, *Carassius auratus*. Annot Zool Jpn 36: 133–139
Foster KW (1933) Color changes in *Fundulus* with special references to the color change of the iridosomes. Proc Natl Acad Sci USA 19: 535–540
Fox DL (1979) Biochromy. University California Press, Berkeley
Fries EFB (1931) Color change in *Fundulus*, with special consideration of the xanthophores. J Exp Zool 60: 384–426
Fries EFB (1958) Iridescent white reflecting chromatophores (antaugophores, iridoleucophore) in certain teleost fishes, particularly in *Bathygobius*. J Morphol 103: 203–242
Fujii R (1959a) Mechanism of ionic action in the melanophore system of fish I. Melanophore-concentrating action of potassium and some other ions. Annot Zool Jpn 32: 47–59
Fujii R (1959b) Mechanism of ionic action in the melanophore system of fish II. Melanophore-dispersing action of sodium ions. J Fac Sci Univ Tokyo Sect IV 8: 371–380
Fujii R (1960) The seat of atropine action in the melanophore-dispersing system of fish. J Fac Sci Univ Tokyo Sect IV 8: 643–657

Fujii R (1961) Demonstration of the adrenergic nature of transmission at the junction between melanophore-concentrating nerve and melanophore in bony fish. J Fac Sci Univ Tokyo Sect IV 9: 171–196
Fujii R (1969) Chromatophores and pigments. In: Hoar WS, Randall DJ (eds) Fish physiology, vol 3. Academic Press, New York, pp 307–353
Fujii R (1993a) Coloration and chromatophores. In: Evans DH (ed) The physiology of fishes. CRC Press, Boca Raton, pp 535–562
Fujii R (1993b) Cytophysiology of fish chromatophores. Int Rev Cytol 143: 191–255
Fujii R, Miyashita Y (1975) Receptor mechanisms in fish chromatophores. I. Alpha nature of adrenoceptors mediating melanosome aggregation in guppy melanophores. Comp Biochem Physiol 51C: 171–178
Fujii R, Miyashita Y (1976a) Beta adrenoceptors, cyclic AMP and melanosome dispersion in guppy melanophores. In: Riley V (ed) Pigment cell, vol 3. Karger, Basel, pp 336–344
Fujii R, Miyashita Y (1976b) Receptor mechanisms in fish chromatophores. III. Neurally controlled melanosome aggregation in a siluroid (*Parasilurus asotus*) is strangely mediated by cholinoceptors. Comp Biochem Physiol 55C: 43–49
Fujii R, Miyashita Y (1978) Receptor mechanisms in fish chromatophores. IV. Effects of melatonin and related substances on dermal and epidermal melanophores of the siluroid, *Parasilurus asotus*. Comp Biochem Physiol 59C: 59–63
Fujii R, Miyashita Y (1979) Photoelectric recording of motile responses of fish leucophores. Annot Zool Jpn 53: 87–94
Fujii R, Miyashita Y (1980) Action of melanophore-stimulating hormone on dermal and epidermal melanophores of siluroid, *Parasilurus asotus*. Yale J Biol Med 5: 422
Fujii R, Miyashita Y (1982) Receptor mechanisms in fish chromatophores. V. MSH disperses melanosomes in both dermal ad epidermal melanophores of catfish (*Parasilurus asotus*). Comp Biochem Physiol 71C: 1–6
Fujii R, Novales RR (1968) Tetrodotoxin: Effects on fish and frog melanophores. Science 160: 1123–1124
Fujii R, Novales RR (1972) Nervous control of melanosome movements in vertebrate melanophores. In: Riley V (ed) Pigmentation: its genesis and biologic control. Appleton-Century-Crofts, New York, pp 315–326
Fujii R, Oshima N (1986) Control of chromatophore movements in teleost fishes. Zool Sci 3: 13–47
Fujii R, Taguchi S (1969) The responses of fish melanophores to some melanin-aggregating and dispersing agents in potassium-rich medium. Annot Zool Jpn 42: 176–182
Fujii R, Nakazawa T, Fujii Y (1973) Effects of ultraviolet radiation on melanophore system of fish. In: McGovern VJ, Russel P (eds) Pigment cell, vol 1. Karger, Basel, pp 195–201
Fujii R, Miyashita Y, Fujii Y (1982) Muscarinic cholinoceptors mediate neurally evoked pigment aggregation in glass catfish melanophores. J Neural Transm 54: 29–39
Fujii R, Oshima N, Miyashita Y (1985) Receptor mechanisms in fish chromatophores. VIII. Mediated by beta adrenoceptors, catecholamines always act to disperse pigment in siluroid melanophores. Comp Biochem Physiol 81C: 1–6
Fujii R, Kasukawa H, Miyaji K, Oshima N (1989) Mechanism of skin coloration and its changes in blue-green damselfish, *Chromis viridis*. Zool Sci 6: 477–486
Fujii R, Hayashi H, Toyohara J, Nishi H (1991a) Analysis of the light reflection from motile iridophores of the dark sleeper, *Odontobutis obscura obscura*. Zool Sci 8: 461–470
Fujii R, Wakatabi H, Oshima N (1991b) Inositol 1,4,5-trisphosphate signals the motile response of fish chromatophores. I. Aggregation of pigment in the tilapia melanophore. J Exp Zool 259: 9–17
Fujii R, Sugimoto M, Oshima N (1992) Blanching at night of denervated bands in teleostean tail fins is due to pigment aggregation in melanophores by melatonin. Comp Biochem Physiol 101A: 29–32
Gray EG (1956) Control of the melanophores of the minnow (*Phoxinus phoxinus* (L.)). J Exp Biol 33: 448–459
Hanyu I, Niwa H (1970) Pineal photosensitivity in three teleosts, *Salmo irideus*, *Plecoglossus altivelis* and *Mugil cephalus*. Rev Can Biol 29: 133–140

Hayashi H, Fujii R (1991) Muscarinic cholinoceptors mediate pigment aggregation within melanophores in some cyprinid fish, *Zacco*. Zool Sci 8: 1055
Hayashi H, Sugimoto M, Oshima N, Fujii R (1993) Motile response of erythrophores in the red abdominal skin of tetra fishes and its possible significance in chromatic adaptation. Pigm Cell Res 6: 37–44
Healey EG, Ross DM (1966) The effects of drugs on the background response of the minnow *Phoxinus phoxinus* L. Comp Biochem Physiol 19: 545–580
Iga T (1968) Action of catecholamines on the melanophores in the teleost. Zool Mag (Tokyo) 77: 19–26 (in Japanese)
Iga T (1979) Alpha adrenoceptors: pigment aggregation in *Oryzias* leucophores. Mem Fac Sci Shimane Univ 13: 87–95
Iga T (1983) Electric stimulation experiments on leucophores of a freshwater teleost, *Oryzias latipes*. Comp Biochem Physiol 74C: 103–108
Iga T, Matsuno A (1986) Motile iridophores of a freshwater goby, *Odontobutis obscura*. Cell Tissue Res 244: 165–171
Iga T, Takabatake I (1982) Action of melanophore-stimulating hormone on melanophores of the cyprinid fish *Zacco temmincki*. Comp Biochem Physiol 73C: 51–55
Iga T, Takabatake I (1983) Melanophores of *Zacco temmincki*. (Teleostei) are light sensitive. J Exp Zool 227: 9–14
Iga T, Takabatake I (1986) Local light stimulation of melanophores of a teleost, *Zacco temmincki*. J Exp Zool 238: 385–391
Iga T, Yamada K, Iwakiri M (1977) Adrenergic receptors mediating pigment dispersion in leucophores of a teleost, *Oryzias latipes*. Mem Fac Lit Sci Shimane Univ Nat Sci 11: 63–72
Imaki-Lamer H, Chavin W (1975) Ultrastructure of the integumental melanophores of the coelacanth, *Latimeria chalumnae*. Cell Tissue Res 163: 383–394
Iwata KS, Fukuda H (1973) Central control of color changes in fish. In: Chavin W (ed) Responses of fish to environmental changes. Thomas, Springfield, pp 316–341
Iwata KS, Watanabe M, Kurihara T (1959) Changes of state and response of the fish scale melanophore during continuous immersion in Ringer's solution. Biol J Okayama Univ 5: 185–194
Iwata KS, Takahashi T, Okada Y (1981) Nervous control in chromatophores of the medaka. In: Seiji M (ed) Pigment cell 1981: phenotypic expression in pigment cells. University Tokyo Press, Tokyo, pp 433–438
Kamada T, Kinosita H (1944) Movements of granules in fish melanophores. Proc Imp Acad Tokyo 20: 484–492
Kasukawa H, Fujii R (1984) Potassium ions act to release transmitter from "cholinergic" sympathetic postganglionic fiber to the glass catfish melanophore. Zool Sci 1: 553–558
Kasukawa H, Fujii R (1985) Receptor mechanisms in fish chromatophores. VII. Muscarinic cholinoceptors and alpha adrenoceptors, both mediating pigment aggregation, strangely coexist in *Corydoras* melanophores. Comp Biochem Physiol 80C: 211–215
Kasukawa H, Oshima N, Fujii R (1986a) Control of chromatophore movements in dermal chromatic units of blue damselfish. II. The motile iridophore. Comp Biochem Physiol 83C: 1–7
Kasukawa H, Oshima N, Fujii R (1986b) A comparative survey on the type of sympathetic neuro-melanophore transmission in catfishes. Comp Biochem Physiol 85C: 115–120
Kasukawa H, Oshima N, Fujii R (1987) Mechanism of light reflection in blue damselfish motile iridophore. Zool Sci 4: 243–257
Kawai I (1989a) Light sensitive response of the scale xanthophores of a teleost, *Oryzias latipes*. Med Biol 118: 93–97 (in Japanese)
Kawai I (1989b) Relation between light sensitive response of carotenoid droplet and calcium on a scale xanthophores of the medaka, *Oryzias latipes*. Med Biol 118: 141–145 (in Japanese)
Kawauchi H, Kawazoe I, Tsubokawa M, Kishida M, Baker BI (1983) Characterization of melanin-concentrating hormone in chum salmon pituitaries. Nature 305: 321–323
Kohda Y, Watanabe M (1982) Agonistic behavior and color pattern in a Japanese freshwater serranid fish, *Coreoperca kawamebari*. Zool Mag (Tokyo) 91: 61–69

Komatsu K, Yamada K (1982) Autordiographic visualization of beta adrenergic receptors in fish melanophores. J Exp Zool 223: 185–188
Kumazawa T, Fujii R (1984) Concurrent releases of norepinephrine and purines by potassium from adrenergic melanosome-aggregating nerve in tilapia. Comp Biochem Physiol 78C: 263–266
Kumazawa T, Fujii R (1986) Fate of adenylic co-transmitter released from adrenergic pigment-aggregating nerve to tilapia melanophore. Zool Sci 3: 599–603
Kumazawa T, Oshima N, Fujii R, Miyashita Y (1984) Release of ATP from adrenergic nerves controlling pigment aggregation in tilapia melanophores. Comp Biochem Physiol 78C: 263–266
Land MF (1972) The physics and biology of animal reflectors. Prog Biophys Mol Biol 24: 75–106
Lerner AB, Case JD, Takahashi Y, Lee TH, Mori W (1958) Isolation of melatonin; the pineal gland factor that lightens. J Am Chem Soc 80: 2587
Levina S, Gordon R (1983) Methionine enkephalin-induced changes in pigmentation of zebrafish (Cyprinidae, *Brachydanio rerio*) and related species and varieties, measured videodensitometrically. Gen Comp Endocrinol 51: 370–377
Luby-Phelps K, Porter KR (1982) The control of pigment migration in isolated erythrophores of *Holocentrus ascensionis* (Osbeck). II. The role of calcium. Cell 29: 441–450
Lythgoe JN, Shand J (1982) Changes in spectral reflexions from the iridophores of the neon tetra. J Physiol 325: 23–34
Lythgoe JN, Shand J (1983) Diel colour changes in the neon tetra, *Paracheirodon innesi*. Environ Biol Fishes 8: 249–254
Lythgoe JN, Shand J, Foster RG (1984) Visual pigment in fish iridocytes. Nature 308: 83–84
Marsland DA (1944) Mechanism of pigment displacement in unicellular chromatophores. Biol Bull 87: 252–261
Marsland DA, Meisner D (1967) Effects of D_2O on the mechanism of pigment dispersal in the melanocytes of *Fundulus heteroclitus*. A pressure-temperature analysis. J Cell Physiol 70: 209–216
Matsumoto J (1979) Morphological and physiological events associated with pigment translocation in swordtail fish erythrophores. In: Klaus SN (ed) Pigment cell, vol 4, Karger, Basel, pp 56–63
Matsumoto J, Watanabe Y, Obika M, Hadley ME (1978) Mechanisms controlling pigment movements within swordtail (*Xiphophorus helleri*) erythrophores in primary culture. Comp Biochem Physiol 61A: 509–517
Matsumoto J, Akiyama T, Hirose Eu, Nakamura M, Yamamoto H, Takeuchi T (1991) Expression and transmission of the melanogenic phenotypes in the orange-colored mutant of medaka fish introduced with mouse tyrosinase gene. Zool Sci 9: 1081
Miyashita Y, Fujii R (1973) Responses of guppy melanophores to 5-hydroxytryptamine. J Pre-Med Course Sapporo Med Coll 14: 39–44
Miyashita Y, Fujii R (1975) Evidence for beta adrenoceptors mediating melanosome dispersion in guppy melanophores. Comp Biochem Physiol 51C: 179–187
Miyashita Y, Fujii R (1977) Evidence for dopamine receptors mediating pigment dispersion in the melanophore of the siluroid. *Parasilurus asotus*—a preliminary note. J Pre-Med Course Sapporo Med Coll 18: 67–70
Miyashita Y, Kumazawa T, Fujii R (1984) Receptor mechanisms in fish chromatophores. VI. Adenosine receptors mediate pigment dispersion in guppy and catfish melanophores. Comp Biochem Physiol 77C: 205–210
Morishita F (1987) Responses of the melanophores of the medaka, *Oryzias latipes*, to adrenergic drugs: Evidence for involvement of $alpha_2$ adrenergic receptors mediating melanin aggregation. Comp Biochem Physiol 88C: 69–74
Morishita F, Yamada K (1989) subtype of alpha adrenoceptors mediating leucosome aggregation in medaka leucophore. J Sci Hiroshima Univ Ser B Div 1, 33: 99–112
Morishita F, Katayama H, Yamada K (1985) Subtypes of beta adrenergic receptors mediating pigment dispersion in chromatophores of the medaka, *Oryzias latipes*. Comp Biochem Physiol 81C: 279–285

Muske L, Fernald RD (1987) Control of a teleost social signals: neural basis for differential expression of a color pattern. J comp Physiol 160: 89–97
Nagai M, Oshima N, Fujii R (1986) A comparative study of melanin-concentrating hormone (MCH) action on teleost melanophores. Biol Bull 171: 360–370
Nagaishi H, Oshima N (1989) Neural control of motile activity of light-sensitive iridophores in the neon tetra. Pigm Cell Res 2: 485–492
Nagaishi H, Oshima N (1992) Ultrastructure of the motile iridophores of the neon tetra. Zool Sci 9: 65–75
Nagaishi H, Nishi H, Fujii R, Oshima N (1989) Correlation between body color and behavior in the upside-down catfish, *Synodontis nigriventris*. Comp Biochem Physiol 92A: 323–326
Nagaishi H, Oshima N, Fujii R (1990) Light-reflecting properties of the iridophores of the neon tetra, *Paracheirodon innesi*. Comp Biochem Physiol 95A: 337–341
Nagaishi H, Oshima N, Fujii R (1992) Effects of atropine on the melanophores and light-reflecting chromatophores of some teleost fishes. Comp Biochem Physiol 103C: 363–368
Naitoh T, Iwata KS (1976) Anomalous colour response of *Odontobutis* to intense illumination. Biol J Okayama Univ 17: 41–45
Naitoh T, Morioka A, Omura Y (1985) Adaptation of a common freshwater goby, yoshinobori, *Rhinogobius brunneus* Temminck et Schlegel to various backgrounds including those containing different sizes of black and white checkerboard squares. Zool Sci 2: 59–63
Namoto S, Yamada K (1983) Effects of monovalent cations on denervated fish melanophores, with special reference to the action of lithium ions. J Sci Hiroshima Univ Ser B Div 1 31: 107–115
Namoto S, Yamada K (1987) Effects of forskolin, isoproterenol and lithium ions on leucophores of a teleost, *Oryzias latipes*: evidence for involvement of adenylate cyclase in pigment-dispersion response. Comp Biochem Physiol 86C: 91–95
Naora H, Iga T (1989) Light response of cultured melanophores of a freshwater teleost, *Zacco temmincki*. Cell Struct Funct 14: 113–120
Naora H, Takabatake I, Iga T (1988) Spectral sensitivity of melanophores of a freshwater teleost, *Zacco temmincki*. Comp Biochem Physiol 90A: 147–149
Needham AE (1974) Significance of zoochromes. Springer, Berlin Heidelberg New York
Negishi S (1985) Light response of cultured melanophores of a teleost adult fish, *Oryzias latipes*. J Exp Zool 236: 327–333
Negishi S (1988) The involvement of microtubules in the light response of medaka melanophores. Zool Sci 5: 951–957
Negishi S, Obika M (1980) The effects of melanophore-stimulating hormone and cyclic nucleotides on teleost fish chromatophores. Gen Comp Endocrinol 42: 471–476
Negishi S, Obika M (1985a) The role of calcium and magnesium on pigment translocation in melanophores of *Oryzias latipes*. In: Bagnara J, Klaus SN, Paul E, Schartl M (eds) Pigment cell 1985. University Tokyo Press, Tokyo, pp 233–239
Negishi S, Obika M (1985b) Effects of hexylene glycol and nocodazole on microtubules and melanosome translocation in melanophores of the medaka *Oryzias latipes*. J Exp Zool 235: 55–63
Negishi S, Fernandez HRC, Obika M (1985) The effects of dynein ATPase inhibitors on melanosome translocation within melanophores of the medaka, *Oryzias latipes*. Zool Sci 2: 469–475
Nishi H, Fujii R (1993) Novel receptors for melatonin that mediate pigment dispersion are present in some melanophores of the pencil fish (*Nannostomus*). Comp Biochem Physiol 103C: 263–268
Novales RR, Fujii R (1970) A melanin-dispersing effect of cyclic adenosine monophosphate on *Fundulus* melanophores. J Cell Physiol 75: 133–136
Obika M (1976) An analysis of the mechanism of pigment migration in fish chromatophores. In: Riley V (ed) Pigment cell, vol 3. Unique properties of melanocytes, Karger, Basel, pp 254–265
Obika M (1986) Intracellular transport of pigment granules in fish chromatophores. Zool Sci 3: 1–11

Obika M (1988) Ultrastructure and physiological response of leucophores of the medaka *Oryzias latipes*. Zool Sci 5: 311–321
Obika M, Meyer-Rochow VB (1990) Dermal and epidermal chromatophores of the Antarctic teleost *Trematomus bernacchil*. Pigm Cell Res 3: 33–37
Obika M, Lo SJ, Tchen TT, Taylor JD (1978) Ultrastructural demonstration of hormone-induced movement of carotenoid droplets and endoplasmic reticulum in xanthophores of the goldfish, *Carassius auratus* L. Cell Tissue Res 190: 409–416
Odiorne JM (1957) Color changes. In: Brown ME (ed) The physiology of fishes, vol 2, Academic Press, New York, pp 387–401
Ohta T (1983) Melanosome dispersion in direct response to light in melanophores of *Rhodeus ocellatus* fry. Annot Zool Jpn 56: 155–162
Ohta T. Muramatsu K (1988) Spectral sensitivity of melanophoress in the primary color response of the rose bitterling, *Rhodeus ocellatus*. Jpn J Ichthyol 34: 483–487
Ohta T, Sugimoto S (1980) Leucosome dispersion under light in medaka leucophores. Jpn J Ichthyol 27: 72–76 (in Japanese)
Olivereau M, Olivereau J (1985) Effects of 17a-methyltestosterone on the skin and gonads of freshwater male silver eels. Gen Comp Endocrinol 57: 64–71
Oshima N (1989) Adenosine inhibits the release of neurotransmitters from melanosome-aggregating nerves of fish. Comp Biochem Physiol 93C: 207–211
Oshima N, Fujii R (1984) A precision photoelectric method for recording chromatophore responses in vitro. Zool Sci 1: 545–552
Oshima N, Fujii R (1985) Calcium requirement for MSH action on non-melanophoral chromatophores of some fish. Zool Sci 2: 127–129
Oshima N, Fujii R (1987) Motile mechanism of blue damselfish (*Chrysiptera cyanea*) iridophores. Cell Motil Cytoskeleton 8: 85–90
Oshima N, Nagaishi H (1992) Study of the motile mechanism in neon tetra (*Paracheirodon innesi*) iridophores. Comp Biochem Physiol 102A: 273–278
Oshima N, Fujii R, Kasukawa H (1984) Simultaneous recording of motile responses of light-absorbing and reflecting chromatophores in vitro. Zool Sci 1: 711–717
Oshima N, Kasukawa H, Fujii R, Wilkes BC, Hruby VJ, Castrucci AML, Hadley ME (1985) Melanin concentrating hormone (MCH) effects on teleost (*Chrysiptera cyanea*) melanophores. J Exp Zool 235: 175–180
Oshima N, Kasukawa H, Fujii R, Wilkes BC, Hruby VJ, Hadley ME (1986a) Action of melanin-concentrating hormone (MCH) on teleost chromatophore. Gen Comp Endocrinol 64: 381–388
Oshima N, Yamaji N, Fujii R (1986b) Adenosine receptors mediate pigment dispersion in leucophores of the medaka, *Oryzias latipes*. Comp Biochem Physiol 85C: 245–248
Oshima N, Furuuchi T, Fujii R (1986c) Cyclic nucleotide action is mediated through adenosine receptors in damselfish motile iridophores. Comp Biochem Physiol 85C: 89–93
Oshima N, Suzuki M, Yamaji N, Fujii R (1988) Pigment aggregation is triggered by an increase in free calcium ions within fish chromatophores. Comp Biochem Physiol 91A: 27–32
Oshima N, Kasukawa H, Fujii R (1989) Control of chromatophore movements in the blue-green damselfish, *Chromis viridis*. Comp Biochem Physiol 93C: 239–245
Oshima N, Hayakawa M, Sugimoto M (1990a) The involvement of calmodulin in motile activities of fish chromatophores. Comp Biochem Physiol 97C: 33–36
Oshima N, Inagaki H, Manabe T (1990b) Evidence for involvement of dynein-tubulin system in pigment aggregation within tilapia melanophores. Comp Biochem Physiol 96A: 517–523
Oshima N, Kitta K, Bern HA (1990c) Reexamination of the effect of prolactin on the teleost fish chromatophores. Zool Sci 7: 1019
Oshima N, Sugimoto M, Fujii R (1992) Effects of choline chloride on the pigment translocation within fish melanophores. Comp Biochem Physiol 102C: 11–15
Pandey AK, Shukla L, Fujii R, Miyashita Y (1981) Effects of sublethal malathion exposure on melanophores of a cichlid, *Sarotherodon mossambicus*. J Lib Arts Sci Sapporo Med Coll 22: 77–81

Parker GH (1948) Animal colour changes and their neurohumours. Cambridge University Press, Cambridge
Pickford G, Atz JW (1957) The physiology of the pituitary gland of fishes. New York Zoological Soc, New York
Pye JD (1964) Nervous control of chromatophores in teleost fishes. III. Local temperature responses in the minnow (*Phoxinus phoxinus* (L.)). J Exp Biol 41: 543–552
Rasquin P (1958) Studies in the control of pigment cells and light reactions in recent teleost fishes. Bull Am Mus Nat Hist 115: 1–68
Reed BL (1968) The control of circadian pigment changes in the pencil fish: a proposed role for melatonin. Life Sci Part II 7: 961–973
Reed BL, Finnin BC (1972) Adrenergic innvervation of melanophores in a teleost fish. In: Riley V (ed) Pigmentation: its genesis and biologic control. Appleton-Century-Crofts, New York, pp 285–294
Robertson OH (1949) Production of the silvery smolt stage in rainbow trout by intramuscular injection of mammalian thyroid extract and thyrotropic hormone. J Exp Zool 110: 337–355
Robentson OH (1951) Factors influencing the state of dispersion of the dermal melanophores in rainbow trout. Physiol Zool 24: 309–323
Rodrigues KT, Sumpter JP (1984) Effects of background adaptation on the pituitary and plasma concentrations of some pro-opiomelano-cortin-related peptides in the rainbow trout (*Salmo gairdneri*). J Endocrinol 101: 227–284
Rohrlich ST (1974) Fine structural demonstration of ordered arrays of cytoplasmic filaments in vertebrate iridophores. J Cell Biol 62: 295–304
Sage M (1970) Control of prolactin release and its role in color change in the teleost *Gillichthys mirabilis*. J Exp Zool 173: 121–127
Satake N (1980) Effects of methionine-enkephalin on xanthophore aggregation. Peptides 1: 73–75
Scheline RR (1963) Adrenergic mechanisms in fish: chromatophore pigment concentration in cuckoo wrasse, *Labrus ossifagus* L. Comp Biochem Physiol 9: 215–227
Schliwa M (1984) Mechanisms of intracellular organelle transport. Cell Muscle Motil 5: 1–82
Schliwa M (1987) Review article: Permeabilized cell models for the study of granule transport in pigment cells. Pigm Cell Res 1: 65–68
Schliwa M, Bereiter-Hahn J (1973) Pigment movements in fish melanophores: morphological and physiological studies. II. Cell shape and microtubules. Z Zellforsch 147: 107–125
Scott GT (1965) Physiology and pharmocology of color change in the flounder *Scopthalmus aquosus*. Limnol Oceanogr 10: R230–R246
Smith DC (1928) The effects of temperature on the melanophores of fishes. J Exp Zool 52: 183–234
Smith DC (1931) The effect of temperature changes upon the pulsations of isolated scale melanophores of *Fundulus heteroclitus*. Biol Bull 60: 269–287
Spaeth RA (1913) The physiology of the chromatophores of fishes. J Exp Zool 15: 527–585
Sugimoto M, Oshima N, Fujii R (1985) Mechanisms controlling motile responses of amelanotic melanophores in the medaka, *Oryzias latipes*. Zool Sci 2: 317–322
Thaler CD, Haimo LT (1990) Regulation of organelle transport in melanophores by calcineurin. J Cell Biol 111: 1939–1948
Thaler CD, Haimo LT (1992) Control of organelle transport in melanophores: regulation of Ca^{2+} and cAMP levels. Cell Motil Cytoskeleton 22: 175–184
Tomita G (1936) Melanophore reactions to light during the early stages of the paradise fish, *Macropodus opercularis*. J Shanghai Sci Inst Sec IV 2: 237–264
von Frisch K (1911) Beiträge zur Physiologie der Pigmentzellen in der Fischhaut. Pflügers Archiv Gesamte Physiol Menschen Tiere 138: 319–387
Wakamatsu Y (1978) Light-sensitive fish melanophores in culture. J Exp Zool 204: 299–304
Wakamatsu Y, Kawamura S, Yoshizawa T (1980) Light-induced pigment aggregation in cultured fish melanophores: Spectral sensitivity and inhibitory effects of theophylline and cyclic adenosine-3′,5′-monophosphate. J Cell Sci 41: 65–74
Waring H (1963) Color change mechanisms of cold-blooded vertebrates. Academic Press, New York

Watanabe M (1960) The mode of action of atropin on melanophores in the isolated scale of a crucian crap. Biol J Okayama Univ 6: 114–123
Watanabe M (1961) Pulsations of the scale melanophore in the crucian carp. Biol J Okayama Univ 7: 65–85
Wikswo MA, Novales RR (1969) The effect of colchicine on migration of pigment granules in the melanophores of *Fundulus heteroclitus.* Biol Bull 137: 228–237
Wilkes BC, Hruby VJ, Castrucci AML, Sherbrooke WC, Hadley ME (1984) Synthesis of a cyclic melanotropic peptide exhibiting both melanin-concentrating and -dispersing activities. Science 224: 1111–1113
Woodhead AD (1966) Effects of thyroid drugs on the larvae of the brown trout, *Salmo trutta.* J Zool (Lond) 149: 394–413
Yamada K (1980) Actions of sympathomimetic amines on leucophores in isolated scales of a teleost fish with special reference to beta-adrenoceptors mediating pigment dispersion. J Sci Hiroshima Univ Ser B Div 1 28: 95–114

Chapter 2

Epithelial Transport of Heavy Metals

E.C. Foulkes[1]

Contents

1 Introduction

There is no rigorous chemical definition of what constitutes a heavy metal. In the biological context the classification generally includes polyvalent metals present in aqueous solution as complex ions or chelate compounds. The heavy

[1] Dept. Environ. Health, College of Medicine, University of Cincinnati, Cincinnati, Ohio 45267–0056, USA

Advances in Comparative and Environmental Physiology, Vol. 20

metals as a class include many frankly toxic elements such as lead and cadmium for which no convincing biological functions has yet been described. Other heavy metals, while toxic at higher concentrations, may play an essential role at low levels, as has been suggested, for instance, for Cr (Mertz 1969).

One of the important chemical properties of all these metals is their high affinity for reactive groups in proteins and other biological molecules. As a result the half-life of unbound metals in biological solutions is very short. The reactivity of heavy metals helps explain their extensive biological effects. It also leads to the prediction that the metals will be strong though relatively non-specific enzyme inhibitors in vitro. Any protein, for instance, which contains -SH groups is likely to react with Hg. However, the biological effects of the metals are much more specific than might be expected from their high reactivity in vitro, and the toxicological significance of such in vitro reactions is often unclear.

The absorption, distribution and excretion of metals in the intact organism, as well as their toxic effects, depend on many factors including the relative metal specificity of different binding sites, the species of the metal administered, the route and rate of exposure and physiological variables such as diet, age, gender and others. If, for instance, a metal is injected into the renal artery of an experimental animal, it will instantaneously become bound to plasma protein and be unable to reach reactive sites in the renal parenchyma. Thus, Cd, Hg or Zn administered as chloride salts are completely recovered in renal venous plasma, without loss by glomerular filtration, after a mean transit time across the kidney from artery to vein similar to that of albumin (Foulkes 1974). If sequestration by plasma protein is prevented, as for instance in the case of Cd by addition of a suitable chelator such as cysteine or EDTA, then the metal will remain diffusible and filterable. However, its ability to react with cells now depends on the relative metal affinities of the exogenous chelator and the cellular binding sites. These statements may be illustrated by the fact that Cd EDTA is completely inert and is handled in the kidney like inulin, whereas Hg chelated to EDTA reacts readily with cells (Foulkes 1991b); EDTA similarly prevents the Cd inhibition of jejunal amino acid absorption, but does not influence the inhibitory action of Hg.

The fact that, except perhaps very transiently, none of the metals exist in the body in the free, ionized form implies that the study of their behavior in biological systems must deal with the properties of metal complexes rather than of the metals themselves. As expected, the valency of a metal, determining the nature of its compounds in the circulation, may exert a major influence on cellular uptake. Reference may be made in this regard to the difference between the relatively ready uptake of anionic compounds of Cr(VI) and the generally very slow uptake of Cr(III) (Mertz 1969).

Extensive work has been performed on the pharmacokinetics of different metals. They are transported in blood and taken up by many tissues, varying with the chemical species involved, presence of high affinity binding sites and other factors. Three epithelial cell layers play major roles in metal metabolism, as targets of toxic action and/or as site of metal uptake and excretion:

(1) Outside of the occupational environment, the major source of the body burden of most metals is intestinal absorption; uptake, retention or rejection by intestinal epithelium are therefore important determinants of total body burden (see, e.g., Page et al. 1986). (2) The epithelium of the respiratory tract and alveoli represents the second major site of entrance of metals into the body. The pulmonary load of metals may be derived from smoking, or from more general air pollution, especially in an industrial setting. (3) A major target tissue in the body for many metals such as Cd, Pb, Ni, UO_2 and Hg, is the proximal tubular epithelium of the renal cortex (see, e.g., Clarkson and Kench 1956). The kidneys are also largely responsible for excretion of absorbed metals, although biliary and fecal excretion can also contribute under certain conditions.

To a significant extent, therefore, it is the reaction of metals with these epithelial tissues which determines their fate in the body; in addition, the tissues are also the target of much of the toxic action of heavy metals. Of course, metals also react with other cell types, such as vascular endothelium, smooth muscle, hepatocytes, etc. This review, however, will focus primarily on transport of heavy metals by intestinal, renal and pulmonary epithelium. It will also consider some of the toxicological implications of this transport.

2 General Problems

2.1 Need for Specific Carrier Systems

In general, inorganic compounds of the heavy metals tend to be more soluble in water than in lipids; as a result, they do not readily penetrate lipid membranes. Specific mechanisms under homeostatic control have been invoked for intestinal absorption of such essential elements as Zn (Cotzias et al. 1961). At the same time it seems a priori unlikely that specific membrane carrier systems should have evolved to mediate uptake of what in many instances are non-essential and highly toxic elements (Foulkes 1986). This dilemma would be resolved if one could assume that transmembrane movement of non-essential heavy metals utilizes mechanisms designed for the transport of essential elements such as Ca or Zn. A good example of such relatively non-specific transport is that of Cd uptake by secretory cells from the pituitary (Hinkle et al. 1987); here transmembrane movement of the toxic metal appears to be facilitated by voltage-gated Ca channels. Further, in the ferret erythrocyte, Cd and Mn are reported competitively to inhibit the Ca influx mediated by the Na–Ca exchanger (Frame and Milanick 1991).

Clearly, mechanisms of metal transport may differ between various tissues. In addition, different metals may themselves follow separate pathways even in the same tissue. This can be seen, for instance, in the work of Blazka and Shaikh (1991) on Cd and Hg influx into rat hepatocytes. About one third of total Cd uptake in this system involves Ca channels. The remaining portion of Cd movement is not sensitive to Ca channel blockers and therefore presumably depends on other mechanisms. Uptake of Hg, according to the same argument, appears to be entirely independent of Ca channels. At least in part, therefore, cadmium is transported in the hepatocyte by a mechanism different from that responsible for transfer of Hg. In jejunal epithelium, available evidence suggests that these two toxic metals, as well as other non-essential metals, may be transported by similar mechanisms (Foulkes and Bergman 1993). The problem now is how to explain cellular transport of polyvalent and water-soluble ions in absence of specific carrier systems.

2.2 Role of Complex Formation in Metal Pharmacodynamics

As already mentioned, the biological reactions of heavy metals are greatly influenced by the chemical form in which they are administered, or secreted, or absorbed. For instance, inorganic Hg primarily accumulates in renal cortex, with little metal reaching the central nervous system; in contrast, methylmercury is a strong neurotoxin (Berlin et al. 1973).The difference here lies in part in the physico-chemical properties of the compounds, especially their lipid-solubility. The role of lipid permeability of metal complexes and their ability to cross membranes is also illustrated by the findings of Hider et al. (1990) on the stimulation of Zn uptake by human erythrocytes in presence of ethylmaltol. There is usually little need to invoke special mechanisms to explain cellular uptake of lipid-soluble metal compounds.

However, lipid solubility of metal complexes is not the only factor explaining their movement into and across cell layers. Thus, Hirayama (1975) found evidence for carrier mediation in the intestinal absorption of methylmercury. Kerper et al. (1992) proposed that methylmercury is transported by the L (leucine-preferring) amino acid carrier across the blood-brain barrier in the form of a cysteine complex. The covalent mercurial diuretic Chlormerodrin is actively secreted in the renal tubule of the dog (Borghgraef et al. 1956). Another instance of carrier-mediated transport of metal complexes is the absorption of anionic proteins such as myoglobin or Cd-metallothionein at the apical cell membrane of the proximal tubule in the rabbit kidney (Foulkes 1978). Metallothionein is the low-molecular weight, cysteine-rich protein discovered by Kagi and Vallee (1960) which, among other functions, contributes to Cd transfer from liver to kidney (Piscator 1964). Renal uptake of metallothionein will be further discussed in Sect. 4.4.

The extent to which various metal compounds are sequestered by plasma protein represents an additional factor controlling metal distribution in the body. Such an effect has been repeatedly observed, for instance with Cd. Thus, administration of this metal as the chloride salt leads to immediate binding to plasma protein (Foulkes 1974) and subsequent accumulation primarily in liver and other tissues but not in the kidney (see e.g., Gunn and Gould 1957). However, when bound to dimercaptopropanol (BAL) in a moderately lipid-soluble complex, Cd readily reaches the renal parenchyma where it is retained (Tepperman 1947). Other low-molecular weight sulfhydryl-containing compounds, including metallothionein (Squibb et al. 1984), mercaptoethanol (Foulkes 1974) and glutathione (Foulkes unpubl. results), exert a similar action.

In general, low-molecular weight chelators prevent metal sequestration by plasma protein; protein-bound metal in the kidney would essentially be restricted to the vascular space. However, uptake of the metal across cell membranes, other than by simple diffusion in a lipid-soluble form, usually appears to require metal interaction with the cell membrane (see Sects. 3.4, 4.4). As a result, very tightly chelated metals such as Cd-EDTA are completely inert and can actually be used as extracellular markers (Stacy and Thornburn 1966) (see also Sect. 1).

The tendency of some metals to form anionic complexes with bicarbonate or chloride (see e.g., Clarkson and Cross 1961) may permit them to enter certain cells through anion channels. For instance, the stilbene derivative DIDS, a well-known inhibitor of anion channels, interferes with uptake of Pb (Simons 1986), Zn (Torrubia and Garay 1989) and Cd (Lou et al. 1991) by human erythrocytes. In rat jejunum, the reaction of inorganic mercury with the brush border involves an anionic chloride complex of the metal (Foulkes and Bergman 1993), but its further transfer into the cells is not sensitive to DIDS (see Sect. 3.7). Contribution of anionic complexes to cadmium uptake by epithelial cells is unlikely to be significant in the intestine in light of the findings on metal-metal interaction described in Sect. 3.4.

2.3 Cellular Trapping and Release of Metals

Intracellular heavy metals do not exist in appreciable amounts in the free form but rapidly bind to a variety of cell constituents (see, e.g., Endo et al. 1986); the biological effect of such binding is not always clear. It is likely, however, that metals in some of these complexes are relatively inert, so that they can accumulate in tissues to a significant level. As a result of binding, their biological half-life in the tissue may be relatively long, ranging from weeks for Hg (Miettinen 1973) to 20 years or longer for Cd (Kjellstrom 1971). Many instances of sequestration and possibly consequent detoxification of metals have been reported, including binding of Cd to metallothionein in kidneys (Piscator 1964), the formation of lead-containing nuclear inclusion bodies in renal tubules (Carroll et al. 1970) and the presence of lead-binding proteins in erythrocytes (Raghavan et al. 1981) and kidney (Oskarsson et al. 1982). To the extent that lead is sequestered in the kidney on a normal lead intake, it does not appear to participate in lead turnover, i.e., it is indeed inert. This conclusion can be drawn from the work of Rabinowitz et al. (1976) who found no evidence for any intermediate lead pool between blood and urine.

Sequestration of toxic metals, as mentioned, is one of the important and plausible functions ascribed to metallothionein. Synthesis of this protein is induced in many cell types, including renal and intestinal epithelium, stressed by exposure to various metals and other agents. In the intestinal mucosa, this results in significant depression of transepithelial Cd absorption from the lumen (Foulkes and McMullen 1986a). Sloughing of these metal-containing cells will then ultimately lead to loss of the metal from the body; in this manner a high concentration of metallothionein in mucosal epithelium can help protect the body against absorption of excess metals (cf. Richards and Cousins 1975). The extent of the protection will depend on the relative affinities of the metal for metallothionein and other cell constituents, and on the degree to which the metal-binding capacity of the cells approaches saturation.

The assumption underlying the suggested protective role for metallothionein is that its metal complexes are inert. This is, however, not true under all circumstances, as in the case of the nephrotoxicity of injected metallothionein (Nordberg et al. 1975). The toxic effect here may be related to preferential lysosomal breakdown of exogenous Cd metallothionein. This interpretation is supported by the report of Cain and Holt (1983) who, 4 h following injection of Cd metallothionein, found only 30% of renal Cd bound to metallothionein. In any case, although synthesis of metallothionein does increase cellular accumulation and retention of metals, its role is probably much more complex than simply that of metal sequestration.

An exclusive role of metallothionein in general trapping of metals is also incompatible with the observation that the renal half-life of Cd greatly exceeds that of Hg (see above), in spite of the much tighter binding of Hg than of Cd to metallothionein. The explanation recently suggested for this apparent discrepancy is based on the fact that, unlike in the case of Cd, the strong affinity of Hg for metallothionein is relatively non-specific and therefore does not greatly exceed that for other thiol compounds (Foulkes 1993). As a result, diffusible thiols like glutathione can compete with metallothionein for Hg but not Cd.

Cell damage may lead to release of protein-bound metals, as discussed in Sect. 4.5 for the kidney. Similarly, infliction of hepatic damage causes release of Cd metallothionein from the liver and permits its increased accumulation by the kidney (Cain and Griffiths 1980). Little is known about the normal mechanism of metal release from various tissues, as further considered in Sect. 3.5 below in reference to the jejunal mucosa.

2.4 Some Further Toxicological Implications

The role of metallothionein in metal metabolism, as discussed in the two preceding sections, clearly helps determine the toxic effects of metals. In general, the pharmacodynamics of metal compounds in the body are obviously important for an understanding of metal toxicology. The underlying mechanisms controlling distribution of metals in the body are, however, not completely understood. For instance, although proximal tubular epithelium is the major target of Hg (Zalups and Barfuss 1990), its high affinity for Hg has not yet been explained (see Sect. 4.2).

The intracellular distribution of metals is also likely to influence their toxic effects. Compartmentation of cellular metal content can be readily shown by fractionation of subcellular constituents and cytosol. It was also demonstrated for Cd in vivo by external monitoring of the rate of isotopic equilibration of kidney with plasma Cd (Foulkes 1985b). The question then arises which of these metal pools is/are toxic. For instance, does only non-metallothionein Cd exert adverse effects, while Cd bound to that protein is inert (Nomiyama and Nomiyama 1982)? This problem is particularly significant in attempts to establish critical concentrations of metals in their target organ, an essential step in assessing risks associated with metal exposure. The critical concentration in a target organ is here defined as that at which the incidence of a given adverse effect in an exposed population begins significantly to exceed background values. Section 4.4 will further consider this problem.

3 Metal Transport in Intestine

3.1 Technical Considerations

In general, net fractional absorption of heavy metals from the intestine is small, although occasional higher values have been reported as, for instance, in the case of lead following administration of vitamin D (C.M. Smith et al. 1978). Absorption is here defined as uptake into the body, not simply transient retention in mucosal cells. Values reported for adults usually lie below 10% of the ingested load. Nevertheless, intestinal absorption normally contributes the major fraction of the body burden of heavy metals. Much of the information available on metal absorption, however, sheds little light on the mechanisms involved in this process. Several factors are responsible for this gap in our knowledge. Thus, in the intact organism, absorption of a metal has at times been equated to net retention following its oral administration (cf. Kostial et al. 1979); such an experimental design does not allow for biliary excretion or enteric secretion. Further, rate and extent of absorption may be greatly influenced by various exogenous as well as endogenous factors (see Sect. 3.8), including especially the contents of the intestinal lumen (Foulkes 1984). Obviously, binding to non-absorbed fibers would prevent uptake of a metal from the intestine, as has been well documented by the production of Zn deficiency on diets high in phytic acid (Pecoud et al. 1975).

Influence of diet on metal absorption was studied in detail especially by Kostial and collaborators, using the intact rat (Kostial et al. 1979). For instance, Cd retention on a milk diet was observed to exceed that on rat chow (Rabar and Kostial 1981). On the other hand, Foulkes (1980) reported that milk strongly inhibits Cd uptake from the rat jejunum perfused with saline in situ. This apparent discrepancy is readily explained by the fact that the baseline for comparison in the work of Rabar and Kostial was the rat gut containing a complex mixture of fibers and compounds capable of binding the metal and/or directly interfering with the interaction between Cd and brush border membranes. In contrast, the investigation of Cd uptake from the perfused intestine started with saline in the lumen.

In much of the experimental work on the mechanism of metal absorption the intestines were exposed to inorganic salts of the metals, rather than to the organic complexes normally present in the lumen. However, except for a small number of specific compounds, or in the case of lipophylic complexes, there is little evidence to suggest that bound metals cross membranes as such (Foulkes 1991; see also Sect. 4.4). Instead, the complexes appear to dissociate in competition with metal-binding sites on the membranes (Foulkes and Blanck 1990); to the extent that ligands reduce the availability of a metal, they will of course slow down its absorption. In general, however, use of inorganic metal salts provides an appropriate model for the study of intestinal metal absorption.

As always, results and their interpretation critically depend on details of the experimental protocol. One very important source of variability, at times difficult to control, is the dependence of metal absorption on physiological factors such as age, gender, pregnancy, diet, bile flow, etc, as further discussed in Sect. 3.8. Other problems may arise from application of inappropriate techniques; this is particularly true of the use of everted sacs for the quantitative study of metal absorption. There are several reasons to question the validity of everted sacs in this connection. One such problem arises from the submucosal trapping of heavy metals, as described above. Endo et al. (1986), for instance, found that removal of the epithelium from everted sacs exerted little effect on uptake of Hg from a 10 μM solution. Another difficulty is posed by the observation that everted sacs exhibit a much higher incidence of polar (intercellular?) channels than does the parent tissue (Foulkes and Bergman 1993); as a result, solute transport by sacs is not likely to possess more than qualitative significance. Limitations of everted sacs for the study of metal absorption are further discussed in Sect. 3.2, and were in part described in an earlier review (Foulkes 1984).

3.2 Absorption Kinetics

Transepithelial flux of heavy metals from the mature intestine in situ has been shown to follow transcellular pathways (Foulkes and McMullen 1987; see also Sect. 3.3). While no evidence was found for significant metal diffusion along paracellular channels in intact jejunum, this is probably not so in everted sacs (see Sect. 3.1). As a consequence, it is only in the intact organ that metal absorption can be described by a series model in which cellular uptake of the metal at the brush border constitutes the first step (step 1), followed by step 2, the basolateral extrusion of the metal. The necessity for this precaution has not always been heeded (e.g., Sahagian et al. 1967; Ohta et al. 1989). As a result, much of the earlier work is difficult to interpret. For instance, the report that EDTA reduces Cd uptake by everted sacs while increasing its transmural movement (Kojima and Kiyozumi 1974) is not readily reconciled with the series model of transport. The most likely explanation, instead, is that the chelator under these conditions opened paracellular leak pathways, as described by Tidball (1964) and Foulkes and Bergman (1993).

Different heavy metals may be absorbed by similar mechanisms (Foulkes and Bergman 1993). There is, however, little consensus as yet on the nature of steps 1 and 2 of metal absorption. To some extent this is likely to be related to experimental limitations. Thus, only step 1 can conveniently be studied, at least qualitatively, in everted sacs (Foulkes and McMullen 1987), while measurement of step 2 in that preparation is not likely to yield physiologically meaningful results.

A much better preparation for the study of metal absorption is obtained by vascular perfusion of isolated segments of intestine (K.T. Smith et al. 1978). However, even in a normally blood-perfused segment in situ, the rate of appearance of luminal metals in portal venous blood is relatively slow, with the mean transit time of Cd, for instance, exceeding 20 min. As illustrated in Fig. 1, Zn and Ni are absorbed more quickly across the jejunal barrier in the rat.

A good inverse correlation was observed between the peak transit time of a metal and its affinity for metallothionein, as illustrated in Table 1; presumably, intracellular trapping by metallothionein under these conditions constitutes a major determinant of the rate of transepithelial transit (see also Sect. 2.2). The

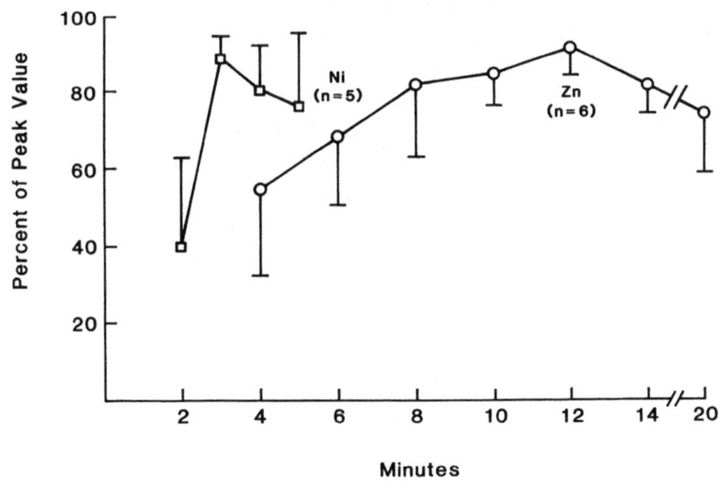

Fig. 1. Transepithelial metal movement in rat jejunum in vivo. Metal chlorides (20 μM) were placed into the jejunum for 10 s. Portal venous blood was collected at times shown. Results represent mean ± 1 SD for n animals. (Foulkes and McMullen 1987)

Table 1. Affinity of metals for metallothionein and their rate of transfer from lumen to blood in rat jejunum. (Based on Foulkes and McMullen 1987)

Metal	Relative affinity for MT	Peak transit time (min)	Predicted mean transit time (min)[a]
Ni	1	3–5	5 ± 5
Zn	59	10–15	9 ± 2
Cd	483	> 20	> 20

[a] Predicted from a 3-compartment series model excluding paracellular movement.

long transmural passage times of the metals may be contrasted with transit times of less than 1 min for sugars and amino acids, solutes unlikely to be extensively trapped in mucosal or submucosal tissue.

A role of mucosal metallothionein in trapping of all heavy metals, however, remains in question. Indeed, induction of metallothionein synthesis, while increasing retention of Cd (see Sect. 2.3), does not affect that of Hg.

3.3 Transcellular and Paracellular Absorption

Transepithelial movement of solutes across more or less tight junctions between cells has been reported in several epithelia, including gall bladder, intestine, choroid plexus and proximal renal tubules of various species (see, e.g., Fromter and Diamond 1972). A role of junctional complexes in the movement of La^{3+} across rabbit gall bladder and intestine was described by Machen et al. (1972), and Whittembury and Rawlins (1971) found evidence for penetration of La^{3+} along paracellular pathways in the proximal tubule of the toad kidney; this movement of La^{3+} may, however, be relatively slow, as exposure periods of the order of 1 h or longer were employed in both cases. Although there is additional evidence for a reaction between divalent cations and tight intercellular junctions (e.g., Powell 1981), a significant contribution of such pathways to transmural movement of heavy metals is unlikely.

This is especially true of the much shorter exposure periods used by Foulkes and McMullen (1987) for the study of transepithelial passage times in the rat jejunum in vivo. In that system, actually, available evidence argues against a role of leaky junctions in absorption of heavy metals. This conclusion is based on two sets of observations: Firstly, as discussed in Sect. 2.3, synthesis of metallothionein decreases transmural movement of Cd (Foulkes and McMullen 1986a). Because metallothionein is normally a cytoplasmic protein, its ability to trap transported metals implies that these metals follow transcellular absorption paths.

In the second place the assumption was tested that the kinetics of unidirectional transmural metal movement can be described by a simple series model, not allowing paracellular movement (see Sect. 3.2). At steady state, S, the tissue content of the metal, which is assumed to represent the pool with which transported metal equilibrates, is given by the expression S = k.t. Here, K represents the rate of disappearance of the metal from the lumen, and t stands for its mean transit time across the tissue, from lumen to blood; actually, peak transit times rather than mean transit times were usually measured. As already shown in Table 1, predictions from the equation fall into the range of experimental observations. If the metal were assumed to pass around the cells, it would be necessary to postulate the unlikely existence of an extracellular metal transport pool in the tissue.

As pointed out above, the recognition that metals follow transcellular pathways during absorption implies that the first event in this process (step 1A) must be the interaction (binding?) between metals and apical cell membranes; the subsequent step 1B, as further defined in the next section, describes the internalization of membrane-bound metal.

3.4 Step 1 of Cd Absorption

In attempting to explain the nature of step 1 of metal absorption it is necessary to account for several important observations; these will here be discussed in particular reference to Cd. It must be emphasized that although different heavy metals may be absorbed by similar mechanisms, these mechanisms may not be identical (Foulkes and Bergman 1993). Further, what may serve as adequate explanation for metal transfer across intestinal epithelium may not necessarily be valid for other epithelial tissues, let alone non-epithelial cells.

The first observations requiring explanation are the apparent saturability (Foulkes 1980) and temperature dependence (Foulkes 1991a) of step 1 of Cd absorption, i.e., the removal of the metal from the lumen. Both of these properties are characteristic of carrier-mediated transport. In light of the a priori rejection of specific carriers for toxic, non-essential metals (see Sect. 2.1). the apparent mediation of Cd uptake might be assumed to reflect competition of the metal with essential elements like Ca and Zn for their specific carriers.

Such competition would have to imply, however, that the homeostatically increased rate of Zn uptake in Zn-deficient rats should be accompanied by an equivalent increase also in Cd uptake; this was found not to be the case (Foulkes and Voner 1981). In addition, the hypothesis that the toxic metal Cd is taken up by carrier systems for the essential metals Ca or Zn predicts that Ca and Zn should act as competitive inhibitors of Cd uptake. In other words, a given concentration of Ca or Zn would be expected to exert a smaller fractional inhibition of Cd uptake at high then at low Cd concentrations. Such an influence of Cd concentration could not be demonstrated in the perfused rat jejunum in vivo (Foulkes 1985a): in this preparation, as illustrated in Fig. 2, Ca, Zn and La inhibit Cd uptake in a non-competitive manner.

In another series of studies in vitro, 5 mM Zn inhibited uptake of Cd by everted sacs of rat jejunum in the cold by $40 \pm 14\%$ at 5 μM Cd, and by

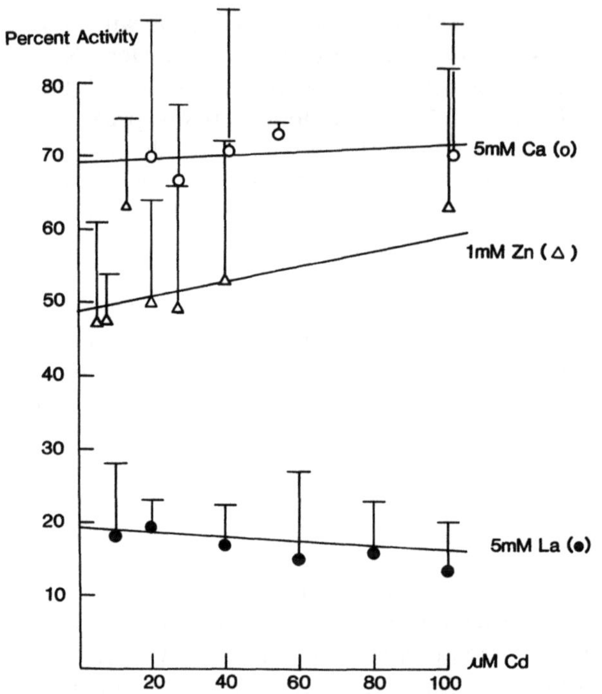

Fig. 2. Non-competitive inhibition of Cd absorption by Ca, Zn and La. Each value represents the mean and 1 SD for 5–12 rats. Each animal served as its own control, with Cd uptake first measured in absence of inhibitory metals (activity 100%), then in presence of Ca, Zn or La. Statistically, the fractional inhibition exerted by these metals is independent of the Cd concentration. (Foulkes 1985a)

40 ± 15% at 150 μM (unpubl. results). Again, the finding that Zn inhibition of Cd uptake is independent of the Cd/Zn ratio is also incompatible with competition between these two metals. In additon, other cations, including Pb and polylysine, also interfere with Cd uptake, and it would not be reasonable to attribute this to competition between Cd and these cations for specific Pb or polylysine carriers.

The interaction between various polyvalent metal cations, as well as polylysine, suggested that the reaction of cadmium with the mucosal brushborder involves first a step 1A, the electrostatic binding of the metals to fixed anionic charges on the membrane (Foulkes 1988). This would exclude participation of anion channels, such as discussed in Sect. 2.2, in the intestinal uptake of Cd. Support for this interpretation is provided by the additional observation that Hg^{2+}, known for its tendency to exist in salt solution as anionic chloride complex, does not behave like the other polyvalent cations tested, and exerts no immediate inhibiton on Cd uptake (see Sect. 3.7).

The hypothesis of electrostatic interactions at the cell membrane is further strengthened by the fact that the apparent saturability of Cd uptake could be abolished in experiments in which, with changing Cd concentrations, the total divalent metal concentration was maintained constant by appropriate addition of Ni or Zn (Foulkes 1991a). There is thus considerable evidence for the suggestion that the first reaction in the cellular uptake of Cd (and presumably certain other metals) by mucosal epithelial cells consists of its binding to the membrane. On this basis, the apparent saturability of step 1 can be ascribed to non-specific charge neutralization.

Once the metal has become bound to the membrane, it may be internalized by step 1B, i.e., steps 1A and 1B act in series to transfer the metal from lumen into cell. The two steps are conveniently separated with suitable chelating compounds, such as EDTA in the case of Cd (Foulkes and McMullen 1987). Provided the chelators may be assumed to remain restricted to the extracellular space during short exposure in the cold, only metal externally bound to the membrane will be removed. Further, the chelators must obviously possess a higher affinity for the metal than do the membrane binding sites (Foulkes 1991b). Step 1A of Cd absorption is fast and temperature-independent (Foulkes 1991a), as expected from an electrostatic interaction; it is further unaffected by blocking of sulfhydryl groups.

Unlike step 1A, step 1B is temperature-sensitive, with a constant Q_{10} of around 1.4 over the range of 0–37 °C; it is not affected by anoxia or metabolic inhibitors, and proceeds much more slowly than step 1A (Foulkes 1991a). These characteristics led to the proposal that membrane fluidity may be a determinant of step 1B. Such fluidity is known to vary with temperature, and the fact that the immature gut possesses a greater intrinsic permeability to Cd than does the mature organ (Foulkes 1991b) can be correlated with changes in membrane fluidity during maturation (Israel et al. 1987). In addition, we have shown that n-butanol, an agent known to increase membrane fluidity, stimulates Cd uptake by brush border membrane vesicles (Bevan and Foulkes 1989).

It is important to emphasize that the mechanism here proposed to explain jejunal Cd uptake requires neither the existence of saturable Cd carriers, nor the involvement of systems responsible for uptake of physiologically important metals like Ca and Zn. Evidence for participation of Ca channels in Cd uptake by other cell types was reviewed in Sect. 2.1. As far as the rat jejunum is concerned, it is likely that the mechanism of cellular uptake of some other heavy metals, such as Pb, Ni and Cu, resembles that here described for Cd; the only exception found so far is Hg (see Sect. 3.7).

3.5 Step 2 of Cd Absorption

The basolateral extrusion (step 2) of Cd from the mucosal epithelial cells is technically more difficult to approach than is step 1. Obviously, the magnitude of step 2 will vary inversely with the degree of intracellular trapping of the metal. This is illustrated in Fig. 3 (Foulkes and McMullen 1986a) by the extent to which step 2 of Cd absorption remains very small compared to step 1 until

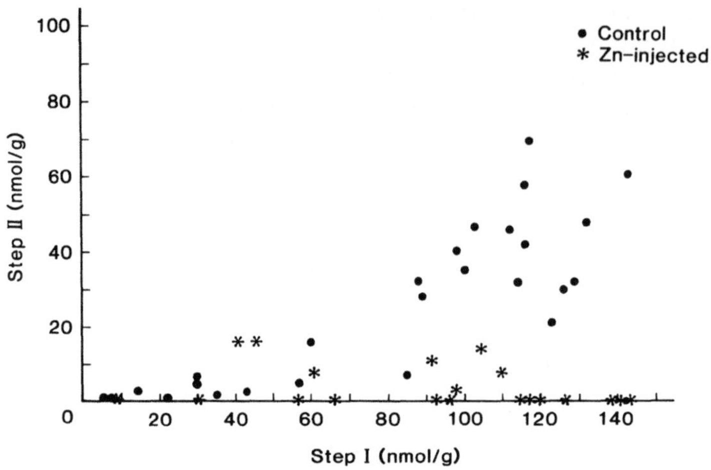

Fig. 3. Step 2 of Cd absorption as function of step 1. Each *symbol* represents results from one rat, perfused for 40 min with 20–150 μM $CdCl_2$; Zn-treated rats received an I.P. injection of 40 μmol $ZnSO_4$ 18 h before the experiment. (Foulkes and McMullen 1986a)

apparently the intracellular capacity to bind Cd becomes exhausted (cf. Sect. 2.3). The study shown compared Cd absorption in jejunum of control animals with animals in which mucosal metallothionein synthesis had been induced with Zn.

The same study showed that increasing the intracellular concentration of Zn does not acutely affect step 2 of Cd absorption, i.e., no evidence could be obtained for Cd–Zn interaction at the basolateral membrane similar to that described at the brush border. It is likely, of course, that any intracellular Cd which is not bound to protein will be ligated by smaller molecules such as glutathione and amino acids, and that step 2 therefore describes extrusion of such complexes (Foulkes 1993). Extrusion of free metal ions in any case would have to proceed against large electrochemical gradients and would therefore require active transport mechanisms.

3.6 Absorption of Zn

The mechanism of intestinal Zn absorption has been studied by many investigators; the comprehensive review by Cousins (1985) provides a good summary of the field. It is important to realize that Zn is an essential metal, whose absorption, unlike that of non-essential toxic metals, is homeostatically controlled. For instance, Zn uptake from the lumen of the Zn-deficient rat exceeds that seen in control animals (Becker and Hoekstra 1971). At the same time, a high Zn diet also increases the metallothionein levels in the intestinal mucosa. It was this inverse correlation between metallothionein and the rate of Zn absorption which led Richards and Cousins (1975) to postulate that mucosal trapping of Zn by metallothionein helps determine the rate of Zn absorption (see also Sect. 2.3). The results of Sugawara (1982) appear to be at variance with this interpretation: pretreatment of rats with Cd increased mucosal trapping of Zn but did not alter its absorption into the body. Other factors, such as the rate of Zn backflow into the lumen (Evans et al. 1979) may be more important in the control of net absorption than is the metallothionein content of the mucosa.

While some movement of Cd from enteric cells into the lumen has been reported (Kiyozumi and Kojima 1978; Foulkes 1980), enteric Zn secretion is much more significant (Pekas 1966; Methfessel and Spencer 1973). Backflow was taken into account in the kinetic model of Zn absorption developed by Bonewitz et al. (1983); this also permits the analysis of drug effects on Zn fluxes into and out of the cells across brush border and basolateral membranes. Mathematical simulation correctly predicted, for instance, that transfer of Zn from lumen into the body should be enhanced by adrenalectomy and retarded by dexamethasone. Regulation of Zn absorption by adrenocortical hormones is not mediated by changes in metallothionein synthesis, indicating the possibility of direct humoral control of Zn metabolism. Clearly, a variety of hormonal and other endogenous as well as exogenous factors can influence metal absorption. This fact has frequently been described and reviewed (e.g., Foulkes 1984). The interesting proposal that the tryptophane metabolite picolinic acid might play a mandatory role in Zn transport by mucosal epithelium (Evans and Johnson 1980) has not been confirmed (Bonewitz et al. 1982).

3.7 Absorption of Other Heavy Metals

Intestinal absorption of some other non-essential heavy metals has also been extensively investigated and will be shortly reviewed here. Much of the background information on absorption of these metals was collected in a useful volume edited by Skoryna and Waldron-Edwards (1970). In general, it is likely that mechanisms similar to those described for Cd may also effect transport of some other metals. Kinetics of Ni absorption for instance, resemble in several ways those of Cd (Foulkes and McMullen 1986b), except for the relatively small intracellular trapping of Ni and consequently relatively short transit time into the body (Foulkes and McMullen 1987). Retention of Ni in the mucosa is not affected by synthesis of metallothionein, in agreement with its low affinity for this protein; milk depresses uptake of both metals. Step 2 of Ni absorption appears to follow first order kinetics over a broad concentration range; it is not influenced by intracellular Zn, a fact previously observed also for Cd (see Sect. 3.5)

Absorption of lead has also been studied by many investigators. Conrad and Barton (1978), for instance, found that as in the case of Cd, mucosal storage, presence of ligands, and interaction with other metals all influence the rate of Pb absorption; further, and as expected from a non-essential metal, no evidence could be found for feedback mechanisms controlling absorption. Barton (1984) reported movement of Pb against the electrochemical gradient in sacs of rat duodenum. Aungst and Fung (1981) determined the kinetic constants of Pb absorption on the basis of the Michaelis-Menten model. While other evidence supports the applicability of Michaelis-Menten kinetics to the study of metal removal from the lumen (step 1) (e.g., Foulkes 1980), usefulness of such kinetics for describing transmural transport is uncertain. It must be emphasized that all these investigations, as well as others not cited here, utilized everted sacs. The physiological significance of the findings therefore remains in doubt (see Sect. 3.1), and the mechanism of Pb absorption requires further exploration.

On the basis of fecal excretion of isotopically labeled Pb, the fractional uptake of an ingested load in the human was found to be as low as that of other toxic metals, in the range of 6–8% (Rabinowitz et al. 1980). Not unexpectedly, absorption of Pb, like that of other metals, is also strongly influenced by diet. Thus, deficiencies of Ca (Quarterman and Morrison 1975) or Fe (Six and Goyer 1972) increase Pb retention. Lactose stimulates absorption of Pb as well as of some other metals (Bushnell and DeLuca 1981), perhaps by a mechanism involving a change in the transmural electrical potential of the intestine.

The importance of non-absorbable ligands in controlling net metal absorption is illustrated by the use of keratin powder (Takahashi and Hirayama 1971) or a synthetic polythiol resin (Clarkson et al. 1973) not only to reduce but actually to reverse absorption of methyl mercury. Extensive work has been performed on the intestinal absorption of methyl mercury or of divalent inorganic mercury. To the extent that this work utilized the everted sac technique, the full significance of these results to the mechanism of epithelial mercury transport is not always clear (Sect. 3.1); this is especially true in light of the high affinity of Hg for -SH groups in the submucosa. Past investigations of Hg uptake in intestinal segments in vivo (e.g., Sasser et al. 1978) also permit only few conclusions to be drawn on the mechanism of this process.

Step 1 of Hg absorption differs in several important aspects from that for Cd (Foulkes and Bergman 1993). For instance, high concentrations of Cd, Zn or Ca inhibit uptake of labelled Cd but not that of Hg; 2 mM La strongly depresses uptake of Cd while stimulating that of Hg. This difference in the behavior of the two metals may be related to the fact that the major fraction of divalent Hg in saline solution is present in anionic form (see Clarkson and Cross 1961). Binding of anionic mercury to the membrane is obviously not subject to the same influence of neutralization of membrane charges as is Cd. The observation of Endo et al. (1986) that increasing the pH of the mucosal solution decreases uptake of Hg by everted sacs of rat jejunum can perhaps be similarly attributed to increased net negative charges on the membrane.

Mercury bound to the membrane by its first reaction with the tissue (step 1A) has been defined by its lability to 1 mM dimercaptosuccinate at 0 °C during 30 s exposure, in full analogy with the use of EDTA for studying Cd absorption (Foulkes and Bergman 1993). Step 1A of either metal is little affected by temperature, while step 1B in both cases is characterized by a temperature quotient (Q_{10}) of 1.4 (Foulkes 1991a; Foulkes and Bergman 1993). Once internalized, Hg becomes largely bound to high-molecular weight protein(s) (cf. Endo et al. 1986). Step 2 of inorganic Hg absorption, as in the case of Pb and Cd, is much slower than step 1, in contrast to the generally smaller trapping of nutritionally essential metals (Neathery and Miller 1975; Bonewitz et al. 1982). Unlike for Cd, trapping of Hg is not increased upon induction of metallothionein synthesis (Foulkes 1993), a finding possibly related to the high affinity of Hg for other sulfhydryl compounds such as glutathione.

The extensive literature on intestinal iron absorption lies beyond the scope of the present review. In summary, the fact that Pb, Ni, CrIII and other polyvalent metal cations tested, with the exception of Hg (see Sect. 3.4), all depress step 1 of Cd absorption, suggests that they exert similar electrostatic effects on brush border membranes. This interpretation provides a simple and general mechanism for intestinal uptake of several of the non-essential toxic metals. Little can be said at the moment about the nature of step 2 of metal absorption.

3.8 Physiological Variables

A number of physiological variables influence movement of the metals cited here out of the intestinal lumen. These include exogenous factors such as the nature of the diet, as discussed in Sect. 3.1. In addition, endogenous variables such as bile flow (Foulkes and Voner 1981) may influence the rate and extent of metal absorption. Finally, hormonal and other mechanisms are presumably involved in the homeostatic control of uptake of essential metals.

Metal absorption by intestinal epithelium is also governed by factors such as sex (e.g., Kello et al. 1979) and age (e.g., Kostial et al. 1978). Thus, the neonatal gut absorbs metals (and other solutes) significantly more readily than can the mature organ. Around the time of weaning, the initial high metal uptake begins to fall towards adult levels. It is difficult to evaluate the contribution of altered variables such as luminal contents, luminal transit time, presence of unstirred layers and mucus, surface area available for absorption, concentration of microvilli and possible other factors to these changes in metal absorption with age. Comparison of function in immature and mature intestine is further complicated by the facts that thickness of the intestinal wall, its internal diameter and its length all change with development. As a result, absorbing area per unit length or weight will vary with age, and neither weight nor length of the tissue can provide a basis for comparing the intrinsic metal permeability at different ages. The term *intrinsic* here refers to the properties of the intestinal absorbing surface itself, independently of composition of luminal contents, unstirred layers, and other external factors.

A new approach to the problem of how to compare the intrinsic properties of the absorbing surface area in the intestine at different stages of development has recently been suggested (Foulkes et al. 1991; Foulkes and Bergman 1993); it uses the rate of alcohol absorption as measure of surface area. Alcohol is absorbed purely passively, and because of its high lipid and water solubility, the exponential rate of its absorption should depend on the size, not the composition of the absorbing area. As a result, the ratio of the rate of absorption of Cd or other solutes to that of ethanol can be directly compared in animals at different stages of development, and provides a relative measure of the intrinsic permeability of the mucosa to these solutes.

These experiments led to the conclusion that shortly after weaning, the intrinsic Cd permeability of the brush border membrane in rat jejunum still remains twice as high as in the mature organ. A similar result was obtained with Hg (Foulkes and Bergman 1993). The fall in the intrinsic permeability of jejunum to these metals during post-weaning development parallels the decrease in the number of "polar" (intercellular?) diffusion pathways.

4 Metal Transport by Renal Epithelium

4.1 Introduction

Renal tubular epithelium is involved in transport of very large amounts of various solutes. Because of the large blood supply of the renal cortex, its very high rate of metabolism, the ability to accumulate high concentrations of many solutes, and other factors, tubules in the renal cortex are the target of many toxic substances, including heavy metals. Within the context of epithelial transport of heavy metals the important points to be considered here are (1) how are the metals taken up by the tubule cells, (2) what is their intracellular disposition, and (3) how does renal epithelium participate in metal excretion. Aspects of these topics, and their toxicological implications, have recently been presented elsewhere in a review which may be consulted for further detail (Foulkes 1992).

4.2 The Renal Tubule as Target Site of Heavy Metals

Whereas metabolism of tissue preparations in vitro is generally sensitive to heavy metals, specific target sites for several of these metals have been identified in vivo. For instance, following chronic exposure to Cd, or acute administration of Hg or uranium, a major fraction of the total body load of these metals accumulates in the renal cortex. This is illustrated by the finding of Friberg (1956) that one day after cessation of Hg injections into rats the concentration of the metal in kidney exceeded that in liver by a factor of about 50.

Reasons for such preferential high accumulation, and its mechanism, will be discussed in Sect. 4.3. A not unexpected consequence of the accumulation is the inhibition of normal tubular function as reflected in decreased solute transport. Indeed, one of the characteristic signs of metal intoxication frequently is a syndrome resembling the congenital Fanconi syndrome (Chisolm et al. 1955). This condition is characterized by faulty reabsorption of solutes normally extracted from proximal tubular urine, with consequent aminoaciduria, phosphaturia, glucosuria, etc.

Other than the well-known general susceptibility of the kidney to toxic insult, the reason for the preferential accumulation of metals in renal cortex is not well understood. Binding of metals to tissue sulfhydryl groups might play a role in this process, and a fall in the concentration of renal sulfhydryl groups has been reported following exposure e.g., to cis-platin (Weiner and Jacobs 1983) or Hg (Johnson 1982). A problem with such an explanation for the high metal affinity of renal cortex arises from the obvious fact that other tissues, like liver, contain equally high concentrations of -SH groups, yet do not accumulate the metals to the same extent. In addition, the fall in free sulfhydryl concentration of renal cortex during exposure to metals is not stoichiometrically related to the amount of metal accumulated.

Another interesting suggestion is that renal accumulation of metals results from their trapping specifically by metallothionein; partial support for this view was presented in Sect. 2.3 above. This hypothesis, however, again fails to take into account that other tissues, including liver, also contain or have the ability to synthesize large amounts of metallothionein.

A potentially important factor contributing to the preferential renal uptake of heavy metals is that these can enter tubule cells from both tubular urine and peritubular blood. Berlin and Gibson (1963), for instance, found that 50% of a dose of Hg infused into rabbits over 4 h was accumulated by normal as well as by non-filtering kidneys. These processes will be discussed further in Sect. 4.4. Much effort has also been spent on determining whether metal intoxication of the kidney results from uptake of heavy metals to some threshold or critical level above which the incidence of malfunction in an exposed population begins to increase (Kjellstrom et al. 1984). This question is considered in the next section (4.3).

4.3 The Concept of Critical Levels of Toxic Metals in Renal Cortex

Especially for the purposes of risk assessment and regulation, many attempts have been made to define a so-called critical level for toxic heavy metals in the kidney. A theoretical analysis of the field was presented in a recent review (Foulkes 1990) and concluded that the concept can be useful, but only within well-defined limitations. Many facts contribute to these limitations.

For instance, metals are known to be distributed between several pools or compartments in the renal cortex (Sect. 2.4), and there is no reason why they should all be equally toxic. Reference was made in this connection to the presumed differences in the properties of Cd metallothionein (MT) and non-MT Cd. Another problem arises if the toxic action of the metal cannot readily be reversed. In such a case, removal of the metal by chelation treatment or because of its short renal half-life, may not alter the toxic response. Renal damage would accordingly not remain correlated with renal concentration of the metal. Such a dissociation between metal accumulation and toxic changes was described by Zhao et al. (1990).

When the renal half-life of a metal is relatively short, as in the case of Hg or U, levels of these metals at time of autopsy may be significantly lower than at the time when the lesion was initiated. This becomes especially true when the overt action of the metal is preceded by a possibly prolonged initiation delay, or when onset of renal damage leads to increased loss of metal from the tissue. For these and additional reasons, only Cd among the metals considered can be characterized by a critical level. Even for Cd, however, no unique critical level may exist; its value, empirically established in a specific population exposed in a defined manner, should not be automatically extrapolated to different populations exposed under different conditions.

4.4 Mechanism of Renal Metal Uptake

Study of metal uptake by renal epithelium in vivo is complicated by the need to administer the metals in various complexed forms, as already discussed in Sects. 1 and 2.2. A method extensively used for this purpose is the double indicator dilution technique as applied to the kidney by Silverman et al. (1970). This permits determination of metal uptake both from post-glomerular peritubular blood as well as from tubular urine. Alternatively, protein-free metal solutions may be directly introduced into the tubular lumen, as for instance in the work of Felley-Bosco and Diezi (1987) with Cd, and Zalups and Barfuss (1990) with Hg. Both these groups observed significant metal uptake in the direction of reabsorption.

Transfer of metals across tubular cell membranes in the direction of secretion has also been reported and can, in selected cases, be attributed to specific transport mechanisms (see also Sect. 2.2). As further discussed in Sect. 4.5, the appearance of metals in urine ahead of simultaneously administered glomerular markers, the so-called urinary precession phenomenon, cannot necessarily be equated to net secretion. Evidence for secretion has been presented by Borghgraef et al. (1956) for the organic mercurial diuretic Chlormerodrin, and by Weiner and Jacobs (1983) for cis platin compounds. Interaction between certain organic mercury complexes and the specific organic anion carrier system at the basolateral membrane of the rat renal tubule was reported by Ullrich et al. (1989).

Another specific means for mediating renal metal uptake is the carrier mechanism specifically transporting anionic proteins such as Cd metallothionein out of the proximal tubular lumen (Foulkes 1978); in contrast, no basolateral metallothionein uptake could be demonstrated. Metal complexes of metallothionein are generally freely filterable at the glomerulus, and reabsorption of filtered Cd metallothionein by the normal kidney is an efficient process

(Nomiyama and Foulkes 1977). The original proposal (Piscator 1964) that metallothionein serves as a transport protein to carry Cd from liver to kidney was, of course, predicated upon the existence of an efficient mechanism of renal uptake of Cd metallothionein from circulating plasma. The proposal is strengthened by the demonstration that Cd metallothionein, infused with osmotic minipumps at a rate approaching that of its likely physiological release from liver, accumulates preferentially in renal cortex (Johnson and Foulkes 1980).

There is no reason to assume that all, or even a major fraction, of plasma Cd circulates as a complex with metallothionein. It follows that a significant portion of renal cortical Cd is likely to have been carried to the kidney in some other complexed form, as for instance with glutathione. Arterial injection of Cd or Hg together with excess glutathione leads to their basolateral uptake by the rabbit kidney (unpubl. obser.) (see Sect. 2.2). A role of gamma-glutamyltranspeptidase in renal Hg uptake was suggested by Tanaka et al. (1990). Renal uptake of metals from both sides of the tubular epithelium may help explain their high renal accumulation.

Considerable effort has accordingly been made to explore the mechanism of renal metal uptake from various low-molecular weight complexes in plasma. The role of the metal ligands in these processes has already been considered in Sects. 1 and 2.2. The question of major interest is whether metals cross cell membranes of renal epithelium by mechanisms similar to those discussed in Sect. 3 in relation to metal uptake by jejunal epithelium.

Evidence in favor of the conclusion that similar metal uptake mechanisms are indeed at work have been discussed in some detail elsewhere (Foulkes and Blanck 1990). One of the pertinent findings is illustrated in Fig. 4, which shows that basolateral renal uptake of Cd is a barrier or diffusion-limited process. The panel on the left describes disappearance of Cd from plasma in presence of the low-molecular weight chelator mercaptoethanol during a single passage through the rabbit kidney. The metal, bound to excess chelator, and inulin had been injected intraarterially in a single bolus, and their recoveries were determined in rapidly collected sequential samples of renal venous blood. Net uptake of Cd under these conditions was relatively slow, judging from the fact that recovery of Cd approached that of non-filtered inulin; the Cd deficit after occlusion is equivalent to the area between the recovery curves for Cd and inulin.

The right panel shows that when the contact time between Cd-containing plasma and renal parenchyma was lengthened by transient trapping of the bolus in the kidney, uptake of Cd was approximately doubled. Now if this uptake resulted from a slow permeation of the Cd-mercaptoethanol complex out of the inulin space of the tissue, one would not expect it to be sensitive to other cations. However, as in jejunum (see Sect. 3.4), addition of Zn to the bolus significantly reduced uptake of Cd.

On the apical side, several metals including Cd (Bevan and Foulkes 1989), Zn (Menard and Cousins 1983) and Pb (Victery et al. 1984) have been shown to bind to the renal tubular cell membrane. Presumably this binding at least in part

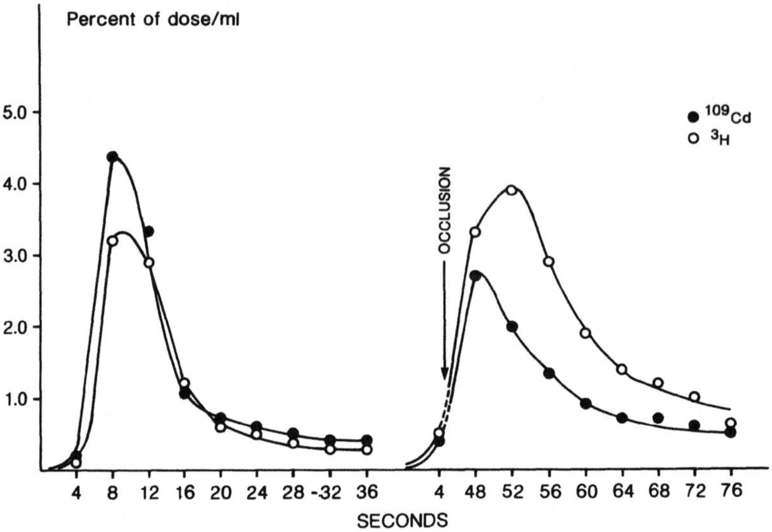

Fig. 4. Diffusion limitation of renal Cd uptake. Bolus containing mercaptoethanol together with labeled Cd and inulin was injected at time zero into the aorta of anesthetized rabbits, and renal venous blood collected continuously, as shown for the control (*left*), or with a 40 s interruption by aortic occlusion (*right*). Results are expressed as percent of injected dose/ml renal venous plasma. (Foulkes 1989)

is the equivalent of step 1A of metal uptake in jejunum. Removal from the lumen of Cd microinjected into the renal tubule is reduced by cysteine (Felley-Bosco and Diezi 1987). The amino acid here, again in analogy with jejunal Cd transport, presumably reduces the availability of Cd to the membrane binding sites. Such an effort could also account for the report that excess cysteine or glutathione reduce both apical and basolateral Cd uptake by LLC-PK1 cells (Brueggeman et al. 1990). All these facts point to significant similarities between membrane transport of metals by renal and jejunal epithelium.

4.5 Role of Epithelium in Renal Metal Excretion

Once heavy metals have been absorbed into the body, a major route for their excretion is in the form of various complexes in urine. The excretion may not always reflect total body burden, as for instance in the case of a bone-seeking element such as Pb. The final urinary concentration is, as usual, determined by the processes of filtration, reabsorption and secretion. Diffusibility and filterability of the complexes has already been referred to in Sect. 1. Thus, the free filterability of PbEDTA, and its low reabsorption, provide the basis for the chelate therapy of Pb poisoning. The present section will review evidence for tubular reabsorption and secretion of metals or metal compounds.

Quantitation of tubular reabsorption of filtered metals is complicated by questions about the glomerular filterability of plasma metals (Foulkes 1977) and by the likely presence of filterable plasma metals in a variety of low-molecular weight complexes. Anticoagulants themselves may bind metals, and the charge distribution in artificial filtration membranes may differ greatly from that in the glomerular membrane. Nevertheless, the clearance studies of Vander et al. (1977) clearly point to tubular reabsorption of lead, a conclusion subsequently reinforced by the stop-flow analysis of Victery et al. (1979). Direct demonstration of tubular metal reabsorption required the development of suitable microinjection or perfusion procedures, as referred to in the preceding section (4.4). Reabsorption of metals from the tubular lumen could readily be demonstrated with these micro techniques. As mentioned, the properties of such apical metal uptake resemble in several aspects the details of epithelial metal transport described for mucosal epithelium.

Evidence of tubular secretion of metals has also been reported by several investigators (see Sect. 4.4). It is important to emphasize, however, that urinary

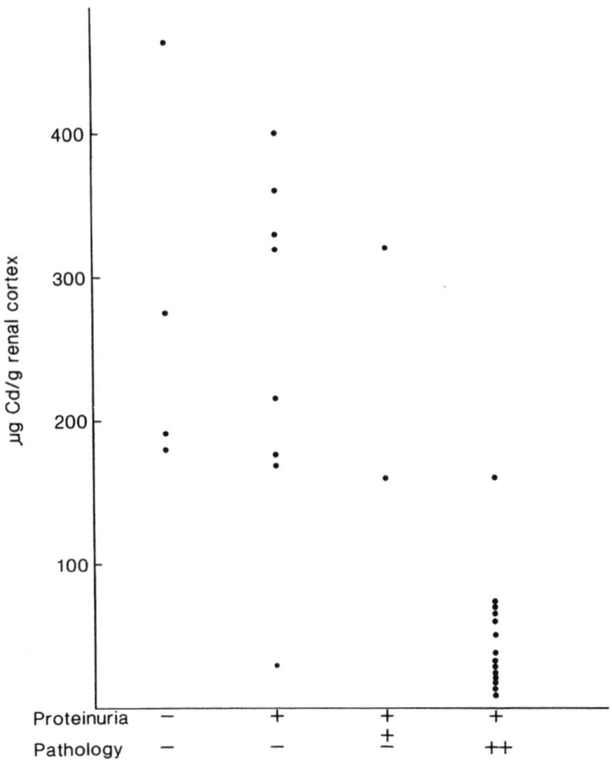

Fig. 5. Changes in renal cortical Cd levels during chronic exposure. Each *point* represents one individual from a population of workers exposed to Cd; the resultant pathological changes are indicated *at the bottom.* (Nomiyama 1977)

precession of a metal ahead of simultaneously injected inulin cannot be equated to net urinary secretion. Such precession was described, for instance, by Vostal and Heller (1968) and Bronner and Thompson (1961). Foulkes and Banks (1968) attributed urinary precession of various cations and anions to transtubular equilibration between tubular urine and plasma at the level of the inner medulla; whether such equilibration involves transcellular or paracellular pathways is not known.

A special case is the loss of renal metal by leakage from damaged kidneys into the urine. The resulting decrease in renal cadmium level has repeatedly been observed (see review by Bernard and Lauwerys 1986), and is illustrated in Fig. 5 for a small number of industrially exposed workers (Nomiyama 1977); note that severe renal damage in these workers was accompanied by a fall in tissue Cd level.

A steady state Cd level is approached only after relatively long exposures, in agreement with the prolonged renal half-life of the metal. Subsequently, renal Cd levels tend to fall, with an increase in urinary Cd levels. The urinary increase may result in part from depression of normal reabsorption of, for instance, Cd metallothionein; indeed, tubular proteinuria is a typical sign of Cd intoxication (Piscator 1966). The net loss of kidney Cd, however, must result from its movement out of the tissue. This presumably, reflects cellular leakage, as in the case of some urinary enzymes, rather than active secretion by tubular epithelium.

5 Metal Transport by Pulmonary Epithelium

In an occupational environment, pulmonary uptake of metals may become a major source of increased metal burdens, especially as it is more complete than that in the intestine. Morrow et al. (1980), for instance, found that 50% of small respirable particles of $PbCl_2$ or $Pb(OH)_2$ deposited in the human lung were absorbed within 14 h. Rabinowitz et al. (1977) reported that normal man absorbs daily over 90% of deposited Pb. In the studies of Hadley et al. (1980), removal of Cd from a slurry of respirable particles instilled intratracheally into rats approached 80% in 24 h; a contribution of the gastrointestinal tract was believed unlikely because of the absence of fecal Cd during the first days following exposure. The efficiency of pulmonary Cd uptake is also reflected in the significantly higher blood Cd of smokers than of non-smokers (Elinder et al. 1983).

The mechanism of the relatively high metal permeability of alveolar epithelium has not yet been clarified. Important contributing factors to this process may be the length of contact of metal with epithelium, absence of the luminal contents which normally affect metal uptake from the intestine, availability of paracellular diffusion pathways, a relatively high membrane fluidity, and other factors such as discussed in Sects. 3 and 4. Little definite information is as yet available on this question.

6 Activity of Some Other Epithelial Tissues

Work from several laboratories supports the likelihood that olfactory epithelium and the primary sensory neurones may provide a direct route of access to the central nervous system for various toxic agents including metals. Thus, Perl and Good (1987) in their study of Al accumulation and Alzheimer's disease concluded that the metal is taken up via the rabbit olfactory epithelium. A contribution of this epithelium to metal uptake was also reported by Hastings and Sun (1987); in their studies, high Cd concentrations following unilateral nasal exposure of rats were found in the corresponding olfactory epithelium and olfactory bulb. The possibility exists, of course, that other metals can also find direct access to the central nervous system in this manner. Little is known about the mechanism of metal uptake by olfactory epithelium.

Extensive work on epithelial solute transport has been carried out with frog skin and amphibian or reptilian bladder. While in such preparations effects of heavy metals on transport especially of sodium have repeatedly been observed, not much information is available on how they react with these tissues. Takada and Hayashi (1980) reported that Cd is taken up by frog skin from both epidermal and dermal sides. Not surprisingly, such uptake leads to changes in normal sodium transport, as reflected in the short circuit current. The relevance of this work for the analysis of Cd effects in mammalian epithelia is somewhat obscured by the use of millimolar as compared to micromolar concentrations. Addition of Cd to the mucosal side at these high concentrations stimulates transcellular sodium transport (Scholtz and Zeiske 1988). Hillyard and Gonick (1976) attributed the stimulation of short circuit current in frog skin by Cd to an action of the metal on apical membranes. Depolarization of the apical membrane by polyvalent cations (La, Co, Ni, Cd) is the explanation offered by Grinstein et al. (1978) for the observed increase in transepithelial sodium transport. This interpretation of the effects of polyvalent cations recalls the proposed role of anionic membrane sites in Cd uptake (see Sect. 3.4). These non-mammalian epithelia may well prove useful in further work on the mechanism of epithelial transport of heavy metals.

Acknowledgment. Preparation of this review was supported in part by NIH grants ES-04840 and ES-02453.

References

Aungst BJ, Fung HL (1981) Kinetic characterization of in vitro lead transport across the rat small intestine: mechanism of intestinal lead transport. Toxicol Appl Pharmacol 61: 39–47

Barton JC (1984) Active transport of lead-210 by everted segments of rat duodenum. Am J Physiol 247: G193–G198

Becker WM, Hoekstra WG (1971) The intestinal absorption of zinc. In: Skoryna SC, Waldron-Edwards D (eds) Intestinal absorption of metal ions, trace elements and radionuclides. Pergamon Press, Oxford, pp 229–256

Berlin M, Gibson S (1963) Renal uptake retention and excretion of mercury. Arch Environ. Health 6: 617–625

Berlin M, Nordberg G, Hellberg J (1973) The uptake and distribution of methyl mercury in the brain of *Saimiri sciurens* in relation to behavioral and morphological changes. In: Miller MW, Clarkson TW (ed) Mercury, mercurials and mercaptans. CC Thomas, Springfield, IL, pp 187–208

Bernard A, Lauwerys R (1986) Effects of cadmium exposure in humans. In: Foulkes EC (ed) Cadmium. Handbook of experimental pharmacology, vol 80. Springer, Berlin Heidelberg New York, pp 135–177

Bevan C, Foulkes EC (1989) Interaction of cadmium with brush border membrane vesicles from the rat small intestine. Toxicology 54: 297–309

Blazka ME, Shaikh ZA (1991) Differences in cadmium and mercury uptakes by hepatocytes: role of calcium channels. Toxicol Appl Pharmacol 110: 355–363

Bonewitz RF, Voner C, Foulkes EC (1982) Uptake and absorption of zinc in perfused rat jejunum: the role of endogenous factors in the lumen. Nutr Res 2: 301–307

Bonewitz RF, Foulkes EC, O'Flaherty EJ, Hertzberg VS (1983) Kinetic of zinc absorption by rat jejunum: effects of adrenalectomy and dexamethasone. Am J Physiol 244: G314–G320

Borghgraef RRM, Kessler RH, Pitts RF (1956) Plasma regression, distribution and excretion of radiomercury in relation to diuresis following the intravenous administration of ^{203}Hg-labelled chlormerodrin to the dog. J Clin Invest 35: 1055–1066

Bronner F, Thompson DD (1961) Renal transtubular flux of electrolytes in dogs with special reference to calcium. J Physiol (Lond) 157: 232–250

Brueggeman IM, Van Bladeren PJ, Temmink JHM (1990) Transport and toxicity of $CdCl_2$ in LLC-PK1 cells. Toxicol Lett 53: 167–168

Bushnell PJ, DeLuca HF (1981) Lactose facilitates the intestinal absorption of lead in weanling rats. Science 211: 61–63

Cain K, Griffiths BL (1980) Transfer of liver cadmium to the kidney after aflatoxin-induced liver damage. Biochem Pharmacol 29: 1852–1855

Cain K, Holt DE (1983) Studies of cadmium-thionein induced nephropathy: time course of cadmium-thionein uptake and degradation. Chem Biol Int 43: 223–237

Carroll KG, Spinelli FR, Goyer RA (1970) Electron probe microanalyzer localization of lead in kidney tissue of poisoned rats. Nature 227: 1056

Chisolm JJ, Harrison HC, Eberlein WR, Harrison HE (1955) Aminoaciduria hypophosphatemia and rickets in lead poisoning: study of a case. Am J Dis Child 89: 159–168

Clarkson TW, Cross A (1961) Studies of the action of mercuric chloride on intestinal absorption. Univ Rochester AEC Rep 588

Clarkson TW, Kench JE (1956) Urinary excretion of amino acids by men absorbing heavy metals. Biochem J 62: 361–372

Clarkson TW, Small H, Norseth T (1973) Excretion and absorption of methyl mercury after polythiol resin treatment. Arch Environ Health 26: 173–176

Conrad ME, Barton JC (1978) Factors affecting the absorption and excretion of lead in the rat. Gastroenterology 74: 731–740

Cotzias GC, Borg DC, Selleck B (1961) Virtual absence of turnover in Cd metabolism: Cd^{109} studies in the mouse. Am J Physiol 201: 927–930

Cousins RJ (1985) Absorption, transport and hepatic metabolism of copper and zinc: special reference to metallothionein and ceruloplasmin. Physiol Rev 65: 238–309

Elinder C-G, Friberg L, Lind B, Jawaid M (1983) Lead and cadmium levels in blood samples from the general population of Sweden. Environ Res 30: 233–253

Endo T, Nakaya S, Kimura R, Murata T (1986) Gastrointestinal absorption of inorganic mercury compounds in vitro. Toxicol Appl Pharmacol 83: 187–196

Evans GW, Johnson EC (1980) Zinc absorption in rats fed a low-protein diet supplemented with tryptophan or picolinic acid. J Nutr 110: 1076–1080

Evans GW, Johnson EC, Johnson PE (1979) Zinc absorption in the rat determined by radioisotope dilution. J Nutr 109: 1258–1264
Felley-Bosco E, Diezi J (1987) Fate of cadmium in rat renal tubules: a microinjection study. Toxicol Appl Pharmacol 91: 204–211
Foulkes EC (1974) Excretion and retention of cadmium, zinc and mercury by rabbit kidney. Am J Physiol 227: 1356–1360
Foulkes EC (1977) Mechanism of renal excretion of environmental agents. In: Lee DHK (ed) Handbook of physiology, section 9: reaction to environmental agents. Am Physiol Soc, Bethesda, MD, pp 495–502
Foulkes EC (1978) Apparent competition between myoglobin and metallothionein for renal reabsorption. Proc Soc Exp Biol Med 159: 321–323
Foulkes EC (1980) Some determinants of intestinal cadmium transport in the rat. J Environ Pathol Toxicol 3: 471–481
Foulkes EC (1984) Intestinal absorption of heavy metals. In: Csaky TZ (ed) Pharmacology of intestinal permeation. Handbook experimental pharmacology, vol 70. Springer, Berlin Heidelberg New York, pp 543–565
Foulkes EC (1985a) Interactions between metals in rat jejunum: implications on the nature of cadmium uptake. Toxicology 37: 117–125
Foulkes EC (1985b) Critical concentration of cadmium in human renal cortex: an update. Trace substances in environmental health, vol 119. University of Missouri, Columbia, pp 129–133
Foulkes EC (1986) Absorption of cadmium. In: Foulkes EC (ed) Cadmium. Handbook of experimental pharmacology, vol 80. Springer, Berlin Heidelberg New York, pp75–100
Foulkes EC (1988) On the transfer of heavy metals across cell membranes. Toxicology 52: 263–272
Foulkes EC (1989) On the mechanism of cellular cadmium uptake. Biol Trace Elem Res 21: 195–200
Foulkes EC (1990) The concept of critical levels of toxic heavy metals in target tissues. Crit Rev Toxicol 20: 327–339
Foulkes EC (1991a) Further findings on the mechanism of cadmium uptake by intestinal mucosal cells. Toxicology 70: 261–270
Foulkes EC (1991b) Nature of Cd and Hg effects on epithelial amino acid transport in vivo and role of chelators. Toxicology 69: 177–185
Foulkes EC (1992) Renal transport of heavy metals. In: Anders MW, Dekant W, Henschler D et al. (eds) Renal disposition and nephrotoxicity of xenobiotics. Academic Press, New York, pp 117–134
Foulkes EC (1993) Metallothionein and glutathione as determinants of cellular retention and extrusion of cadmium and mercury. Life Sci 52: 1617–1620
Foulkes EC, Banks RO (1968) Significance of urinary precession of sodium over simultaneously injected inulin. Am J Physiol 215: 574–581
Foulkes EC, Bergman D (1993) Inorganic mercury absorption in mature and immature rat jejunum: transcellular and intercellular pathways in vitro and in everted sacs. Toxicol Appl Pharmacol 120: 89–95
Foulkes EC, Blanck S (1990) Acute cadmium uptake by rabbit kidneys: mechanism and effects. Toxicol Appl Pharmacol 102: 464–473
Foulkes EC, McMullen DM (1986a) Endogenous metallothionein as determinant of intestinal cadmium absorption: a reevaluation. Toxicology 38: 285–291
Foulkes EC, McMullen DM (1986b) On the mechanism of nickel absorption in the rat jejunum. Toxicology 38: 35–42
Foulkes EC, McMullen DM (1987) Kinetics of transepithelial movement of heavy metals in rat jejunum. Am J Physiol 253: G134–G138
Foulkes EC, Voner C (1981) Effects of Zn status, bile and other endogenous factors on jejunal cadmium absorption. Toxicology 22: 115–122
Foulkes EC, Mort T, Buncher R (1991) Intestinal cadmium permeability in mature and immature rats. Proc Soc Exp Biol Med 197: 477–481
Frame MDS, Milanick MA (1991) Mn and Cd transport by the Na-Ca exchange of ferret red blood cells. Am J Physiol 261: C467–C475

Friberg L (1956) Studies on the accumulation, metabolism and excretion of inorganic mercury after prolonged subcutaneous administration to rats. Acta Pharmacol Toxicol 12: 411–427
Fromter E, Diamond J (1972) Route of passive ion permeation in epithelia. Nature (New Biol) 235: 9–13
Grinstein S, Candia O, Erlij D (1978) Nonhormonal mechanisms for the regulation of transepithelial sodium transport: the roles of surface potential and cell calcium. J Membr Biol 40 (Spec No): 261–280
Gunn SA, Gould TC (1957) Selective accumulation of Cd^{115} by cortex of rat kidney. Proc Soc Exp Biol Med 96: 820–823
Hadley JG, Conklin AS, Sanders CL (1980) Rapid solubilization and translocation of ^{109}CdO following pulmonary deposition. Toxicol Appl Pharmacol 54: 156–160
Hastings L, Sun TJ (1987) Effects of cadmium on rat olfactory system. Olfaction and taste IX. Ann NY Acad Sci 510: 355
Hider RC, Ejim L, Taylor PP, Gale R, Huehns E, Porter JB (1990) Facilitated uptake of zinc into human erythrocytes. Biochem Pharmacol 39: 1005–1012
Hillyard SD, Gonick HC (1976) Effects of Cd^{++} on short-circuit current across epithelial membranes. I. Interactions with Ca^{++} and vasopressin on frog skin. J Membr Biol 26: 109–119
Hinkle PM, Kinsella PA, Osterhoudt KC (1987) Cadmium uptake and toxicity via voltage-sensitive calcium channels. J Biol Chem 262: 16333–16337
Hirayama K (1975) Transport mechanism of methyl mercury. Intestinal absorption, biliary excretion and distribution of methyl mercury. Kumamoto Med J 28: 151–163
Israel EJ, Pang KY, Harmatz PR, Walker WA (1987) Structural and functional maturation of rat gastrointestinal barrier with thyroxine. Am J Physiol 252: G762–G767
Johnson DR (1982) Role of renal cortical sulfhydryl groups in development of mercury-induced renal toxicity. J Toxicol Environ Health 9: 119–126
Johnson DR, Foulkes EC (1980) On the proposed role of metallothionein in the transport of cadmium. Environ Res 21: 360–365
Kagi JHR, Vallee BL (1960) Metallothionein: a cadmium and zinc containing protein from equine renal cortex. J Biol Chem 235: 3460–3465
Kello D, Dekanic D, Kostial K (1979) Influence of sex and dietary calcium on intestinal cadmium absorption in rats. Arch Environ Health 34: 30–33
Kerper LE, Ballatori N, Clarkson TW (1992) Methylmercury transport across the blood-brain barrier by an amino acid carrier. Am J Physiol 262: R761–R765
Kiyozumi M, Kojima S (1978) Studies on poisonous metals. V. Excretion of cadmium through bile and gastrointestinal mucosa and effect of chelating agents on its excretion in cadmium-pretreated rats. Chem Pharm Bull (Tokyo) 26: 3410–3415
Kjellstrom T (1971) A mathematical model for the accumulation of cadmium in human kidney cortex. Nord Hyg Tidskr 52: 111–119
Kjellstrom T, Ellinder C-G, Friberg L (1984) Conceptual problems in establishing the critical concentration of cadmium in human kidney cortex. Environ Res 33: 284–295
Kojima S, Kiyozumi M (1974) Studies on poisonous metals I. Transfer of cadmium chloride across rat small intestines in vitro and effect of chelating agents on its transfer. Yakugaku Zasshi 94: 695–701
Kostial K, Kello D, Jugo S, Rabar I, Maljkovic T (1978) Influence of age on metal metabolism and toxicity. Environ Health Perspect 25: 81–86
Kostial K, Rabar I, Blanusa M, Landeka M (1979) Effect of age on heavy metal absorption. Proc Nutr Soc 38: 251–256
Lou M, Garay R, Alda JO (1991) Cadmium uptake through the anion exchanger in human red blood cells. J Physiol 443: 123–136
Machen TE, Erlig D, Wooding FBP (1972) Permeable junctional complexes: the movement of lanthanum across rabbit gall bladder and intestine. J Cell Biol 54: 302–312
Menard MP, Cousins RJ (1983) Zinc transport by brush border membrane vesicles from rat intestine. J Nutr 113: 1434–1442
Mertz W (1969) Chromium occurrence and function in biological systems. Physiol Rev 49: 163–239

Methfessel AH, Spencer H (1973) Zinc metabolism in the rat. I. Intestinal absorption of zinc. J Appl Physiol 34: 58–62
Miettinen J (1973) Absorption and elimination of dietary mercury or methylmercury in man. In: Clarkson TW, Miller MW (eds) Mercury, Mercurials and Mercaptans. CC Thomas, Springfield, pp 233–240
Morrow PE, Beiter H, Amato F, Gibb FR (1980) Pulmonary retention of lead: an experimental study in man. Environ Res 21: 373–384
Neathery MW, Miller WJ (1975) Metabolism and toxicity of cadmium, mercury and lead in animals: a review. J Dairy Sci 58: 1767–1781
Nomiyama K (1977) Does a critical concentration of cadmium in human renal cortex exist? J Toxicol Environ Health 3: 607–609
Nomiyama K, Foulkes EC (1977) Reabsorption of filtered metallothionein in the rabbit kidney. Proc Soc Exp Biol Med 156: 97–99
Nomiyama K, Nomiyama H (1982) Tissue metallothioneins in rabbits chronically exposed to cadmium, with special reference to the critical concentration of cadmium in the renal cortex. In: Foulkes EC (ed) Biological roles of metallothionein. Elsevier North Holland, Amsterdam, pp 47–67
Nordberg GF, Goyer R, Nordberg M (1975) Comparative toxicity of cadmium-metallothionein and cadmium chloride on mouse kidney. Arch Pathol 99: 192–197
Ohta H, DeAngelis MV, Cherian MG (1989) Uptake of cadmium and metallothionein by rat everted intestinal sacs. Toxicol Appl Pharmacol 101: 62–69
Oskarsson A, Squibb KS, Fowler BA (1982) Intracellular binding of lead in the kidney: the partial isolation and characterization of post-mitochondrial lead binding components. Biochem Biophys Res Commun 104: 290–298
Page AL, El-Amamy MM, Chang AC (1986) Cadmium in the environment and its entry into terrestrial food chain crops. In: Foulkes EC (ed) Cadmium. Handbook of experimental pharmacology, vol 80. Springer, Berlin Heidelberg New York, pp 33–74
Pecoud A, Donzel P, Schelling JL (1975) Effect of foodstuffs on the absorption of zinc sulfate. Clin Pharmacol Ther 17: 469–474
Pekas JC (1966) $Zinc^{65}$ metabolism: gastrointestinal secretion by the pig. Am J Physiol 211: 407–413
Perl DP, Good PF (1987) Uptake of aluminum into central nervous system along nasal-olfactory pathway;s. Lancet May 2: 1028
Piscator M (1964) Cadmium in the kidneys of normal human beings and the isolation of metallothionein from liver of rabbits exposed to cadmium. Nord Hyg Tidskr 45: 76–82
Piscator M (1966) Proteinuria in chronic cadmium poisoning 3. Electrophoretic and immunoelectrophoretic studies on urinary proteins from cadmium workers, with special reference to the excretion of low molecular weight proteins. Arch Environ Health 12: 335–344
Powell D (1981) Barrier function of epithelia. Am J Physiol 241: G275–G288
Quarterman J, Morrison JN (1975) The effects of dietary calcium and phosphorus on the retention and excretion of lead in rats. Br J Nutr 34: 351–362
Rabar I, Kostial K (1981) Bioavailability of cadmium in rats fed various diets. Arch Toxicol 47: 63–66
Rabinowitz MB, Weatherill GW, Kopple JD (1976) Kinetic analysis of lead metabolism in healthy humans. J Clin Invest 58: 260–270
Rabinowitz MB, Wetherill GW, Kopple JD (1977) Magnitude of lead intake from respiration by normal man. J Lab Clin Med 90: 238–248
Rabinowitz MB, Kopple JD, Weatherill GW (1980) Effect of food intake and fasting on gastrointestinal lead absorption in humans. Am J Clin Nutr 33: 1784–1788
Raghavan SRV, Culver BD, Gonick HC (1981) Erythrocyte lead-binding protein after occupational exposure IV. Influence of lead inhibition of membrane Na^+, K^+-adenosinetriphosphatase. J Toxical Environ Health 7: 561–568
Richards MP, Cousins RJ (1975) Mammalian zinc homeostasis: requirement for RNA and metallothionein synthesis. Biochem Biophys Res Commun 64: 1215–1223
Sahagian BM, Harding-Barlow I, Perry HM (1967) Transmural movements of zinc, manganese, cadmium and mercury by rat small intestine. J Nutr 93: 291–300

Sasser LB, Jarboe GE, Walter BK, Kelman BJ (1978) Absorption of mercury from ligated segments of the rat gastrointestinal tract. Proc Soc Exp Biol Med 157: 57–60
Scholtz E, Zeiske W (1988) A novel synergistic stimulation of Na^+-transport across frog skin (*Xenopus laevis*) by external Cd^{2+}- and Ca^{2+}-ions. Pfluegers Arch Eur J Physiol 413: 174–180
Silverman M, Aganon MA, Chinard FP (1970) Specificity of monosaccharide transport in the dog kidney. Am J Physiol 218: 743–750
Simons TJ (1986) The role of anion transport in the passive movement of lead across the human red cell membrane. J Physiol (Lond) 378: 287–312
Six KM, Goyer RA (1972) The influence of iron deficiency on tissue content and toxicity of ingested lead in the rat. J Lab Clin Med 79: 128–136
Skoryna SC, Waldron-Edwards D (eds) (1970) Intestinal absorption of metal ions, trace elements and radionuclides. Pergamon Press, New York
Smith CM, DeLuca HF, Tanaka Y, Mahaffey KR (1978) Stimulation of lead absorption by vitamin D administration. J Nutr 108: 843–847
Smith KT, Cousins RJ, Silbon BL, Failla ML (1978) Zinc absorption and metabolism by isolated vascularly perfused rat intestine. J Nutr 108: 1849–1857
Squibb K, Pritchard JB, Fowler BA (1984) Cadmium-metallothionein nephropathy: relationships between ultrastructural/biochemical alterations and ultracellular cadmium binding. J Pharm Exp Ther 229: 311–321
Stacy BD, Thornburn GD (1966) Chromium-51 ethylene-diaminetetraacetate for estimation of glomerular filtration rate. Science 152: 1076–1077
Sugawara N (1982) Role of metallothionein in zinc uptake from rat jejunum. In: Foulkes EC (ed) Biological roles of metallothionein. Elsevier, New York, pp 155–162
Takada M, Hayashi H (1980) Effect of cadmium on active sodium transport by the abdominal skin and the isolated epidermis of the bullfrog. Jpn J Physiol 30: 257–269
Takahashi H, Hirayama K (1971) Accelerated elimination of methyl mercury from animals. Nature 232: 201–202
Tanaka T, Naganuma A, Imura N (1990) Role of gamma-glutamyltranspeptidase in renal uptake and toxicity of inorganic mercury in mice. Toxicology 60: 187–198
Tepperman HM (1947) Effect of BAL and BAL-glucoside therapy on excretion and tissue distribution of injected cadmium. J Pharmacol Exp Ther 89: 343–349
Tidball CS (1964) Magnesium and calcium as regulators of intestinal permeability. Am J Physiol 206: 243–246
Torrubia JO, Garay R (1989) Evidence for a major route for Zn uptake in human red blood cells: $[Zn(HCO_3)_2Cl]^-$ influx through the Cl^-/HCO_3^- anion exchanger. J Cell Physiol 138: 316–322
Ullrich KJ, Rumrich G, Wieland T, Dekant W (1989) Contraluminal para-aminohippurate (PAH) transport in the proximal tubule of the rat kidney. Pfluegers Arch Eur J Physiol 415: 342–350
Vander AJ, Taylor DL, Kalitis K, Mouw DR, Victery W (1977) Renal handling of lead in dogs: clearance studies. Am J Physiol 233: F532–F538
Victery W, Vander AJ, Mouw DR (1979) Renal handling of lead in dogs: stop-flow analysis. Am J Physiol 237: F408–F414
Victery W, Miller CR, Fowler BA (1984) Lead accumulation by rat renal brush border membrane vesicles. J Pharmacol Exp Ther 231: 589–596
Vostal J, Heller J (1968) Renal excretory mechanisms of heavy metals. I. Transtubular transport of heavy metal ions in the avian kidney. Environ Res 2: 1–10
Weiner MW, Jacobs C (1983) Mechanism of cisplatin nephrotoxicity. Fed Proc 42: 2974–2978
Whittembury G, Rawlins FA (1971) Evidence for a paracellular pathway for ion flow in the kidney proximal tubule. Pfluegers Arch Eur J Physiol 330: 302–309
Zalups RK, Barfuss D (1990) Accumulation of inorganic mercury along the renal proximal tubule of the rabbit. Toxicol Appl Pharmacol 106: 245–253
Zhao JY, Foulkes EC, Jones M (1990) Delayed nephrotoxic effects of cadmium and their reversibility by chelation. Toxicology 64: 235–243

Chapter 3

Heavy Metal Cytotoxicity in Marine Organisms: Effects on Ca^{2+} Homeostasis and Possible Alteration of Signal Transduction Pathways

A. Viarengo[1]

Contents

Summary

The data reported show that heavy metals, such as Hg^{2+}, Cu^{2+}, Cd^{2+}, and Zn^{2+}, are able to affect the mechanisms of Ca^{2+} homeostasis. It has been demonstrated that heavy metals, probably interacting with the sulphydrylic groups of the membrane proteins involved in the structure of the Ca^{2+} channels, can enhance the flux of Ca cations through the plasma membrane, the membrane of calciosomes, and mitochondria.

Moreover, the high affinity of heavy metals for SH residues could explain the high sensitivity of Ca^{2+} ATPases to these metals. In fact, Ca^{2+} translocases contain SH groups critical for the enzyme activity.

In the case of Cu^{2+}, this cation, probably through the production of oxygen radicals, stimulates the process of lipid peroxidation of cell membranes, possibly contributing to the alteration of structural and enzymatic proteins involved in Ca^{2+} homeostasis through the oxidation of the sulphydrylic residues. In addition, heavy metals are also able to inhibit the Na^+/K^+ ATPase, a well-known SH-containing enzyme present in the plasma membrane. This is of particular

[1] Istituto di Fisiologia Generale, Università di Genova, Corso Europa 26, 16132 Genova, Italy

Advances in Comparative and Environmental Physiology, Vol. 20

interest considering that Na^+/K^+ translocase regulates a Na^+/Ca^{2+} exchanger, whose function in Ca^{2+} homeostasis has been demonstrated in the plasma membrane of many different cells.

The role of GSH and metallothioneins as important components of the cell mechanisms to reduce the cytotoxic effects of heavy metals has here also been described.

Data are also presented concerning the effects of heavy metals on the mechanism of signal transduction. It seems possible that the cations affecting Ca^{2+} homeostasis could render the cell no more sensitive to Ca-mediated hormones and external stimuli. In addition, heavy metals could exert a direct effect on different aspects of signal transduction pathways.

1 Introduction

Metals play fundamental roles in the cells of living organisms, a large number of them being recognized as essential for biological processes. Among these, Na^+ and K^+ (alkaline metals, group I) are involved in the formation of transmembrane cell potential (Hille 1984); Mg^{2+} (alkaline metal, group II) acts as a cofactor in many enzymatic reactions and it is involved in the maintenance of subcellular structures such as ribosomal assembly (Aikawa 1976); Ca^{2+} (alkaline metal, group II) plays an important role as a second messenger in the signal transduction pathways (Rasmussen and Barret 1984); Zn^{2+} and Cu^{2+} (subgroup B I and II) are important in membrane structure and function and are involved in the activation of more than 100 important enzymes (Evans 1973); Co^{2+} and Fe^{3+} (transition metals) are fundamental in the structure of biological compounds such as vitamins, cytochromes and respiratory pigments, etc. (Elinder 1986; Elinder and Friberg 1986).

However, when no established biological function is known for a metal or if an essential metal is present in excess in the environment, all these elements would usually be considered "pollutants" since they are accumulated in the cells where they can exert toxic effects.

In particular, elements of IB and IIB subgroups include both essential metals such as Zn^{2+} and Cu^{2+}, and hazardous cations such as Cd^{2+} and Hg^{2+} which are widely recognized as being among the more toxic environmental contaminants (Unep 1989).

These elements, due to their electron configuration, exhibit similar chemical characteristics (Eichorn 1973; Vouk 1986). They typically exhibit a strong affinity for nucleophilic groups of biological compounds, notably SH residues ($S > O > N$).

Interestingly, the more toxic metal cations are those showing greater affinity for SH groups ($Zn < Cd < CuI < Hg$). Therefore, although the importance of their biochemical interactions with oxydryl residues and imidazole groups of biological compounds should not be underestimated (Eichorn 1973), the effects of heavy metal cations on biological structures can be generally ascribed to their tendency to bind to SH residues of aminoacids, peptides and proteins. Therefore, in this paper, I will attempt to give a general interpretation, at molecular and cellular level, of the different cytotoxic effects of these cations based on their similar chemical characteristics.

In past years, many authors have described heavy metal effects on different components of the cells of both terrestrial and marine organisms. These observations have included inhibition of Na^+/K^+ ATPases (Bowler and Duncan 1970; Bouquegneau 1977; Bouquegneau and Gilles 1979; Webb 1979) and aminoacid transport (Viarengo et al. 1980; Farmanfarmaian et al. 1981), alteration of mitochondrial metabolism (George 1972; Scott and Major 1972; MacInnes and Thurberg 1973; Zaba and Harris 1978; Gould et al. 1985) and of the ATP/AEC values (Zaroogian et al. 1982; Viarengo et al. 1985), inhibition of total protein synthesis (Viarengo et al. 1980), alteration of the lysosomal physiology (Viarengo et al. 1981, 1985, 1987), and of various important metabolic pathways (Webb 1979; Viarengo 1989), depolymerization of cytoskeleton structures such as microtubule (Viarengo et al. 1989) or microfilaments (Moore and Viarengo unpubl. data), alteration of nuclear DNA or RNA metabolism (Viarengo et al. 1982; Dixon 1983; Accomando et al. 1990).

However, not all the molecular mechanisms responsible for these alterations are fully understood, nor has an attempt been made to find out a possible common trait in the action of heavy metal cations on cellular metabolism. In this regard, there is increasing evidence that heavy metals, due to their aforementioned chemical characteristics, can affect the mechanisms of Ca^{2+} homeostasis. In the cells, calcium homeostasis is realized by the concerted operation of different mechanisms which maintain the cytosolic Ca^{2+} level in the physiological range of 10^{-7} M. Specific Ca^{2+} ATPases are responsible for active Ca^{2+} extrusion and compartmentalization at the plasma membrane and at the endoplasmic or sarcoplasmic reticulum level, respectively.

Ca^{2+} flux is also regulated by different voltage (VOCs) and receptor operating (ROCs) Ca^{2+}-channels. Moreover, Na^+/Ca^{2+} exchangers are present in both plasma membrane and mitochondrial membranes. In these organelles, a Ca^{2+} uniporter seems to be involved in Ca^{2+} uptake from the cytosol (Carafoli 1987; Pietrobon et al. 1990).

Because the proteins involved in calcium translocation across cellular membranes often contain one or more SH groups critical to their activity (e.g., Ca^{2+} ATPases, Ca^{2+}-channels), they are a potential specific target for heavy metal cations.

The importance of the fact that heavy metal cations could impair Ca^{2+} homeostasis is that, as mentioned above, Ca^{2+} plays a fundamental role as second messenger in regulating various and important metabolic pathways. Consequently, an uncontrolled rise in the cytosolic Ca^{2+} concentration could compromise the physiological status of the cells and the correct cellular responses to Ca-mediated hormones and environmental stimuli.

This review provides a general description of how heavy metal toxicity in marine organisms appears to be mediated through alteration of Ca-dependent processes. Marine organisms have been chosen because they are continuously exposed to varying concentrations of metals in sea water, this fact being particularly evident near the coast where the metal contamination is generally higher.

2 Effects on Plasma Membranes

The mechanisms of cellular uptake of heavy metals are still not entirely understood. Metal cations could enter the cell through Ca^{2+} channels, as demonstrated in the case of Cd^{2+} (Hinkle et al. 1987; Verbost et al. 1989b), in the form of $[MeCl_2]^0$, through the lipid bilayer as proposed by Simkiss (1983), bound to high-molecular weight molecules by active endocytosis (Viarengo, unpubl. data), by carrier mediated mechanisms (Cousins 1985) as demonstrated for Zn, etc. (Fig. 1).

However, independently from the route of heavy metal entry into the cells, the cations crossing the external cell membrane can interact with the membrane components; therefore, the plasma membrane can be considered the first site of action of heavy metals. As noted above, heavy metal cations, due to their high affinity for SH groups, will tend to interact with sulphydrylic residues of structural and enzymatic proteins; among these, the Ca^{2+} translocase present in the plasma membrane could represent the primary target of heavy metal cations (Fig. 2).

The interaction of Cd^{2+} with the Ca^{2+} translocase present in the plasma membrane of fish gills has been recently studied (Verbost et al. 1988). It has been found that ATP-driven Ca^{2+} transport into the basolateral plasma membrane vesicles of the gill cells is extremely sensitive to the metal, being inhibited by 3 nM free Cd^{2+} in the presence of 250 μM free Ca^{2+}. Although the Ca^{2+} pump is calmodulin-dependent, Cd^{2+} does not seem to interfere with the calmodulin-mediated activation of the enzyme. Rather, the effects of Cd^{2+} are likely related to a direct interaction of the metal with the Ca^{2+} pump at the Ca^{2+} translocation

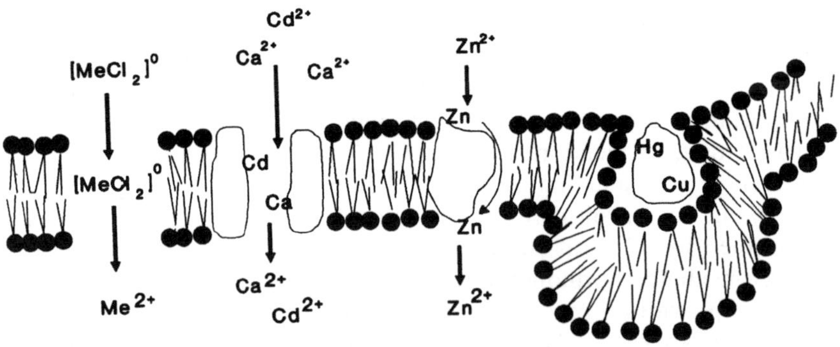

Fig. 1. Proposed mechanisms for heavy metal transport through the plasma membrane

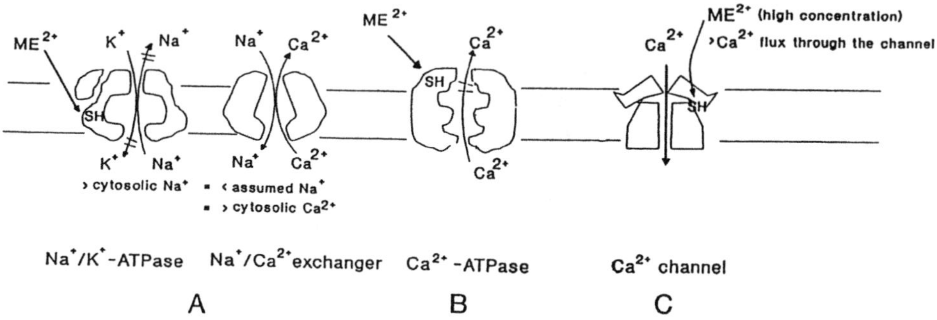

Fig. 2. Schematic representation of heavy metal effects on the enzymatic and structural proteins involved in Ca^{2+} transport through the plasma membrane

site, probably by binding to SH groups important for the catalytic activity of the enzyme (Verbost et al. 1988).

Similar results were obtained studying the effects of Cd^{2+} on transmembrane Ca^{2+} transport in human erythrocytes (Verbost et al. 1989a). Ca^{2+} translocase was found to be inhibited at extremely low concentrations ($K_i = 2$ nM free Cd^{2+}), whereas $Cd^{2+} < 10^{-4}$ M did not alter the membrane permeability to Ca^{2+}. Also, in this case, the metal seemed to inhibit the enzyme by direct interaction at the Ca^{2+} binding site and not to affect calmodulin stimulation.

Additionally, Cd^{2+} was found to competitively inhibit the ATP-driven Ca^{2+} transport ($I_{50} = 1.6$ nM free Cd^{2+} at 1 μM Ca^{2+}) in rat intestinal epithelial cells (Verbost et al. 1987). These results were obtained using specific buffers for Ca^{2+} (EGTA, ethylene glycol bis-aminoethylether-N,N′-tetraacetic acid) and for Cd^{2+} (HEEDTA, N-2-hydroxyethylethylenediamine-N,N′,N′-triacetic acid).

Micromolar Hg^{2+} concentrations were found to inhibit Ca^{2+} transport through rat corpus-luteum plasma membranes (Minami and Penniston 1987).

Similar results have been obtained in recent studies on the effects of heavy metals on Ca^{2+}-stimulated ATPase activity present in the plasma membrane of mussel gill cells (Viarengo and Nicotera 1991).

These studies demonstrate that low concentrations of heavy metals inhibit in vitro the Ca^{2+} ATPase activity associated with plasma membrane preparations from mussel gill cells. In fact, when 10 μM Hg^{2+} (or Pb^{2+}) was present in the reaction mixture, a significant decrease of the Ca^{2+} ATPase activity was observed. The presence of Cu^{2+}, Cd^{2+} and Zn^{2+} also decreased the enzymatic activity, albeit at higher concentrations (Viarengo and Nicotera 1991; Viarengo et al. 1991). In addition, it has been shown that heavy metals (10–50 μM) strongly inhibit the formation of the phosphorylated enzyme (Viarengo and Nicotera 1991; Viarengo et al. 1991) that represents a fundamental step in Ca^{2+} transmembrane transport (Carafoli 1991). These data indicate that heavy metals may reduce the Ca^{2+}-stimulated ATPase activity by affecting the formation of the phosphorylated intermediate.

Moreover, it has been demonstrated that the simultaneous addition to the reaction mixture of different metal cations has cumulative inhibiting effects on the Ca^{2+} ATPase activity as well as on the formation of the phosphorylated enzyme (Viarengo and Nicotera 1991; Viarengo et al. 1991). Clearly, these data could have important environmental implications. In fact, these results indicate that in field studies it is not important to consider the concentration of a single contaminant metal but, on the contrary, the "total charge" of metals present in the sea water (both essential metals such as Cu^{2+}, Zn^{2+} and pollutants such as Cd^{2+}, Hg^{2+}, Pb^{2+}) which, as demonstrated in our experiments, could interact with the Ca^{2+} ATPase, thus altering the enzyme activity.

Taken together, these results imply that in mussel gill cells, the Ca^{2+} translocase present in the plasma membranes could represent a target for heavy metals.

These observations have been confirmed by in vivo experiments in which mussels were exposed to sublethal concentrations of Cu^{2+} (0.63 μM). It was found that the inhibition of the Ca^{2+} ATPase activity was maximal (−82%) after 4 days of metal exposure, but it tended to return to control values when the period of Cu^{2+} treatment was extended (Viarengo et al. 1992).

Interestingly, exposure of mussels to sublethal Cu^{2+} concentrations also causes a net inhibition (−84%) of Na^+/K^+ ATPase activity in the gill cells (Viarengo et al. 1992). This implies that the function of a putative Na^+/Ca^{2+} exchanger, present in these cells, whose activity depends on the Na^+/K^+ pump, should be affected in the gills of Cu-exposed mussels. Similar results on the inhibitory effects of heavy metals on Na^+/K^+ ATPase have been previously observed in the gill cells of metal-exposed fish (Bouquegneau 1977) and in other marine organisms (Bouquegneau and Gilles 1979).

A Na^+/Ca^{2+} exchanger found by Roer (1980) in *Carcinus maenas* hypodermic cells and in squid axons (Barrit 1982) is thought to be involved in Ca^{2+} homeostasis only when the Ca^{2+} concentration in the cytosol is pathologically high. This finding seems particularly important in the case of metal-exposed organisms in which a possible increase in free Ca^{2+} concentration may not be compensated by the activity of the Ca^{2+}-stimulated ATPase that, as mentioned above, is inhibited by heavy metals.

Although the interaction of heavy metal cations with SH groups of plasma membrane proteins should represent the focal point of their effects on Ca^{2+} homeostasis mechanisms, it is important to point out that these cations can also interact with the phospholipid components of the membrane.

Artificial phospholipid membranes show higher affinity for Cd^{2+} than for Ca^{2+} which is physiologically present in the membranes. Interestingly, Cd^{2+}, at the minimal concentration of 10^{-8} M, expands the phospholipid membranes (Webb 1979).

It is now recognized that heavy metals such as Ca^{2+}, Mg^{2+} and Zn^{2+} are important components physiologically present in the cell membranes, most of them being involved in interactions with phospholipid polar heads. Such interactions could be important for the correct functioning of proteins involved in cation transport (Papahadjopoulos 1968). Therefore, alteration of such metal phospholipid/protein interactions due to the presence of heavy metal cations such as Cd^{2+} (Evtushenko et al. 1986) and Hg^{2+} or high levels of Cu^{2+} could lead to variations in the biochemical membrane characteristics and therefore, in some cases, also to alterations of the Ca^{2+} homeostasis mechanisms.

3 Effects on the Endoplasmic Reticulum Ca^{2+} Stores

Sarcoplasmic reticulum and calciosomes are usually referred to as intracellular calcium stores in muscle and non-muscle cells, respectively. Both these Ca^{2+} stores derive from endoplasmic reticulum, although sarcoplasmic reticulum represents a specialized form present in muscle cells.

These stores are involved in regulating the concentration of free cytosolic Ca^{2+} through their ability to sequester Ca^{2+} and to release the cation, thus allowing for cell activation. Calcium is usually accumulated in the vesicles by the activity of Ca, Mg-ATPases (Carafoli 1987).

In sarcoplasmic reticulum, Ca^{2+} release is activated by membrane depolarization or by Ca-induced Ca^{2+} release. In this case, Ca ion flux is always through ryanodine sensitive channels (Fill and Coronado 1988). However, in non-muscle cells, calciosomes represent a heterogeneous pool of Ca^{2+} stores from which Ca^{2+} release can be initiated by either Ins $(1,4,5)P_3$ (IP_3) formation by receptor-activated phospholipase C (Berridge and Irvine 1989) or ryanodine-sensitive channels as in the sarcoplasmic reticulum (Burgoyne and Cheek 1991; Walton et al. 1991).

Regarding Ca^{2+} accumulation, it is known that SH groups play an important role in Ca^{2+} sequestering in the sarcoplasmic and in the endoplasmic reticulum. In particular, it was demonstrated that Ca^{2+} ATPase contains 26

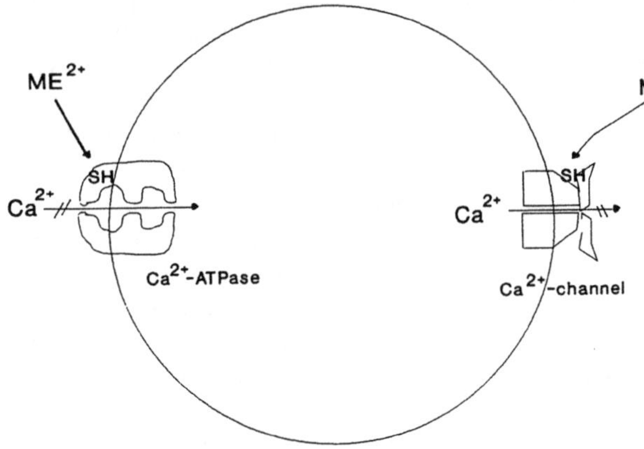

Fig. 3. Possible effects of heavy metals on the Ca-transport system present in the membranes of the endoplasmic reticulum calcium stores. No data are available to clarify the effects of heavy metals on the different types of Ca^{2+} channels in calciosomes

cysteine residues at least two of which are functionally important (Thorley-Lawson and Green 1977; Yamada and Ikemoto 1978). In fact, microsomal Ca^{2+} sequestering capability was found to be highly sensitive to SH reagents (Zhang et al. 1990). Therefore, calciosomes should represent a target for heavy metals that penetrate into the cell (Fig. 3).

In 1978, Shepard and Simkiss were able to demonstrate that in vitro Hg^{2+}, Cu^{2+}, Pb^{2+}, and Zn^{2+}, in a range from 10 to 50 μM, inhibit the Ca^{2+} ATPase present in the microsomal fraction obtained from the gills of the fish *Rutilus rutilus* (Shepard and Simkiss 1978).

Utilizing purified sarcoplasmatic reticulum vesicles obtained from rabbit skeletal muscle, it was found that Hg^{2+} could inhibit the Ca, Mg-ATPase involved in Ca^{2+} sequestration (Abramson et al. 1983).

More recently, it has been shown that Cu^{2+} and Cd^{2+}, at a concentration of 10 μM, decrease by about 50% the activity of the Ca-stimulated ATPase present in the hepatic microsomal vesicles. At a concentration of 50 μM, the enzyme is totally inhibited (Zhang et al. 1990).

Interestingly, heavy metals (Cu^{2+}, Hg^{2+}, Ag^{+}, Cd^{2+}, and Zn^{2+}) in concentration from 10 to 20 μM also evoke a fast release of the Ca^{2+} previously accumulated in the endoplasmic vesicles, although Zn^{2+} is less effective in stimulating the release of Ca^{2+} ions (Abramson et al. 1983). Sulphydrylic reagents such as N-Ethylmaleimide were also found able to cause the release of Ca^{2+} from the vesicles.

The effects of heavy metal cations can be interpreted taking into account that the two metals show high affinity for SH residues of the proteins probably involved in the control (or structure) of Ca^{2+}-channels.

Accordingly, it is important to point out that about 20% (Cavallini et al. 1969) of the copper in the cells is present as CuI, whose affinity for SH groups is extremely high. Moreover, the capacity of Cu^{2+} to generate oxyradicals and promote lipid peroxidation could help to explain the effects of this metal on Ca^{2+} release. It appears, in fact, that the oxidation (or some other perturbation) of critical sulphydrylic groups of the protein involved in the structure of the Ca^{2+}-channels present in the endoplasmic reticulum can lead to fast release of the cation from the calciosomes (Trimm et al. 1986).

Experiments on rabbit skeletal muscle sarcoplasmatic reticulum have confirmed that heavy metals can alter the Ca^{2+}-channels of the cytoplasmic Ca^{2+} stores (Abramson et al. 1983). These experiments demonstrate that heavy metals, at minimal concentration, trigger Ca^{2+} release from the vesicles. It seems that heavy metals increase the membrane permeability to this cation by interacting with SH groups of proteins present in the sarcoplasmatic reticulum (or calciosomes). Also in this case, Cu-mediated Ca^{2+} release seems to be initiated either by the binding of heavy metals to the thiol groups of the Ca^{2+}-channel proteins or by the oxidation of these SH groups to form disulfide bridges (Abramson et al. 1983). Therefore, it seems that in general those metal cations that react more strongly with thiol groups are more effective in promoting Ca^{2+} release, although the capacity of Cu^{2+}

to generate oxyradicals must also be considered. However, in different cells of different organisms, not all the cytoplasmic stores show the same behaviour in terms of susceptibility to heavy-metal-induced Ca^{2+} release, this probably being due to a different kind of channel present in the calciosomes.

For example, the fact that in mammalian cardiac muscle, Hg^{2+} and Ag^{+}, but not Cu^{2+} and Cd^{2+}, are able to stimulate Ca^{2+} release from loaded vesicles (Prabhu and Salama 1990) emphasizes that the phenomenon is not yet fully understood.

4 Effects on Mitochondria

It is well known that heavy metals can affect mitochondrial physiology and ATP production (Webb 1979; Viarengo et al. 1985). As mentioned above, these organelles play an important role in Ca^{2+} homeostasis at least when the Ca^{2+} concentration rises to a pathological level (Carafoli 1987) at which point they probably represent the last defense against Ca^{2+} cytotoxicity. In this regard, it has been found that heavy metals inhibit in vitro mitochondrial Ca^{2+} uptake (Zaba and Harris 1976; Jarvisalo et al. 1980; Chávez et al. 1985). Also in this case, among the possible mechanisms by which heavy metals can alter mitochondrial activity, the interaction with SH groups of the inner membrane proteins has been postulated (Southard et al. 1974) (Fig. 4).

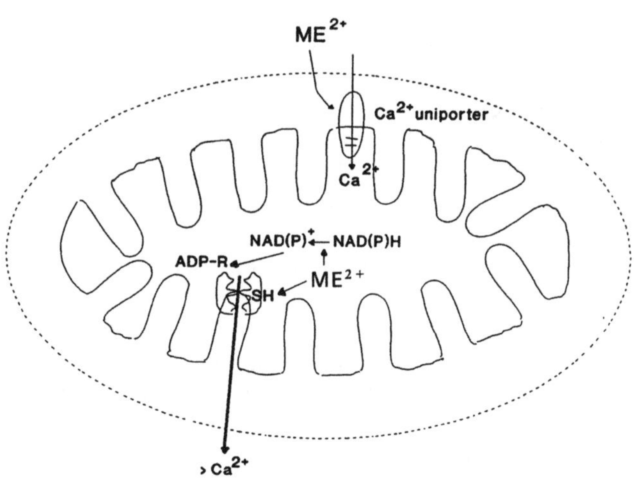

Fig. 4. General scheme illustrating the possible effects of heavy metals on the proteins involved in Ca^{2+} uptake and release in the inner mitochondrial membrane. The Ca^{2+}/Na^{+} exchanger is not shown; its role in heavy metal-induced Ca^{2+} release is not understood and it is not clear from what kind of channels Ca^{2+} is released by intoxicated mitochondria (metals, oxidative stress, etc.)

In general, mitochondrial Ca^{2+} transport seems to be regulated by critical SH groups since SH reagents, such as N-ethylmaleimide, inhibit Ca^{2+} uptake and produce rapid Ca^{2+} release (Rizzuto et al. 1987; Chavez and Holguin 1988).

Recent results have shown that in rat kidney mitochondria 5 μM Hg^{2+} activates Ca^{2+} release. The effects of Hg^{2+} could not be reversed by the addition of cysteine concentrations up to 200 μM. It is important to point out that cysteine removes about 70–80% of the Hg bound to mitochondrial membranes (from 6 to 1.6 nmol/mg prot). These results suggest that only a few critical protein SH groups are involved in triggering the Ca^{2+} release. It has also been found that Hg^{2+} causes a decrease in the NAD(P)H/NAD(P) ratio and a collapse of the inner membrane potential in mitochondria. However, the same experiments clearly demonstrated that the addition of EDTA to the reaction mixture restores the membrane potential. These results demonstrate that Hg^{2+} binding does not affect mitochondrial membrane integrity.

The available data seem to demonstrate that the Hg^{2+} binding proteins, which are most likely to be involved in the control of Ca^{2+} release, have a molecular weight of about 20000–30000 and are located on the external side of the inner mitochondrial membrane. Thirty thousand MW proteins are known to play a role in Ca^{2+} release from mitochondria. The activity of these proteins seems to be regulated by sulphydrylic reagents and NAD binding (Lotscher et al. 1980; Panfili et al. 1980; Hofstetter et al. 1981). It is tempting to speculate that the proteins which bind Hg^{2+} and which are probably involved in Ca^{2+} release in Hg-treated rat kidney mitochondria, could be the same low-molecular weight proteins previously demonstrated to be involved in the regulation of Ca^{2+} release from the mitochondrial matrix.

As for marine organisms, it has been demonstrated that the addition of copper to fish mitochondria can alter the influx of cations, through the inner membrane, finally inhibiting respiration (Zaba and Harris 1976).

In vivo experiments with fish (*Torpedo marmorata*) confirm the deleterious effects of Cu^{2+} on mitochondria. It was, in fact, demonstrated that Cu^{2+} accumulation in neuronal cells results in a significant alteration of mitochondrial membranes due to a considerable increase in lipid peroxidation (Aloj Totaro et al. 1986).

Results obtained on mussels exposed to sublethal concentrations of Cu^{2+} have shown that mitochondrial membranes are greatly damaged in the digestive gland (Viarengo et al. 1985; Viarengo 1989). As in fish, copper induces a large increase in lipid peroxidation in mussel tissues, a process that is probably responsible for the loss of integrity of mitochondrial membranes in metal-exposed animals (Viarengo et al. 1990).

However, in the case of Cu^{2+}-exposed organisms, the possibility that Cu ions could induce an oxidative stress leading to NAD(P)H oxidation, and consequent Ca^{2+} release, cannot be ruled out. With regard to this, it has been demonstrated that in rat liver mitochondria both enzymatic and non-enzymatic oxidation of pyridine nucleotides results in Ca^{2+} release. It has been hypothesized that in such conditions, NAD(P) hydrolysis would allow

ADP-ribosylation of a protein located in the inner mitochondrial membrane which is possibly involved in triggering Ca^{2+} release (Richter and Frei 1985).

Finally, it is important to note that various mitochondrial enzymes are selectively inhibited by heavy metals. Succinic dehydrogenase, for example, contains thiol groups and it is particularly sensitive to heavy metals (Webb 1979). Therefore, an alteration of mitochondrial metabolism could also result in a reduced mitochondrial capacity of calcium storage.

5 GSH and Metallothioneins in Preventing or Reducing Heavy Metal Cytotoxicity

Glutathione, the most abundant non-protein soluble cellular thiol, plays a role in metabolic and transport functions (Meister and Anderson 1983), in the detoxification and excretion of exogenous organic compounds, and in the protection against oxidative damage and radiation (Meister 1983; Meister and Anderson 1983) (Fig. 5).

As for the role of GSH in heavy metal cation homeostasis and cytotoxicity, it has been clearly demonstrated that GSH can form, both in vitro and in vivo, complexes with different metals such as Cu^{2+}, Cd^{2+}, Hg^{2+}, and $HgCH_3$ (Simpson 1961; Norseth and Clarkson 1971; Thomas and Smith 1979; Fayi and George 1985). A closer understanding of the cellular functions of GSH comes from the utilization of compounds able either to deplete or to increase the concentration of this tripeptide (Fig. 6). Among these compounds is buthionine sulfoximine (BSO), which can deplete the cellular GSH content through

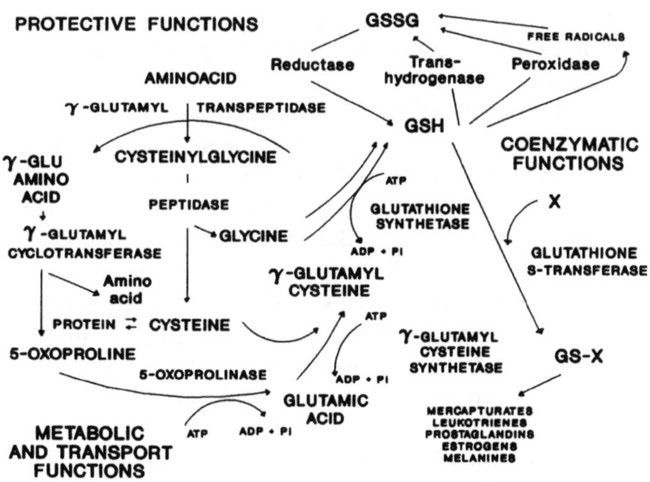

Fig. 5. General scheme of glutathione (*GSH*) metabolism and functions

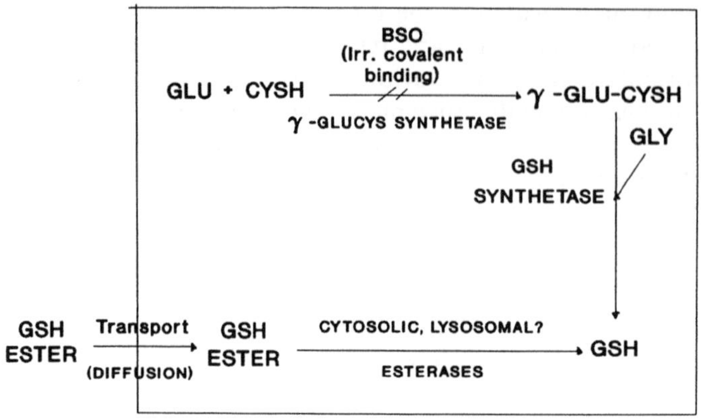

Fig. 6. Modulation of the cellular levels of *GSH* by buthionine sulphoximine (*BSO*) and GSH ester

inhibition by irreversible covalent binding of γ-glutamilcysteine synthetase, the rate-limiting enzyme in GSH biosynthesis (Griffith and Meister 1979; Meister 1983).

Moreover, the GSH concentration can be increased by the administration of glutathione monoester, which (unlike GSH) is rapidly transported through the plasma membranes of the cells into the cytoplasm where it is converted to glutathione (Puri and Meister 1983; Singhal et al. 1987; Meister 1988).

Experiments involving these compounds have proven useful to clarify the role of GSH against heavy metal cytotoxicity by modulating the cellular concentration of GSH from 30 to 400% of its basal level (Naganuma et al. 1990).

It has been found that in mice treated with a micromolar concentration of Hg^{2+}, GSH depletion resulting from buthionine sulphoximine pretreatment leads to a marked enhancement of metal-induced toxicity. Moreover, administration of GSH mono-ester prevented both cell toxicity and animal death (Naganuma et al. 1990).

Similar results were obtained in cultured Chinese Hamster cells (Ochi et al. 1988). Zn^{2+} pretreatment of these cells results in a ninefold increase in the metallothionein level. In Zn-pretreated cells, Cd cytotoxicity was markedly reduced both in control and in GSH-depleted cells (Ochi et al. 1988). This defense is attributable to the role played by metallothioneins in heavy metal homeostasis (Fig. 7). Metallothioneins are a class of soluble, low-molecular weight, cysteine-rich proteins, with an extremely high heavy metal content (7 g atoms/mol). These proteins are always present in the cells in low amounts (physiological Zn, Cu-thionein pool) and their neosynthesis can be induced by heavy metal cations that are assumed into the cells (inducible metallothionein pool).

GSH and metallothionein are also involved in reducing the cytotoxic effects of copper. In this case, most of the CuII in the reducing environment of the cell cytoplasm shifts to CuI which, due to its high affinity to SH groups, is complexed by GSH as CuIGS (Freedman and Peisach 1989). Successively, CuI is transferred to metallothionein (Freedman and Peisach 1989) the concentration of which increases with the time of copper exposure. The main difference between copper and the other heavy metals is that, in the case of Cu, the oxyradical production by the Fenton reaction represents an additional factor in determining cytotoxicity:

$$Cu^{+} + H_2O_2 \rightarrow Cu^{2+} + HO\bullet + OH^{-} .$$

In fact, enhanced oxyradical production leads to peroxidation of cellular membrane lipids and to oxidation of sulphydrylic residues of proteins, particularly of those related to Ca^{2+} homeostasis (Ca^{2+} ATPases, Ca^{2+} channels, SH-containing proteins, etc). In Cu-exposed cells, a net increase in the GSH peroxidase activity would result in the inhibition of the hydroxyl radical production by removing H_2O_2, a substrate in the Fenton reaction (Freedman et al. 1989).

It seems, therefore, that GSH represents per se an initial protection against heavy metal toxicity because it is able to bind metal cations entering the cells (Fig. 7). At this stage, the physiological metallothionein pool that is always present in the cells contributes in tightly binding minimal amounts of metal. In fact, toxic metals (Hg, Cd, Cu, Ag) showing a high affinity for SH are able to displace Zn^{2+} from the preexisting metallothionein. The excess metal can induce the neosynthesis of apothioneins which, in turn, will bind heavy metal cations. This induced metallothionein pool therefore represents a specific secondary defense against heavy metal cytotoxicity.

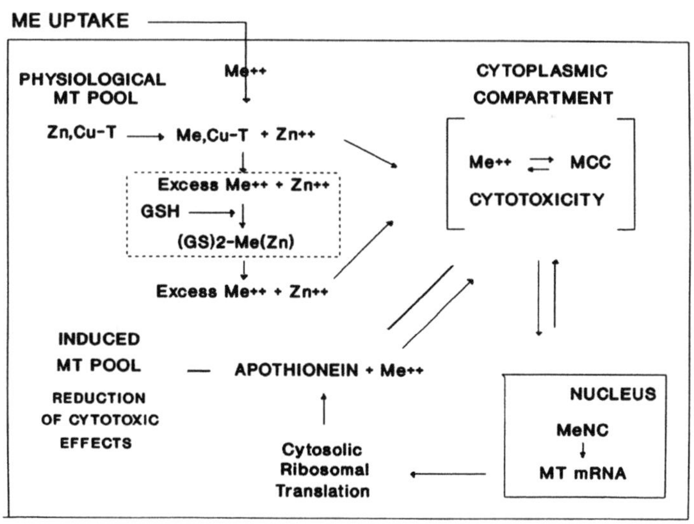

Fig. 7. The theorized role of *GSH* and metallothioneins (*MT*) in the reduction of the cytotoxic effects of heavy metals and in metal cation homeostasis; *MCC* metal cytosolic compounds; *MENC* metal nuclear compounds

It has been recently demonstrated that the GSH conjugates of Cu, Zn and $HgCH_3$ can be excreted from the hepatocytes into bile (Ballatori and Clarkson 1983; Gregus and Varga 1985; Houwen et al. 1990). In this case this excretion process can determine the biological half-life of the metal into the cells.

Interestingly, in mussels exposed to copper and subsequently detoxified, the addition of a mixture of cysteine, glycine and glutamic acid to the sea water facilitates, during the first days of detoxification, the elimination of the metal from the tissues (Viarengo et al. unpubl. data).

Concerning the role of GSH in reducing heavy metal cytotoxicity in marine organisms, data obtained on Hg-exposed clams *Anadara granosa* demonstrate that GSH provides effective protection at least for what concerns heavy metal effects on the lysosomal vacuolar system (Chandy and Patel 1985).

Regarding the effects of heavy metals on the GSH level of the cells of marine organisms, it has been found in marine fish exposed to Cd^{2+} that the level of GSH in the liver increases three- fourfold and that exposure to mercury and lead induces analog GSH variations (Thomas and Wofford 1984; Thomas and Juedes 1992). Similar results have been obtained in Cd-exposed *Euglena gracilis* (Coppellotti 1989). Contrariwise, Cu^{2+} leads to a decrease in the concentration of GSH in the tissues of metal-exposed mussels; this decrease is associated with a marked increase in lipid peroxidation (Viarengo et al. 1990), due to the role of GSH as a scavenger of oxygen and lipid radicals.

Taken together, these results demonstrate the importance of GSH as a first line of defense against some aspects of heavy metal cytotoxicity. However, no current data demonstrate that such a protection is able to prevent the effects of heavy metals on Ca^{2+} homeostasis. Therefore, this aspect represents a primary field for further research.

6 Effects on Cytosolic Free Ca^{2+}

In the light of the results noted so far, it is now possible to consider the consequences of impairing the Ca^{2+} homeostasis mechanisms by heavy metal cations.

It has been recently found that 50 μM Hg^{2+} can dramatically alter Ca^{2+} homeostasis in cultured rat liver hepatocytes (Nieminen et al. 1990). A concentration as low as 2 μM was found to result in a sustained increase of free Ca^{2+} in PC12 cells (Viarengo and Nicotera 1991) (Fig. 8A).

This effect of Hg^{2+} has also been confirmed in the cells of marine invertebrates. In fertilized sea urchin eggs, 10–100 μM Hg^{2+} increases the influx of Ca^{2+} through voltage-dependent channels (Walter et al. 1989).

More detailed studies on the effects of heavy metals on the cytosolic Ca^{2+} level were carried out on mussel haemolymph cells loaded with the Ca-fluorescent dye Fluo3. Fluorescence of the Fluo3/Ca^{2+} complex was evaluated on single cells by confocal laser microscopy (see Fig. 9). The results demonstrate

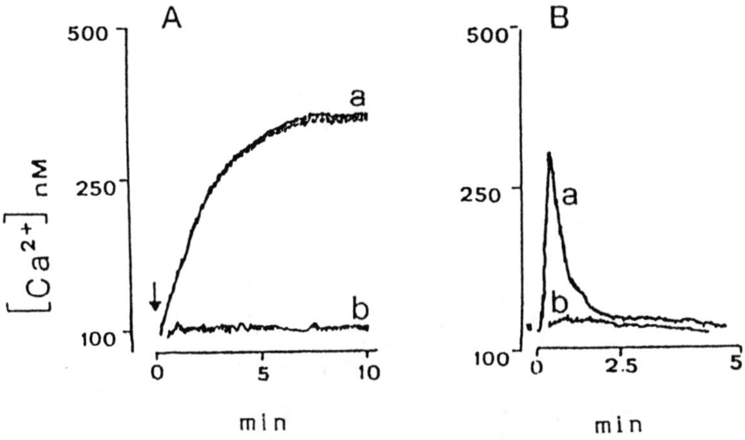

Fig. 8A, B. Effect of Hg^{2+} on cytosolic Ca^{2+} in resting and stimulated neuroadrenergic PC12 cells. **A** PC12 cells were exposed to 2 μM Hg^{2+} (*trace a*) or preincubated with the Ca^{2+}-channel blocker Verapamil (20 μM) and then exposed to Hg^{2+} (*trace b*); **B** PC12 cells were stimulated with 20 μM bradykinin (*trace a*) or (*trace b*) preincubated for 10 min with 0.5 μM Hg^{2+} and then stimulated by bradykinin as in *trace a*. (Modified from Viarengo and Nicotera 1991)

that cells exposed to both Hg^{2+} and Cu^{2+} exhibit a sustained increase in the cytosolic Ca^{2+} level which was heavy metal concentration dependent. In particular, when mussel haemolymph cells were exposed to Hg^{2+} from 2 to 50 μM, a significant increase of free cytosolic Ca^{2+} was observed (from + 200 to + 900%, respectively) within 10 min.

Interestingly, haemocyte pretreatment with Verapamil, a well-known blocker of voltage-dependent channels in many cell types, as well as the utilization of a low Ca^{2+}-containing medium, significantly reduced the effects of higher (50–100 μM) Hg^{2+} concentrations. However, neither lowering the extracellular Ca^{2+} concentration nor Verapamil pretreatment had any significant effect on the Ca^{2+} increase induced by low (1–2 μM) metal concentrations. These data clearly indicate that at higher Hg^{2+} concentrations, the metal exerts its toxic effects in part by enhancing the influx of Ca ions through voltage-dependent Ca^{2+} channels. At low concentrations (0.5–2 μM) Hg^{2+} seems to act on plasma membrane Ca^{2+} pumps and on endocellular calcium stores (Viarengo et al. 1993).

It is also important to point out that minimal concentrations of Cu^{2+}, an essential metal, have been shown to affect Ca^{2+} homeostasis in mussel haemolymph cells. Haemocyte exposure to low (0.5–2 μM) Cu^{2+} concentrations can induce a sustained increase of the cytosolic Ca^{2+} level (from about +40% to +200%). However, higher concentrations of copper were less effective, probably due to quenching of fluorescence of the Fluo3/Ca^{2+} complex by Cu^{2+} ions. As in the case of low Hg^{2+} concentrations, cell pretreatment with Verapamil or the utilization of a low Ca^{2+}-containing medium did not affect the increase in fluorescence induced by low copper concentrations (Viarengo et al. 1993).

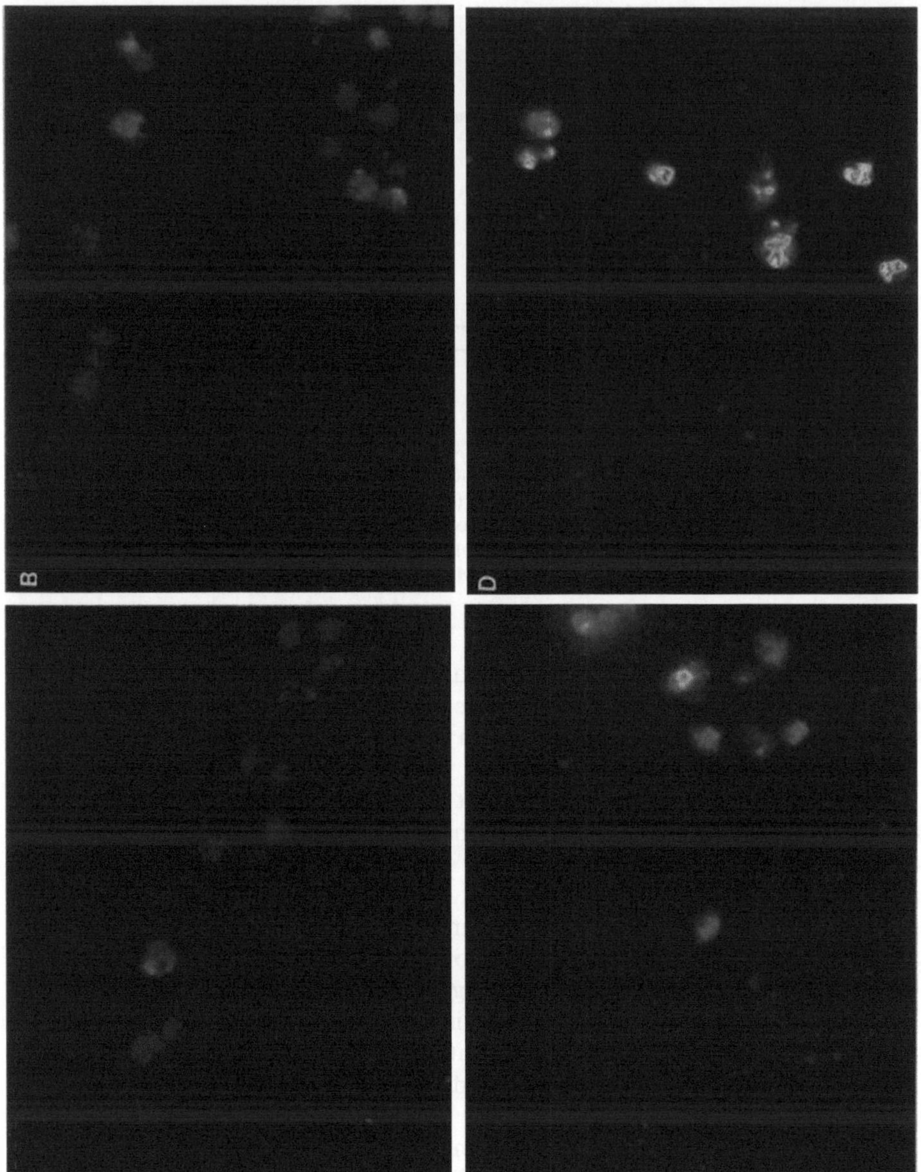

Fig. 9A–D. Fluorescence laser microscopy of single mussel haemocytes loaded with Fluo3 AM; **A** control; **B** 0.5 μM Cu^{2+}; C 1 μM Cu^{2+}; **D** 50 μM Hg^{2+}. Cells were exposed to heavy metals for 10 min

Data regarding the effects of Cd^{2+} and Zn^{2+} on free cytosolic Ca^{2+} concentrations are difficult to interpret, because the fluorescent probes utilized for measuring the free calcium concentrations (such as Fura2 and Fluo3) can form metal complexes which give emission spectra similar to those observed with Ca^{2+}. Therefore, it is difficult to distinguish between the fluorescent response due to a real Ca^{2+} increase and one caused by a Cd^{2+} or Zn^{2+} influx into the cytosol.

The increase in cytosolic Ca^{2+} observed in metal-exposed cells confirms that minimal heavy metal concentrations can impair the mechanisms of Ca^{2+} homeostasis.

Heavy metals cause a marked Ca^{2+} increase (Fig. 8A) that, as reported by all the researchers cited, is persistent for a long period of time (minutes). This sustained increase is distinct from the transient (seconds) Ca^{2+} spike observed following stimulation with certain hormones (Murphy et al. 1980; Rooney et al. 1989). Whether this phenomenon is the initial step that initiates cytotoxicity or the result of an altered cell metabolism, needs to be clarified. However, it is likely that, due to the role of this cation as a second messenger, an uncontrolled Ca^{2+} increase could trigger a sequence of reactions leading to alterations of the physiological status of the cells and, ultimately, to cell death (Schanne et al. 1979).

6.1 Toxicological Implications of Derangement of Ca^{2+} Homeostasis

When the cytosolic Ca^2 level is abnormal for a sustained period, the result is an activation of different Ca-dependent biochemical pathways or cellular functions (Fiskum 1989) such as mitochondrial activity, glycogen metabolism, exocytosis and cytoskeleton disassembly. This last effect has been studied in greater detail. It has been found that cytoskeletal alterations cause plasma membrane blebbing, which leads to cell morphological and functional changes (Jewell et al. 1982).

This effect has been recently confirmed by experimental data demonstrating that in rabbit kidney cells 10 μM Hg^{2+} was able to deregulate cytosolic Ca^{2+}, this causing extensive blebbing (Smith et al. 1991). Moreover, results on copper-exposed mussels have demonstrated that the microtubular structure of gill cilia are highly altered by the metal (Viarengo et al. 1989).

In addition, both direct action of elevated calcium levels and the consequent Ca-dependent calmodulin activation could have summatory effects in stimulating enzyme activities and affecting cellular functions; this could greatly contribute to the alteration of the physiological status of the cell.

Moreover, it has been recently found that Ca-activation of hydrolytic enzymes could directly trigger cytotoxicity.

It has been demonstrated that Ca^{2+} can stimulate the activity of phospholipases. These widely distributed enzymes are able to hydrolyze phospholipids thus leading to membrane damage (Chien et al. 1979; Shier and DuBourdieu 1983).

Furthermore, the hydrolysis of cytosolic proteins due to neutral proteases is known to be stimulated by increased Ca^{2+} concentrations. These Ca-stimulated proteolytic enzymes are involved in the degradation of specific proteins such as membrane receptors and cytoskeletal proteins (Murachi et al. 1981; Pontremoli and Melloni 1986).

Ca^{2+} can also activate nuclear endonucleases, indicating that both DNA and RNA metabolism of this organelle could be severely affected by a heavy metal-induced Ca^{2+} increase (Wyllie et al. 1984).

Finally, it is important to note that most of the aforementioned toxic effects due to increased Ca^{2+} levels could be synergic with the direct effects of heavy-metal binding to SH groups of enzymatic and/or structural proteins.

7 Heavy Metal Effects on Signal Transduction

The fact that heavy metals can enhance the cytosolic Ca^{2+} level implies that the intoxicated cells may be not properly responsive to Ca-mediated hormonal stimuli. This kind of cellular malfunctioning can therefore lead to a general alteration of the physiology of the animal.

Moreover, heavy metals can sometimes exert a direct effect on different aspects of signal transduction pathways.

As mentioned above, when PC12 cells were exposed to 0.5 μM Hg^{2+}, the metal, at least initially, did not appear to cause an increase of free cytosolic Ca^{2+} but rendered the cells insensitive to hormone (bradykinin) dependent Ca^{2+} variations (Viarengo and Nicotera 1991; Fig. 8B). It was speculated that this effect of Hg^{2+} could be due to a depression of inositol triphosphate (IP_3) production following hormonal stimulation.

As known, IP_3 production depends on the activity of phospholipase C, a membrane enzyme which hydrolyzes phosphatidyl inositol diphosphate (PIP_2) to produce IP_3 and diacylglycerol (Lehninger 1970).

The results of a study concerning the effects of metals on this enzyme in skeletal muscle demonstrate that, at least in vitro, heavy metals such as Cu^{2+} and Pb^{2+} inhibit the activity of phospholipase C (Shute and Smith 1985).

Recent data on marine invertebrates further demonstrate that micromolar concentrations of heavy metals (Cu^{2+} and Hg^{2+}) affect the activity of phospholipase C in the homogenate obtained from different mussel tissues (gills and digestive gland) (Panfoli et al. 1993). Moreover, in the digestive gland of mussels exposed to copper (0.65 μM), the metal causes a significant reduction in the phospholipase C activity (Viarengo and Panfoli unpubl. data).

These results indicate that animal exposure to heavy metal cations may, therefore, cause an impaired cell response to certain hormonal stimulation as previously demonstrated for the response to bradykinin of Hg^{2+}-treated PC12 cells.

However, not all heavy metals seem to have the same kind of effect on IP_3 production. In fact, in human skin fibroblasts, minimal concentrations of Cd^{2+} (0.1 μM) produce, in a few seconds, a fourfold increase in IP_3 concentration. Such a variation is associated with a spike in cytosolic free Ca^{2+} similar to what happens in hormone stimulated cells (Smith et al. 1989). At 1 μM concentration, Cd^{2+} initially induces in the cells a large enhancement of Ca^{2+} followed by a sustained Ca^{2+} increase. This latter effect, typical of heavy metal alteration of Ca^{2+} homeostasis mechanisms, should render the cell insensitive to Ca-dependent hormonal stimulation.

The Ca^{2+} spike in Cd-stimulated fibroblasts is independent of the Ca^{2+} concentration in the external medium, demonstrating that the spike is due to activation of endocellular stores.

From the data presented by Smith it appears that Cd^{2+} presumably interacts on the surface of the cell plasma membrane with a component that could be considered a "Cd^{2+} receptor" that stimulates IP_3 production by activation of phospholipase C. This will imply not only IP_3 production but also the formation of diacylglycerol with consequent activation of protein kinase C. Protein kinase C is a cytosolic enzyme that can be activated only after translocation to the internal plasma membrane surface; this translocation is Ca^{2+}-dependent (Nishizuka 1986). Protein kinase C acts on numerous intracellular proteins involved in many basic cell functions, such as mitosis, exocytosis, and control of Ca^{2+} movements across cellular membranes (pumps, channels, ionic exchangers) (Lehninger 1970).

The studies concerning the effects of heavy metal cations on different components of signal transduction pathways demonstrated that metals can also directly affect the activity of protein kinase C. Initial results postulate a biphasic effect of heavy metal cations on this enzyme, i.e., activation and inhibition at low and high metal concentrations, respectively (Mazzei et al. 1984; Murakami et al. 1987). However, it has been finally clarified by in vitro experiments in the presence of Ca^{2+} chelators in the reaction mixture that Cd^{2+}, Cu^{2+}, and Hg^{2+}, in the 10 μM range, inhibit the activity of protein kinase C as well as the binding capacity of regulatory ligands (Speizer et al. 1989).

Calmodulin is generally considered the universal mediator of calcium, acting as a second messenger. This Ca^{2+}-binding protein recognizes changes in the cytosolic free Ca^{2+} levels induced by external stimuli and transmits this information to intracellular proteins under the control of these signals. Calmodulin could represent a potential target for heavy metal cations, Cd^{2+} in particular, that penetrate into the cells. It has been in fact demonstrated that bovine brain calmodulin can bind Cd^{2+} at the Ca^{2+} binding sites and that Cd^{2+} can displace Ca^{2+} even at submicromolar concentrations. This seems to be due to the fact that Cd^{2+} has an ionic radius similar to that of Ca^{2+} and, therefore, the heavy metal cation can mimic Ca^{2+} in activating calmodulin (Flik et al. 1987).

It has been found in bovine brain cells that Cd-calmodulin is able to stimulate the phosphodiesterase activity but it is less effective than the Ca calmodulin complex (Flik et al. 1987).

Concerning marine organisms and particularly in fish gonadal cells, Cd^{2+} was found to be as effective as Ca^{2+} in stimulating calmodulin-dependent phosphorylation of endogenous substrates in the cytosolic fraction (Bhera and Gall 1991). These data do not clarify whether the calmodulin activation modality depends on direct activation of calmodulin through the binding of Cd^{2+} at the Ca^{2+} binding sites, or whether it is due to the displacement of Ca^{2+} from cytoplasmic Ca^{2+} ligands and a consequent increase in the cytosolic Ca^{2+} concentration. This would result in indirect calmodulin activation.

However, also in this case, it appears that Cd^{2+} when present in the cell cytoplasm at micromolar concentrations, could alter per se most of the calmodulin-mediated signal transduction pathways in marine organisms.

It is important to point out, however, that AMPc-dependent protein kinase, an enzyme involved in the AMPc-mediated signal transduction pathways, was also demonstrated to be inhibited by micromolar concentrations of heavy metal cations (Speizer et al. 1989).

Finally, recent data have shown that heavy metals can also act on rat brain cells inhibiting the GABA (γ-aminobutyric acid) binding to GABA A and B receptors (Drew et al. 1991). As known, GABA A receptors are "channel receptors" in which neurotransmitter binding activates the influx of chloride ions through the channel; contrarily, GABA B receptor activation reduces (through a G protein) the activity of adenylate cyclase and, consequently, the cytosolic AMPc level (Bormann 1988). This fact leads to a decreased flux of Ca^{2+} and K^+ across the plasma membrane. Interestingly, the GABA B receptor in the reported experiment was found to be more sensitive (by one order of magnitude) to transition metals than GABA A. This could imply that in neurones Ca^{2+} homeostasis could also be affected.

However, Kardos et al., through in vitro experiments, have demonstrated that micromolar concentrations of Cu^{2+} inhibit GABA-mediated chloride uptake into membrane vesicles obtained from rat cerebral cortex cells (Kardos et al. 1989). It seems that in vertebrate neurons the GABA A receptor complex is modulated by heavy metal cations of the IIB subgroup (Smart and Constanti 1989).

In addition, it has been demonstrated that in lobster muscle Zn^{2+} and Cu^{2+} in the range of micromolar concentrations depressed the GABA-evoked increase in conductance in a reversible way probably by binding to an imidazole group at a specific protein-binding site (Smart and Constanti 1982). Further, Cu^{2+} was found to be able to block the conduction of action potential (Arhem 1980). It has also been demonstrated that the metal inhibits the membrane permeability to chloride in molluscan neurons (Chiarandini et al. 1967; Rózsa and Salánki 1990). Copper ions also alter Na^+ permeability in *Aplysia californina* neurons (Weinrich and Wonderlin 1987). In *Helix pomatia* neurons Cu^{2+} was described to alter both a Cl^- outward current and inward component due to Na^+ flux through Ca-activated non-selective cationic channels (Kiss et al. 1991).

Finally, data on fish demonstrate that Pb^{2+} significantly impairs neurotransmitter functions, increasing, among the other parameters, histamine and serotonin concentrations and reducing monoaminooxidase and acetylcholinesterase activities (Katti and Sathyanesan 1986).

Overall, these data seem to indicate that heavy metals can affect the signal transduction pathways directly, thus rendering the cells unable to correctly respond to environmental stimuli. However, as mentioned above, heavy metals can also impair the mechanisms of calcium homeostasis, leading to a sustained increase of the cytosolic Ca^{2+} level. As a result of this, the cell would become insensitive to Ca-dependent hormones.

These aspects of heavy metal cytotoxicity represent a promising research opportunity to further clarify the effects of heavy metal cations on the physiological status of the organisms.

Acknowledgments. This work was supported by a C.N.R. grant (Target Project "Biotechnology and Bioinstrumentation"). The author thanks Prof. M. Orunesu for the helpful discussion, Dr. M. Pertica, Dr. G. Mancinelli, Dr. R. Accomando and Dr. L. Canesi for their assistance in the preparation of this manuscript. Thanks are also due to Dr. R. Fabbri for her skillful technical assistance.

References

Abramson JJ, Trimm JL, Weden L, Salama G (1983) Heavy metals induce rapid calcium release from sarcoplasmic reticulum vesicles isolated from skeletal muscle. Proc Natl Acad Sci USA 80: 1526–1530

Accomando R, Viarengo A, Orunesu M (1990) In vivo and in vitro effects of heavy metals on DNA polymerase activities in the digestive gland of *Mytilus galloprovincialis* Lam Comp Biochem Physiol 95C: 271–274

Aikawa JK (1976) Biochemistry and physiology of magnesium. In: Ananda S, Prasad M (eds) Trace elements in human health and disease. vol II. Essential untoxic elements. Academic Press, New York, pp 47–77

Aloj Totaro E, Pisanti FA, Glees P, Continillo A (1986) The effect of copper pollution on mitochondrial degeneration. Mar Environ Res 18: 245–253

Arhem P (1980) Effects of some heavy metal ions on the ionic currents of myelinated fibres from *Xenopus laevis*. J Physiol 306: 219–231

Ballatori N, Clarkson TW (1983) Biliary transport of glutathione and methylmercury. Am J Physiol 244: G435–G441

Barrit GJ (1982) Calcium movement across the cell membranes. In: Anghileri LJ, Tuffet-Anghileri AM (eds) The role of calcium in biological systems. CRC Press, Boca Raton, pp 17–30

Behra R, Gall R (1991) Calcium/calmodulin-dependent phosphorylation and the effect of cadmium in cultured fish cells. Comp Biochem Physiol 100C: 191–195

Berridge MJ, Irvine RF (1989) Inositol phosphates and cell signaling. Nature 341: 197–205

Bormann J (1988) Electrophysiology of GABA A and GABA B receptor subtypes. Trends Neurosci 11: 112–116

Bouquegneau JM (1977) ATPase activity in mercury intoxicated eels. Experientia 33: 941–942

Bouquegneau JM, Gilles R (1979) Osmoregulation and pollution of the aquatic medium. In: Gilles R (ed) Metabolism and osmoregulation in animals. Wiley, New York, pp 563–580

Bowler K, Duncan CJ (1970) The effect of copper on membrane enzymes. Biochim Biophys Acta 196: 116–121

Burgoyne R, Cheek TR (1991) Locating intracellular calcium stores. Trends Biochem Sci 16: 319–320
Carafoli E (1987) Intracellular calcium homeostasis. Annu Rev Biochem 56: 395–433
Carafoli E (1991) Calcium pump of the plasma membrane. Physiol Rev 71: 129–153
Cavallini D, De Marco C, Duprè S, Rotilio G (1969) The copper catalyzed oxidation of cysteine to cystine. Arch Biochem Biophys 130: 354–361
Chandy JP, Patel B (1985) Do selenium and glutathione (GSH) detoxify mercury in marine invertebrates? Effects on lysosomal response in the tropical blood clam *Anadara granosa*. Dis Aquat Org 1: 39–47
Chávez E, Holguin JA (1988) Mitochondrial calcium release as induced by Hg^{2+}. J Biol Chem 263: 3582–3587
Chávez E, Briones R, Michel B, Bravo C, Jay D (1985) Evidence for the involvement of dithiol groups in mitochondrial calcium transport. Studies with cadmium. Arch Biochem Biophys 242: 493–497
Chiarandini DJ, Stefani E, Gerschenfeld HM (1967) Inhibition of membrane permeability to chloride by copper in molluscan neurons. Nature 213: 97–99
Chien KR, Abrams J, Ptau RG, Farber JL (1979) Ischemic myocardial cell injury. Prevention by chlorpromazine of an accelerated phospholipid degradation and associated membrane disfunction. Am J Pathol 94: 505–509
Coppellotti O (1989) Glutathione, cysteine and acid-soluble thiol levels in *Euglena gracilis* cells exposed to copper and cadmium. Comp Biochem Physiol 94C: 35–40
Cousins RJ (1985) Absorption, transport, and hepatic metabolism of copper and zinc: special reference to metallothionein and ceruplasmin. Physiol Rev 65: 238–309
Dixon DR (1983) Sister chromatid exchange and mutagens in the aquatic environment. Mar Pollut Bull 14: 282–289
Drew CA, Spence I, Johnston GAR (1991) The effect of metal cations of the first transition series on GABA binding to GABA A and GABA B binding sites in rat brain. Neurochemistry 57: S147
Eichorn GL (ed) (1973) Inorganic Biochemistry, Elsevier, Amsterdam
Elinder CG (1986) Iron. In: Friberg L, Nordberg GF, Vouk V (eds) Handbook on the toxicology of metals, 2nd edn. Elsevier, Amsterdam, pp 276–297
Elinder CG, Friberg L (1986) Cobalt. In: Friberg L, Nordberg GF, Vouk V (eds) Handbook on the toxicology of metals, 2nd edn. Elsevier, Amsterdam, pp 211–227
Evans GW (1973) Copper homeostasis in the mammalian system. Physiol Rev 53: 535–570
Evtushenko ZS, Belcheva NN, Lukyanova ON (1986) Cadmium accumulation in organs of the scallop *Mizuhopecten yessonensis* II. Subcellular distribution of metals and metal-binding proteins. Comp Biochem Physiol 83C: 377–383
Farmanfarmaian A, Socci R, Iannacconne V (1981) Interaction of heavy metals with intestinal transport mechanisms in echinoderms. In: Lawrence JM (ed) Echinoderms. Proc Int Conf Tampa Bay. University of South Florida, Tampa, pp 339–386
Fayi L, George SG (1985) Purification of very low molecular weight Cu-complexes from the European oyster. In: Vernberg FJ, Thurberg FP, Calabrese A, Vernberg WB (eds) Marine pollution and physiology: recent advances. University of South Carolina Press, Columbia, pp 145–156
Fill M, Coronado R (1988) Ryanodine receptor channel of sarcoplasmic reticulum. Trends Neurosci 11: 453–457
Fiskum G (ed) (1989) Cell calcium metabolism. Plenum Press, New York
Flik G, Van de Winkel JGJ, Part P, Bonga SEW, Lock AC (1987) Calmodulin-mediated cadmium inhibition of phosphodiesterase activity, in vitro. Arch Toxicol 59: 353–359
Freedman JH, Peisach J (1989) Intracellular copper transport in cultured hepatoma cells. Biochem Bhiophys Res Commun 164: 134–140
Freedman JH, Ciriolo MR, Peisach J (1989) The role of glutathione in copper metabolism and toxicity. J Biol Chem 264: 5598–5605
George SG (1972) Subcellular accumulation and detoxication of metals in aquatic animals. In: Vernberg WB, Calabrese A, Thurberg FP, Vernberg FJ (eds) Physiological mechanisms of marine pollutant toxicity. Academic Press, New York, pp 3–22

Gould E, Greig RA, Rusanowski D, Marks BC (1985) Metal-exposed sea scallops, *Plactopecten magellanicus* (Gmelin): a comparison of the effects and uptake of cadmium and copper. In: Vernberg FJ, Thurberg FP, Calabrese A, Vernberg WB (eds) Marine pollution and physiology: recent advances. University of South Carolina Press, Columbia, pp 157–186

Gregus Z, Varga F (1985) Role of glutathione and hepatic glutathione S-transferase in the biliary excretion of methyl mercury, cadmium and zinc: a study with enzyme inducers and glutathione depletors. Acta Pharmacol Toxicol 56: 398–403

Griffith OW, Meister A (1979) Potent and specific inhibition of glutathione synthesis by buthionine sulfoximine (S-n-butyl homocysteine sulfoximine). J Biol Chem 254: 7558–7560

Hille B (ed) (1984) Ionic channels of excitable membranes. Sinauer Associates, Sunderland, Mass

Hinkle PM, Kinsella PA, Osterhoudt KC (1987) Cadmium uptake and toxicity via voltage-sensitive calcium channels. J Biol Chem 262: 16333–16337

Hofstetter W, Muhlebach T, Lotscher HR, Winterhalter KH, Richter C (1981) ATP prevents both hydroperoxide-induced hydrolysis of pyridine nucleotides and release of calcium in rat liver mitochondria. Eur J Biochem 117: 361–367

Houwen R, Dijkstra M, Kuipers F, Smit EP, Havinga R, Vonk RJ (1990) Two pathways for biliary copper excretion in the rat. The role of glutathione. Biochem Pharmacol 39: 1039–1044

Jarvisalo JO, Kilpio J, Saris NE (1980) Toxicity of cadmium to renal mitochondria when administered in vivo and in vitro. Environ Res 22: 217–233

Jewell SA, Bellomo G, Thor H, Orrenius S (1982) Bleb formation in hepatocytes during drug metabolism is caused by disturbances in thiol and calcium ion homeostasis. Science 217: 1257–1259

Kardos J, Kovács I, Simonyi M (1989) Copper ion inihibits GABA-mediated chloride uptake into membrane vesicles from rat cerebral cortex. J Neurochem 52: S95

Katti SR, Sathyanesan AG (1986) Lead nitrate induced changes in the brain constituents of the freshwater fish *Clarias batrachus* (L.). Neurotoxicology 7: 47–51

Kiss T, Györi J, Osipenko ON, Maginyan SB (1991) Copper-induced non-selective permeability changes in intracellularly perfused snail neurons. J Appl Toxicol 11: 349–354

Lehninger AL (1970) Biochemistry. Worth, New York

Levesque PC, Atchison WD (1987) Interactions of mitochondrial inhibitors with methylmercury on spontaneus quantal release of acetylcholine. Toxicol Appl Pharmacol 87: 315–324

Lotscher HR, Winterhalter KH, Carafoli E, Richter C (1980) Hydroperoxide-induced loss of pyridine nucleotides and release of calcium from rat liver mitochondria. J Biol Chem 255: 9325–9330

MacInnes J, Thurberg F (1973) Effects of metals on the behaviour and oxygen consumption of the mud snail. Mar Pollut Bull 4: 185–197

Mazzei GJ, Girard PR, Kuo JF (1984) Environmental pollutant Cd^{2+} biphasically and differentially regulates myosin light chain kinase and phospholipid/Ca^{2+} dependent protein kinase. FEBS Lett 173: 124–128

Meister A (1983) Selective modification of glutathione metabolism. Science 220: 471–477

Meister A (1988) Glutathione metabolism and its selective modification. J Biol Chem 263: 17205–17208

Meister A, Anderson ME (1983) Glutathione. Annu Rev Biochem 52: 711–760

Minami J, Penniston JT (1987) Ca^{2+} uptake by corpus-luteum plasma membranes. Biochem J 242: 889–894

Murachi T, Tanaka K, Hatanaka M, Murakami T (1981) Intracellular Ca^{2+}-dependent protease (calpain) and its high molecular weight endogenous inhibitor (calpastatin). Adv Enzyme Regul 19: 407–413

Murakami K, Whiteley MK, Routtenberg A (1987) Regulation of protein kinase C activity by cooperative interaction of Zn^{2+} and Ca^{2+}. J Biol Chem 262: 13902–13906

Murphy E, Coll K, Rich TL, Williamson JR (1980) Hormonal effects on calcium homeostasis in isolated hepatocytes. J Biol Chem 255: 6600–6608

Naganuma A, Anderson ME, Meister A (1990) Cellular glutathione as a determinant of sensitivity to mercuric chloride toxicity. Biochem Pharmacol 40: 693–697

Nieminen AL, Gores GJ, Dawson TL, Herman B, Lemasters JJ (1990) Toxic injury from mercuric chloride in rat hepatocytes. J Biol Chem 265: 2399–2408

Nishizuka Y (1986) Studies and perspectives of protein kinase C. Science 233: 305–312

Norseth T, Clarkson TW (1971) Intestinal transport of Hg-labeled methyl mercury chloride. Arch Environ Health 22: 568–577

Ochi T, Otsuka F, Takahashi K, Ohsawa M (1988) Glutathione and metallothioneins as cellular defence against cadmium toxicity in cultured Chinese hamster cells. Chem Biol Interact 65: 1–14

Panfili E, Sottocasa GL, Sandri G, Liut G (1980) The Ca^{2+} binding glycoprotein as the site of metabolic regulation of mitochondrial Ca^{2+} movements. Eur J Biochem 105: 205–210

Panfoli I, Morelli A, Viarengo A, Orunesu M (1993) Biochemical characterization of a phosphatidylinositol 4,5-bisphosphate-specific phosphalipase C activity in gills and digestive gland of the marine mussel *Mytilus galloprovinciolis* Lam. Comp Biochem Physiol 105B: 139–145

Papahadjopoulos D (1968) Surface properties of acidic phospholipids: interactions of monolayers and hydrated liquid crystals with uni- and bi-valent metal ions. Biochim Biophys Acta 163: 240–254

Pietrobon D, Di Virgilio F, Pozzan T (1990) Structural and functional aspects of calcium homeostasis in eukaryotic cells. Eur J Biochem 193: 599–622

Pontremoli S, Melloni E (1986) Extralysosomal protein degradation. Annu Rev Biochem 55: 455–481

Prabhu SD, Salama G (1990) The heavy metal ions Ag^{+} and Hg^{2+} trigger calcium release from cardiac sarcoplasmic reticulum. Arch Biochem Biophys 277: 47–55

Puri RN, Meister A (1983) Transport of glutathione, as γ-glutamylcysteinylglycyl ester, into liver and kidney. Proc Natl Acad Sci USA 80: 5258–5260

Rasmussen H, Barret PQ (1984) Calcium messenger system: an integrated view. Physiol Rev 64: 938–984

Richter C, Frei B (1985) Ca^{2+} movements induced by hydroperoxides in mitochondria. In: Sies H (ed) Oxidative stress. Academic Press, London, pp 221–241

Rizzuto R, Pitton G, Azzone GF (1987) Effect of Ca^{2+} peroxides, SH reagents, phosphate and aging on the permeability of mitochondrial membranes. Eur J Biochem 162: 239–249

Roer RD (1980) Mechanisms of resorption and deposition of calcium in the carapace of the crab *Carcinus maenas*. J Exp Biol 88: 205–218

Rooney TA, Sass EJ, Thomas AP (1989) Characterization of cytosolic calcium oscillations induced by phenylephrine and vasopressin in single Fura-2-loaded hepatocytes. J Biol Chem 264: 17131–17141

Rózsa KS, Salánki J (1990) Heavy metals regulate physiological and behavioral events by modulating ion channels in neuronal membranes of molluscs. Environ Monitor Assess 14: 363–375

Schanne FAX, Kane AB, Young EE, Farber JL (1979) Calcium dependence of toxic cell death: a final common pathway. Science 206: 700–702

Scott D, Major C (1972) The effect of copper (II) on survival, respiration, and heart rate in the common blue mussel *Mytilus edulis*. Biol Bull 143: 679–684

Shepard K, Simkiss K (1978) The effects of heavy metal ions on Ca^{2+} ATPase extracted from fish gills. Comp Biochem Physiol 61B: 69–72

Shier WT, DuBourdieu DT (1983) Stimulation of phospholipid hydrolysis and cell death by mercuric chloride: evidence for mercuric ion acting as a calcium mimetic agent. Biochem Biophys Res Commun 110: 758–763

Shute JK, Smith ME (1985) Inhibition of phosphatidylinositol phosphodiesterase activity in skeletal muscle by metal ions and drugs which block neuromuscular transmission. Biochem Pharmacol 34: 2471–2475

Simkiss (1983) Lipid solubility of heavy metals in saline solutions. J Mar Biol Assoc UK 63: 1–6

Simpson RB (1961) Association constants of methylmercury with sulphydryl and other bases. J Am Chem Soc 83: 4711–4717

Singhal RK, Anderson ME, Meister A (1987) Glutathione, a first line of defence against cadmium toxicity. FASEB J 1: 220–223

Smart TG, Constanti A (1982) A novel effect of Zn^{2+} on the lobster muscle GABA receptor. Proc R Soc Lond B Biol Sci 215: 327–341

Smart TG, Constanti A (1989) The vertebrate neuronal GABA-A receptor complex is modulated by group IIB divalent cations. Br J Pharmacol 96: 328

Smith JB, Dwyer SD, Smith L (1989) Cadmium evokes inositol polyphosphate formation and calcium mobilization. J Biol Chem 264: 7115–7118

Smith MW, Phelps PC, Trump BF (1991) Cytosolic Ca^{2+} deregulation and blebbing after $HgCl_2$ injury to cultured rabbit proximal tubule cells as determined by digital imaging microscopy. Proc Natl Acad Sci USA 88: 4926–4930

Southard J, Nitisewojo P, Green DE (1974) Mercurial toxicity and the perturbation of the mitochondrial control system. Fed Proc Fed Am Soc Exp Biol 33: 2147–2153

Speizer LA, Watson MJ, Kanter JR, Brunton LL (1989) Inhibition of phorbol ester binding and protein kinase C activity by heavy metals. J Biol Chem 264: 5581–5585

Thomas DJ, Smith JC (1979) Partial characterization of a low-molecular-weight methylmercury complex in rat cerebrum. Toxicol Appl Pharmacol 47: 547–556

Thomas P, Juedes MJ (1992) Influence of lead on the glutathione status of Atlantic croaker tissues. Aquat Toxicol 23: 11–30

Thomas P, Wofford HW (1984) Effects of metals and organic compounds on hepatic glutathione, cysteine, and acid-soluble thiol levels in mullet (*Mugil cephalus* L.). Toxicol Appl Pharmacol 76: 172–182

Thorley-Lawson DA, Green NM (1977) The reactivity of thiol groups of the adenosine triphosphatase of sarcoplasmic reticulum and their location on tryptic fragments of the molecule. Biochem J 167: 739–748

Trimm JL, Salama G, Abramson JJ (1986) Sulfhydryl oxidation induces rapid calcium release from sarcoplasmic reticulum vesicles. J Biol Chem 261: 16092–16098

Unep (1989) State of Mediterranean marine environment. MAP Tech Rep Ser 28. UNEP, Athens

Verbost PM, Senden MHMN, Van Os CH (1987) Nanomolar concentrations of Cd^{2+} inhibit Ca^{2+} transport systems in plasma membranes and intracellular Ca^{2+} stores in intestinal epithelium. Biochim Biophys Acta 902: 247–252

Verbost PM, Flik G, Lock RAC, Bonga SEW (1988) Cadmium inhibits plasma membrane calcium transport. J Membr Biol 102: 97–104

Verbost PM, Flik G, Pang PKT, Lock RAC, Bonga SEW (1989a) Cadmium inhibition of the erythrocyte Ca^{2+} pump. J Biol Chem 264: 5613–5615

Verbost PM, Van Rooij J, Flick G, Lock RAC, Bonga SEW (1989b) The movement of cadmium through freshwater trout branchial epithelium and its interference with calcium transport. J Exp Biol 145: 185–197

Viarengo A (1989) Heavy metals in marine invertebrates: mechanisms of regulation and toxicity at the cellular level. Aquat Sci 1: 295–317

Viarengo A, Nicotera P (1991) Possible role of Ca^{2+} in heavy metal cytotoxicity. Comp Biochem Physiol 100C: 81–84

Viarengo A, Pertica M, Mancinelli G, Capelli R, Orunesu M (1980) Effects of copper on the uptake of aminoacids, on protein synthesis and on ATP content in different tissues of *Mytilus galloprovincialis* Lam. Mar Environ Res 4: 145–152

Viarengo A, Zanicchi G, Moore MN, Orunesu M (1981) Accumulation and detoxification of copper by the mussel *Mytilus galloprovincialis* Lam.: a study of subcellular distribution in the digestive gland cells. Aquat Toxicol 1: 147–157

Viarengo A, Pertica M, Mancinelli G, Palmero S, Orunesu M (1982) Effects of Cu^{2+} on nuclear RNA polymerase activities in mussel digestive gland. Mar Biol Lett 3: 345–352

Viarengo A, Secondini A, Scoppa P, Moore MN, Orunesu M (1985) Effetti del Cu^{2+} e del Cd^{2+} sull'attività mitocondriale e sulla carica energetica adenilica delle cellule della ghiandola digestiva di mitili. Atti Riunione Congiunta SIB, SIF, SINU, Pisa, Settembre

Viarengo A, Moore MN, Mancinelli G, Mazzucotelli A, Pipe RK, Farrar SV (1987) Metallothioneins and lysosomes in metal toxicity and homeostasis in marine mussels: the effect of cadmium in the presence and absence of phenanthrene. Mar Biol 94: 251–257

Viarengo A, Arena N, Pertica M, Canesi L, Martino G, Gaspa L, Orunesu M (1989) Effect of copper on microtubule structure in the gill cells of metal exposed mussels. Mar Environ Res 28: 453

Viarengo A, Canesi L, Pertica M, Poli G, Moore MN, Orunesu M (1990) Heavy metal effects on lipid peroxidation in the tissues of *Mytilus galloprovincialis* (Lam.). Comp Biochem Physiol 97C: 37–42

Viarengo A, Pertica M, Mancinelli G, Orunesu M (1991) Heavy metal effects on the Ca^{2+} ATPase activity present in the gill cell plasma membrane of mussels (*Mytilus galloprovincialis* Lam.). Atti Riunione Congiunta SIBS, SIF, SINU, Sorrento, Settembre

Viarengo A, Pertica M, Mancinelli G, Orunesu M (1992) Effects of Cu^{2+} on the Ca^{2+} ATPase activity present in the plasma membrane of metal exposed mussels. Atti Riunione Congiunta SIBS, SIF, SINU, Roma, Settembre

Viarengo A, Canesi L, Moore MN, Orunesu M (1993) Effects of Cu^{2+} and Hg^{2+} on the cytosolic Ca^{2+} level in molluscan blood cells evaluated by confocal microscopy. Mar Biol (in press)

Vouk V (1986) General chemistry of metals. In: Friberg L, Nordberg GF, Vouk VB (eds) Handbook on the toxicology of metals, vol I. Elsevier, Amsterdam, pp 14–35

Walter P, Allemand D, de Renzis G, Payan P (1989) Mediating effect of calcium in $HgCl_2$ cytotoxicity in sea urching egg: role of mitochondria in Ca^2-mediated cell death. Biochim Biophys Acta 54: 1–8

Walton DW, Airey JA, Sutko JL, Beck CF, Mignery GA, Sudhof TC, Deerinck TJ, Ellisman MH (1991) Ryanodine and inositol trisphosphate receptors coexist in avian cerebellar Purkinje neurons. J Cell Biol 113: 1145–1157

Webb M (1979) Interactions of cadmium with cellular components. In: Webb M (ed) The chemistry, biochemistry and biology of cadmium. Elsevier, North-Holland Biomedical Press, Amsterdam, pp 285–340

Weinreich D, Wonderlin WF (1987) Copper activates an unique inward current in molluscan neurons. J Physiol 394: 429–443

Wyllie AH, Morris RG, Smith AL, Dunlop D (1984) Chromatin cleavage in apoptosis: association with condensed chromatin morphology and dependence on macromolecular synthesis. J Pathol 14: 67–71

Yamada S, Ikemoto N (1978) Distinction of thiols involved in the specific reaction steps of the Ca^{2+} ATPase of the sarcoplasmic reticulum. J Biol Chem 253: 6801–6807

Zaba BN, Harris EJ (1976) Uptake and effects of copper in rat liver mitochondria. Biochem J 160: 707–714

Zaba BN, Harris EJ (1978) Accumulation and effects of trace metal ions in fish liver mitochondria. Comp Biochem Physiol 61C: 89–94

Zaroogian GE, Gentile JH, Heltshe JF, Johnson M, Ivanovici AM (1982) Application of adenine nucleotide measurements for the evaluation of stress in *Mytilus edulis* and *Crassostrea virginica*. Comp Biochem Physiol 71B: 649–655

Zhang GH, Yamaguchi M, Kimura S, Higham S, Kraus-Friedmann N (1990) Effects of heavy metal on rat liver microsomal Ca^{2+} ATPase and Ca^{2+} sequestering. J Biol Chem 265: 2184–2189

Chapter 4

Comparative Pulmonary Morphology and Morphometry: The Functional Design of Respiratory Systems

J.N. Maina

Contents

Department of Veterinary Anatomy, University of Nairobi, P.O. Box 30197, Nairobi, Kenya

Advances in Comparative and Environmental Physiology, Vol. 20

"It is generally agreed that animal life originated in the sea and that adaptive radiation subsequently led to the colonization of other environments – shores and estuaries, streams and lakes, bog, mountain, and desert. In their invasion of these habitats, animals left the equable, relatively stable surroundings of the open sea and subjected themselves to the rigors of temperature fluctuations and extremes, a variety of ionic backgrounds, areas of depleted oxygen or the possibility of aerial exposure and potential desiccation. The spur for this radiation presumably lay in the prize of access to unexploited habitats and sources of energy."

Davenport (1985)

1 Introduction

Though there are assertions that life is possible without oxygen, such states can only exist for definite periods and in simplest of life forms (Hochachka et al. 1973; Herreid 1980; Portner et al. 1985). Intestinal parasites have been said to live without molecular oxygen and intertidal molluscan facultative anaerobes survive for days without it (Ghiretti 1966). Extended life in total anoxia is impossible in more advanced animals. Adaptively, however, in unpropitious ambient conditions, some animals drastically reduce their energy needs by adopting latent (ametabolic) life states, variably termed aestivation, lethargy, hibernation, quiescence and dormancy. Cryptobiosis, a state when life virtually stops is the extreme of such low energy retreats (Hochachka and Guppy 1987). Even in such a state an infinitesimal quantity of energy must, nevertheless, be essential to sustain the integrity of the basic molecular life-support processes such as protein turnover and ion pumping. The presence of oxygen is crucial in maintaining what was appropriately termed by Kleiber (1961) "the fire of life". Metabolic rate is an expression of intensity and speed of life while death is empirically a state of zero cellular energy production. Energy is crucial to all biological events from molecular and biochemical level to ecological and evolutionary processes. To varying extents, the principal factors which appear to determine the metabolic demands of an organism are size, phylogeny, temperature and physical activity (Bennett 1988a). The design and efficiency of the respiratory system hence is not dependent on any single quantum factor or physical law. The definitive form presents the best mutually inclusive design most favourable to life. It is envisaged that by studying a cross-section of representative animals, data may be educed and collated to help integrate and hopefully formulate the fundamental unifying or alternative principles in appreciation of the permutations and trade-offs on which the phenomenology, mechanistics, and regulatory physiologies and organizational attributes of the respiratory systems are entrenched. Paucity of pertinent data has been, and still is, the constraining factor in making firm and wide-ranging conclusions on these aspects. Nevertheless, it is only by assembling the components of this maze that the gaps in this spectrum of events can be identified. Before outlining the structure and design of respiratory organs, the factors which have modulated their ultimate form and function are briefly considered.

2 Metabolic Determinants

2.1 Body Size, Phylogeny and Ethology

The extant animals (from unicellular to multicellular) differ in mass by a colossal factor of 10^{21}, terrestrial animals differ by six orders of magnitude while birds span almost five (Lindstedt and Calder 1981; Schmidt-Nielsen 1984). Body mass is the most important determinant of metabolism and hence sets the ecological opportunities available to an animal, its physiology, behaviour and morphology (Peters 1983; Calder 1984). According to the surface law, metabolism in endotherms is higher in smaller animals than in larger ones. O_2 uptake scales to body mass by 0.75 power, i.e. $VO_2 = aW^{0.75}$ (Heusner 1982). A 100% increase in body mass leads to a 68% increase in O_2 uptake (Bennett and Dawson 1976; Bennett 1988a). A 3–9 shrew has a heart rate of about $\times 1200\ min^{-1}$ and its VO_2 (1.92 mmol $min^{-1}\ kg^{-1}$) is the highest among mammals (Fons and Sicart 1976). Allometric interdependence of metabolism and body mass is reflected in the structure and design of organisms (Else and Hubert 1985a, b): small animals have smaller body cells with a high mitochondrial density and large cristae surface areas. In animals of same size measured under the same conditions, endotherms consume five to ten times more O_2 than ectotherms (Dawson and Hubert 1970). Phylogenetic differences between animals closely correspond with their total VO_2 (Else and Hubert 1985a). Metabolic differentials occur within and without taxa and in some cases clines exist within taxa. In birds, passerine species which are more active and operate at a higher body temperature (42 °C) have a metabolic rate 50–60% greater than nonpasserines of similar body mass (Lasiewski and Dawson 1967; Aschoff and Pohl 1970). In nonpasserines, hawks and owls have a lower metabolic rate than ducks and herons (Zar 1969). In both pigeons and heteromyid rodents, desert-dwelling species have lower basal metabolic rates than mesic varieties (MacMillen 1983). The largest and probably the most important metabolic differential occurs between the endotherms and ectotherms. Birds evolved independently from reptiles and acquired metabolic capabilities which are six to ten times higher than those of their progenitors at the same body temperature (Bennett and Dawson 1976). The aerobic capacities of mammals range from those of the terrestrial groups to the volant bats. The capacity of an animal to take up and transport large amounts of O_2 was a prerequisite to key events like transition to terrestrial life and development of endothermy and homeothermy. Inherent adaptations to these demands should be commensurate with morphophysiological changes. Whereas the actual environment during the incipient stages of the development of life and that in the subsequent epochs the organisms went through may never accurately be known, albeit with onerous caveats comparative studies of respiratory structures and processes should yield useful data which could elucidate the factors that have conspired to mold the respiratory organs and processes. In contrast to features like nutrition, thermoregulation, and reproduction which can easily be adjusted to suit a habitat, respiration is a continuous and an invariable process compatible with physical laws: only means and efficiency can be altered.

3 Symmorphosis

The concept of symmorphosis first advanced by Taylor and Weibel (1981) basically describes the consequences of function on form. It was defined as "a state of structural design commensurate to functional needs resulting from regulated morphogenesis whereby the formation of structural elements is regulated to satisfy but not exceed the requirements of the functional system." There should hence be a stoichiometric and economic match between structure and function and overdesign and redundancy are untenable. This concept corresponds with the longstanding physiological one of homeostasis and the optimality concept described by among others Rosen (1968), Maynard-Smith (1978) and Alexander (1982). The universal validity of symmorphosis has been questioned particularly from phylogenetic and developmental perspectives (Gould and Lewontin 1979; Lewontin 1979). Intuitively, optimization though desirable would curtail responses to selective pressures. Extinctions may partly result from such refinements and inflexibilities. Structural elements largely carry out multiple functions to which they are variably adapted (Garland and Huey 1987; Bennett 1988a) hence inference of a component or an aspect as manifesting "excessive construction" or constituting a "limiting factor" may be highly subjective (Gans 1979). Garland and Huey (1987) on empirically re-analyzing Taylor and Weibel's (1981) and their subsequent data (using mass-corrected residuals) failed to confirm the concept. A recent revisit of the subject (Weibel et al. 1991) noted that of all the structural and functional variables involved in O_2 transfer from the gas exchanger to the mitochondrial sink, it is only the pulmonary gas exchanger which does not conform to symmorphosis. The lung maintains a remarkable excess of diffusing capacity (DLo_2) up to a factor of 2, a feature which enables animals to vary their VO_2 to suit prevailing demands. Jones et al. (1985) observed that goats can maintain VO_{2max} even under hypoxic conditions and only the smallest mammals use all of their DLo_2 for O_2 uptake (Taylor et al. 1989). According to Heusner (1983) redundancy of the lung in respiratory function should increase with body size. As a philosophical definition of structural-functional correlation in organic and tissue design, the concept of symmorphosis would appear to be heuristically satisfactory and as a working hypothesis implicitly useful. Its apparent failure at finer levels goes to show the complexity of the modulation processes of tissue structure and function. Lung design is not entirely determined by gas exchange. Factors like biomechanics, genetics and often imputed endocrine function may variably play important roles. Further, it is tenable that modicum interspecific variations as to what may constitute redundancy or limitation may exist (Garland and Huey 1987). In this account, paradigms of pulmonary structural-functional interdependence will be evident, though refinement to the extent defined in the dictum of symmorphosis cannot be proved due to paucity of essential data.

4 Water Versus Air as Respiratory Media

During the incipient stages of evolution of animal life, the aquatic medium provided a stable environment, the inherent buoyancy enabling the development of the delicate invertebrates. Subsequent modulation of form, size and function lead to the more complex vertebrates and finally the tetrapods. The physicochemical properties of water and air like gas solubility, rheology and thermal capacity are remarkably different. Their attributes, particularly in O_2 and CO_2 physical characteristics, are of paramount importance in understanding the phylogenesis of life, respiration, transition from water to air breathing, the advent of terrestriality, the evolution of endothermy and homoiothermy and in general that of perpetuation of species.

Water is a 1000 times more dense and viscous than air. The O_2 content of water (0.04–9.0 $cm^3\ L^{-1}$) is notably lower than that of air (105–210 $cm^3\ l^{-1}$), a differential factor of about 20, the rate of O_2 diffusion in water ($2.5 \times 10^{-5} cm^2\ s^{-1}$) is remarkably lower (by a factor of 10^5) than in air ($1.98 \times 10^{-1} cm^2\ s^{-1}$) and the capacitance coefficient (increment of concentration per increment in partial pressure) of O_2 in water is only 1.82 $nmol\ min^{-1}\ torr^{-1}$ compared to the much higher one in air of 54.73 $nmol\ min^{-1}\ torr^{-1}$ (Dejours 1976, 1979, 1988). Due to strong convective currents and high diffusion rates, O_2 availability is more reliable in an aerial environment than in an aquatic one. In the open air all over the world, the O_2 concentration (FO_2) is 0.2095 and FCO_2 is 0.0034. This stability enables empiric predictions of the composition of gases (in air) under specific circumstances to be accurately made with as little information as temperature and barometric pressure, an uncertain process in water without detailed physicochemical data, particularly for CO_2. In water, the tensions of these gases are extremely variable and depend on factors like water circulation, dissolved substances, temperature, density and nature of flora and fauna (Carter 1935, 1955). Some waters exhibit remarkable circadian fluctuations in FO_2: during the day, due to photosynthetic activity of algae and other simple plants, the water may even be supersaturated with O_2 while at night it may be virtually anoxic. This cycle was observed in the hyperosmotic and alkaline waters of Kenyan Lake Magadi, which has abundant algal and phytoplankton growth, by Maina (1990a, 1991). Only in very few aerial environments, like the fossorial habitat (McNab 1966; Williams and Rausch 1973) do the O_2 and CO_2 concentrations differ significantly from the normal values. In caves with a large number of bats, the concentration of NH_3 can be as high as 1.4 torr (Mitchell 1964) and in the pouch of a marsupial with a young, Farber and Tenney (1971) determined an average FO_2 of 0.142 and FCO_2 of 0.043.

Due to the disparate physicochemical properties of O_2 and CO_2 in air and water, the advantages conferred by air- over water-breathing are evident. Depending on the design of the respiratory organs, for reasons of lower O_2 concentration and the rheology of water, a water-breather does more work than an air-breather in extracting an equivalent quantity of O_2 from an external medium. The ventilatory rate of aquatic crabs is three times that in air (Herreid and Full 1988). In man, 1–2% of O_2 uptake at rest is utilized in mechanical ventilation of the lung, whereas in fish the value may be ten times higher (Dejours et al. 1970; Jones and Schwarzfeld 1974). In crabs, the cost of ventilation is 30% of the total VO_2 (Herreid and Full 1988). The energetic cost of aquatic ventilation in the crab, *Cancer*, may be as high as 76% (McMahon et al. 1979). Thus air-breathing has the great advantage of energy saving on respiratory work and gives flexibility in adjusting the respiratory rate without undue increase in energetic demands. Such animals adapt easily to wide metabolic demands. At the critical level of aquatic hypoxia at which aerobic metabolism becomes impossible, long-term solutions are essential to meet respiratory demands. These may entail reduction in metabolic needs by adopting definite behavioural and physiological processes, escape from a hostile environment and adoption of appropriate respiratory regulatory processes and morphological adjustments. As a last recourse, adaptations at biochemical and molecular levels occur (Hochachka 1979). The remarkably different respiratory properties of water and air demanded drastic modifications and adjustments of the gills, apparently the earliest of the respiratory organs, and in some cases their total extirpation was essential before air-breathing and terrestrial habitation was possible. Such modifications were subtle and gradual. Whereas CO_2 off-loading is not a problem to a water-breather due to its high solubility, aerial breathers have to evolve an efficient ventilatory mechanism and a greater buffering power of blood to prevent acidosis (Howell 1970; Randall et al. 1981). The PCO_2 of the arterial blood of the air-breather *Electrophorus electricus* is 27.7 mm Hg (Johansen et al. 1968b). *Lepidosiren*, an obligate air-breather, has a better CO_2 buffering power of blood than *Neoceratodus* which has fully functional gills (Johansen and Lenfant 1967). To dissociate the primary functions of the gills (respiration, ionic exchange and pH regulation), a bimodal respiratory stage was imperative. The air-breathing organ was involved in O_2 uptake and the gills or any other aquatic respiratory surface like the skin in CO_2 removal and hence pH regulation and ionic homeostasis. A further relocation of CO_2 excretion was necessary for terrestrial habitation. The lung assumed an important acid-base regulatory function by controlling the CO_2 tension in blood through ventilation, as ionic exchange was taken over primarily by the kidneys and to a lesser extent by the skin. It is quite explicit, however, that oxygenation is the primary purpose of respiration and in case of conflict, CO_2 clearance is subordinate (Dejours 1988).

5 Evolution of Aerial Breathing

Since the primeval stages of the advent of life, cyclic changes in gaseous composition in water and air have occurred. The early anoxic reducing atmosphere essentially consisted of methane, ammonia and hydrogen (Tappan 1974; Schidlowski 1975). The anoxic environment was essential for the formation of DNA and RNA, the carriers of genetic information and hence the depositories of life. O_2 production had to await the evolution of simple photosynthesizing plants. Concentration of O_2 had risen to about 3% by the Precambrian period (Cloud 1974), a time marked by the appearance of Metazoa, with fish appearing in the Cambrian period; 300 million years elapsed before the atmospheric O_2 concentration reached this modicum level, as much of it was taken up oxidizing ferrous iron (which was abundant on the earth's crust) to ferric iron. It is rather paradoxical that life should have started in an O_2-free environment and subsequently become so intricately dependent on it. This is simply because O_2 coincidentally happened to be an ideal terminal electron acceptor (due to its very high electron affinity) facilitating the production of much energy, with water, a neutral factor, being the final product. Cyclic fluctuations of atmospheric O_2, climatic changes, changes in size and spatial distributions of land and those of water and wet masses are evident over the eons of vertebrate evolution. In the mid-Silurian period (425 million years ago), late Cretaceous (60 million) and mid-Tertiary and Quaternary (40 and 10 million, O_2 tensions were above the present levels, while the lowest concentrations occurred about the mid-Permian (250 million) and late Cretaceous (100 million) (Cloud 1974). The mid-Paleozoic hypoxic episode which has been termed the "mid-Paleozoic crisis" (Tappan 1974) saw the extinction of 90% of the fish species by the late Devonian (Inger 1957): a further 50% had succumbed by the end of the Permian. Though aerial O_2 tension in air is always higher than which in water and local and regional variations occur, the changes in atmospheric O_2 tensions correlate with known massive extinction episodes. Aquatic hypoxia hence appears to be the main but probably not the only driving pressure towards evolution of aerial breathing (Packard 1974; Randall et al. 1981). Clear evidence of advantages conferred by air-breathing are obvious when it is appreciated that the primitive extant fish are exclusively represented by air-breathing groups (Lauder and Liem 1983). Such animals extended their ecological range and occupied niches which would otherwise have been alien to them. In bony fish, the lungs may have arisen from the posterior pair of gill pouches and served to retain engulfed air (Spengel 1904). The majority of fish possessed lungs by the Paleozoic era (Szarslo 1977) and all the extant primitive osteichthyes possess lungs (Tenney and Tenney 1970).

6 Transition from Aquatic to Terrestrial Life

From the current concepts of palaeontology, it is widely accepted that life started in water (Satchell 1971; Davenport 1985). Clearly water-breathing is the more ancient mode of respiration. Fish, the most successful vertebrate group with 20,000 species or so have lived in water for over 350 million years. Transition from aquatic to terrestrial life is deemed to be the single most important event to have occurred in the evolution of the extant tetrapods (Schmalhausen 1968). Aerial breathing and terrestriality opened greater ecological opportunities to those animals which endeavoured and succeeded in it. Romer (1967) opined that the major developments necessary for evolution from aquatic to terrestrial life were limbs and lungs though equally, if not more important was the threat of dehydration and the need for appropriate adaptation to curtail it. As to a factor which is dependent on the geometric consequences of size, volume increases as a cube, while area increases as the square of the radius. The theoretical limiting radius of a hypothetical geometrically spherical cell (exhibiting average metabolism in a normal environment) has been estimated to be 0.5 mm if such a cell was to rely entirely on O_2 diffusion (Harvey 1928). The simple unicellular and minute multicellular life forms survived, and the contemporary ones still do, on physical diffusion of O_2 over their thin and proportionately extensive bodies unaided by specialized structures (see Fig. 1a). The need for subsequent functional and structural adaptations called for cellular complexity and diversity, features which resulted in diffusional limitations to cells and tissues internally situated. Such changes variably called for development of (1) system(s) specially charged with bringing cells closer to the respiratory medium, (2) an extensive area to facilitate optimal influx and efflux of respiratory gases, (3) a delivery system of O_2 from the respiratory site to the tissue cells, and (4) presence of a respiratory pigment to enhance O_2 uptake. These adaptations were, however, inadequate in meeting demands from an acutely O_2-depleted environment. Towards the end of the Silurian and the beginning of the Devonian period, the earth's temperature rose accelerating putrefaction of organic plant materials (Romer 1967) and oceans rose flooding much of the rain forests. Aquatic hypoxia and hypercarbia, predisposed by drying of the inland waters, occurred. Arguably, Barrel (1916) proposed that air-breathing started in freshwater during the Devonian time. Due probably to effect on the central control, VO_2 falls drastically with rising PCO_2, particularly when the PO_2 is low (Black et al. 1954; Saxena 1960; Tenney 1979). These changes (increased PCO_2 and reduced PO_2) induced the water-breathers to strive for an alternative source of O_2, which was abundant just above the water-air interface. As expected, the transitional period corresponds with the "mid-Paleozoic crisis" (Tappan 1974) when much of the aquatic animal life succumbed (Sect. 5). Strictly taken, the transition from water- to air-breathing was an adaptive response to changes in the aquatic environment rather than merely a means for terrestrial colonization. This is clearly demonstrated by the air-breathing fish (Sect. 11.1) which even after acquiring preadaptations for terrestriality, mainly by attaining aerial respiration, permanently live in water. Initially the water-breathers would have adopted simple behavioural characteristics like skimming the water surface and using the skin and buccal cavity as respiratory surfaces as they evolved physiological tolerance to hypoxia. On the brink of extinction, air-breathing may have been the only alternative. It required hybridization of diverse stratagems, such as structural reorganization of the respiratory organs, co-option of other organs to supplement gas exchange and development of entirely new gas exchange organs. Concomitantly, changes in the control of respiration from a centrally modulated one to a diffusely coordinated one, a process which entailed attenuation of the central control on the branchial ventilation and detection and integration of both aquatic and aerial breathing was necessary (Shelton and Croghan 1988; Smatresk 1988; Boutilier 1990). In *Protopterus*, Jesse et al. (1967) reported that exposure to 8% O_2 increased lung ventilation threefold but gill ventilation only by half, a finding suggestive of different receptors being involved in the control and coupling of respiration in the air-breather. As an indication of the strict demands for air-breathing, of the about 20,000 species of fishes, only a handful have evolved the capability for aerial respiration (Fange 1976). Some of the earliest air-breathers are the Devonian armoured fish, like the arthrodire *Bothriolepis* (Denison 1941; Moy-Thomas 1971). The primitive air-breathers are arguably thought to have given rise to lungfish, coelacanth and semiaquatic amphibians which subsequently gave rise to terrestrial tetrapods (Romer 1972).

Air-breathing has evolved independently many times among phyletically different teleostean lineages (Gans 1971; Lauder and Liem 1983) and in some taxa, reversion to water-breathing has occurred (Gans 1976; Randall et al. 1981). Correspondingly, extant air-breathing fish possess remarkable diversity of anatomical, physiological, biochemical and behavioural features. Physiological differences attributable to variations in environmental selective pressures were observed between the Panamanian and South American swamp-eel (*Synbranchus marmoratus*) by Graham and Baird (1984). Some contemporary animals are initially perfect aquatic breathers only to subsequently change to perfect terrestrial air-breathers, and amphibians and reptiles daily pass from one medium to the other. Increase in complexity and development of features like endothermy and homoiothermy lead to increased O_2 needs. To meet these requirements, the respiratory organs and processes underwent drastic reorganizations with enhanced efficiency. Modifications and adaptations depended on factors like nature, severity and duration of the external pressure, phylogenetic level of development and the ecological opportunities accessible to the organism. It is perhaps opportune at this juncture to echo Tenney and Boggs' (1985) observation that "the great events in the respiratory system evolution can best be appreciated in lower classes of animals".

7 Endothermy

Subsequent to terrestriality, probably the next most significant event which influenced the form and function of vertebrates was the evolution of endothermy and eventually that of homoiothermy. The interspecific effect of body mass on resting O_2 needs is quantitatively the same in ectotherms and endotherms but the absolute values for similar-sized animals is always less in an ectotherm at its body temperature (Zeuthen 1953; Gleeson 1991). Consequent to high metabolic intensity in an endotherm, resting VO_2 is substantial (White 1978; Bennett 1988a). Compared with the endotherms (birds and mammals), the basal metabolic rate (BMR) of reptiles is generally 80–90% lower at the same body temperature and mass (Lillywhite 1987). The BMR of lizards at similar body mass and temperature is only 0.28th that of eutherian mammals and 0.17th that of passerine birds (Dawson and Hubert 1970). The lizard, *Sceloporus occidentalis* in a natural environment uses 3–4% of the energy a bird or a mammal uses in a day, owing to the fact that the body temperature of a reptile decreases when it is inactive (Bennett and Naggy 1977). Both the resting and active metabolic rates of reptiles are about one-tenth those of mammals (Gans 1978). Diving turtles at 10°C can withstand anoxia for as long as 17 days (Ultsch et al. 1984). Contemporary reptiles are essentially ectothermic (Cowles 1947) and in the animal kingdom only birds and mammals are truly endothermic. The metabolic cost of maintaining endothermic homoiothermy is extremely high, requiring at least 90% of the total energy budget (Heinrich 1977; Bennett 1988a). In the lizard *Iguana iguana*, the cost is over 700% greater at 40°C than at 15°C (Morberly 1968). The selective forces which induced endothermy in mammals and birds are still debatable (Heinrich 1977; Crompton et al. 1978; Bennett and Ruben 1979). Interestingly, evidence adduced from ecological, bone structure, bioenergetic and biomechanical studies suggests that some degree of endothermy may have been possible in the Mezozoic and even the late Paleozoic reptiles (Bakker 1975). Naggy et al. (1972) observed that endothermy, with its high metabolic capacity, may have evolved several times in animals (and even plants). Mammals remarkably vary in their degree of endothermy, body temperature and metabolism. The Metatheria are intermediate between the Prototheria and the Eutheria (Dawson and Hubert 1970; Hubert 1978; McNab 1984, 1986a, b).

Although the scaling of mitochondrial parameters with body size in mammals and reptiles is the same, the total mitochondrial volume density and cristae surface area in mammalian tissue are four times greater than in reptiles (Else and Hubert 1981, 1985a). The phylogenetic metabolic variations between different groups of animals rests in the level of their total organismal VO_2 which is a cumulative feature of metabolism of the individual cells comprising body tissues (Smith 1956; Lasiewski and Calder 1967; Else and Hubert 1985a; Bennett 1988a). At the same body temperature, mammals consume five to ten times more O_2 than similar sized ectotherms (Dawson and Hubert 1970; Bennett and Dawson 1976). Evolution of reptiles to mammals entailed a drastic change from low to high metabolic capabilities. The principal structural factor involved in this change was increased tissue mitochondrial volume density (Else and Hubert 1985a). Low energy expenditure which characterizes ectotherms is a fundamental advantage of the ectothermic condition but it offers fewer opportunities for greater physical activity (Bennett 1988a). A fully endothermic-homeothermic animal is relatively independent of the environmental temperature for its metabolic activities. As a sign of the remarkable advantages gained by endothermy, some insects (Heinrich 1981), tunas and some sharks (Carey et al. 1971) and some snakes (Hutchison et al. 1966) have evolved the capacity of myogenetic endothermy. The majority of insects will not fly until their flight muscle temperature has reached 35–38 °C (Heinrich and Bartholomew 1972). Evolution of endothermy, among other features, called for corresponding development of more efficient respiratory mechanisms, processes and structures as pointed out by Gans (1978) and Benett (1988a) and will be evident in this account. The notable similarities of solutions to challenges posed by endothermy suggest that the process evolved in almost the same way in birds and mammals, or alternatively endothermy is too highly refined to require definite and invariable processes and structures.

8 Design Characteristics of Respiratory Organs

Gas exchange across a respiratory surface takes place by passive physical diffusion and not actively as was once thought when PO_2 in pulmonary capillary blood was envisaged to be higher than in the terminal air-spaces. To function as a gas exchanger, a surface need only be moist and convectively perfused with blood and ventilated with air or water. Diffusive gas exchange across the thin respiratory barriers is completed within 250–500 ms (West 1974). There are no structural elements which are unique or ubiquitous to a respiratory tissue like e.g. hepatocytes are to the liver, neurones to nervous tissue and osteocytes to bone. It is probably on this nonstereotyping that the versatility, diversity and malleability of the respiratory organs (ROs) has resided and enabled animals at remarkably different levels of phylogenetic development to attain air-breathing without need for what could have had to be a common and elaborate limiting distinct RO. Variably, skin, buccal cavity, and GI-tract serve as respiratory sites. The respiratory system has played a central role in the evolutionary changes of the extant animal life as evinced by the fact that to a fair measure the level of organization and sophistication of the system gives an unequivocal picture of the evolutionary status of the organism. It is our conviction that phylogeny presents the raw materials on which selective pressures act and subsequently on which adaptational changes prevail.

The requisite structural attributes of an efficient respiratory organ are an extensive surface area and a thin partitioning between respiratory media. Commensurate functional features are an efficient ventilatory and perfusive

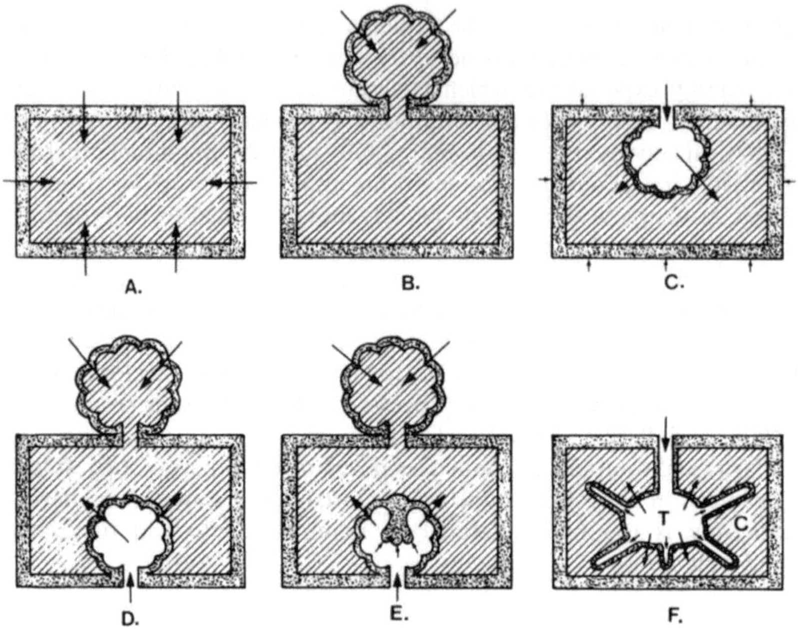

Fig. 1A–F. Design characteristics of respiratory organs. **A** Simple organisms with a large surface-to-volume ratio and thin bodies acquire their O_2 needs by diffusion through the skin (*arrows*). With increased tissue complexity, some parts of the body are specialized for gas exchange by becoming well vascularized and having a large surface area. This is attained by evagination from the body surface as in the gills (**B**) or by invagination as in the lungs (**C**). In hypoxic environments air-breathing organs have evolved to supplement O_2 uptake through the gills (**D**). Further modifications of the air-breathing organs like in the cat-fish, *Clarias mossambicus*, in form of labyrinthine organs are occasionally observed (**E**). The tracheal system of the insects (**F**) is unique in that O_2 is delivered by the trachea (*T*) directly to the tissue cells (*C*)

mechanism to maintain a gaseous pressure gradient across the barrier. In simple animals, the extensive surface area (per body volume) is adequate in gas transfer, while in the larger and more elaborate ones respiration is restricted to localized specialized sites. In the latter, this is achieved by outfolding (evagination) or intucking (invagination) of a part of a body surface (Fig. 1): the gills fall in the first category and the lungs in the second. Surface area is further increased by hierarchical growth and compartmentalization of the respiratory organs (ROs). Gas exchange capacity can be further enhanced by co-option of other organs or parts of the body as accessory respiratory organs (Fig. 1D, E). Transfer of O_2 to the cells (in most animals) is effected by blood vessels and enhanced by carriers like haemocyanin and haemoglobin. Reconstruction of the stages through which the respiratory organs have passed poses a number of problems, the main one being lack of fossil materials on soft tissue like the lung. To understand the evolution of air-breathing and subsequently terrestriality, we have to be contented with examining: (1) animals adapted to atypical habitats, (2) those that manifest putative respiratory transitional features, (3) those at differing phylogenetic levels of development and (4) those which lead unique behavioural life-styles. This account will focus on the diversity of the organization of the ROs, highlighting how different animals extract O_2 from their external milieus and/or have arrived at a solution to air-breathing. It is by no means a treatise either on the evolution of these species, that of the pulmonary systems or their structure, but aims to point out uniformity and variations in the functional design of ROs. We believe that when these factors are considered, it is reasonable to extrapolate and appreciate the design and changes which the ROs have gone through from one level of development and sophistication to another. Evidently, the process has entailed somewhat similar morphological adaptations to comparable physiological requirements. It will be emphasized that : (1) the evolution of the ROs and respiratory processes has largely followed parallel and at times convergent pathways rather than a serial pattern, and in some cases retrogressive changes have occurred; (2) the evolution of respiration is unique in that among the contemporary animals the mammalian lung is by no means the best endowed gas exchanger and hence anthropocentric interpretations of comparative pulmonary physiological and structural features, as often is the case, based on such highly refined group of animals, are largely misleading; (3) respiration appears to be too important for survival for primitive characteristics to be transmitted from one evolutionary level to another, and (4) strictly speaking, but what is rarely appreciated, the so called air-breathing animals have not totally dissociated themselves from water in that gas exchange can still only occur across wet surfaces.

9 Gas Exchange Models

9.1 Physical Models

The efficiency of a gas exchanger depends on extrinsic and intrinsic factors. The extrinsic factors include those which regulate ventilation and perfusion and the intrinsic ones the physical and geometric properties of the components inherent in the gas exchanger. The former have been comparatively well reviewed by Tenney (1979), Randall et al. (1981) and Tenney and Boggs (1985), while the physical (morphological and morphometric) features are outlined in this account. Here, the significance of the spatial organization of the structural components of the ROs, features which though highly important largely go unappreciated at the expense of structural and functional attributes, will be highlighted. Though theoretically there are infinite ways in which the structural components of a gas exchanger can be arranged, in consequence of longstanding developmental and genetic programming, the interaction between the gas exchange media occurs in a definite and consistent pattern. Gas exchange

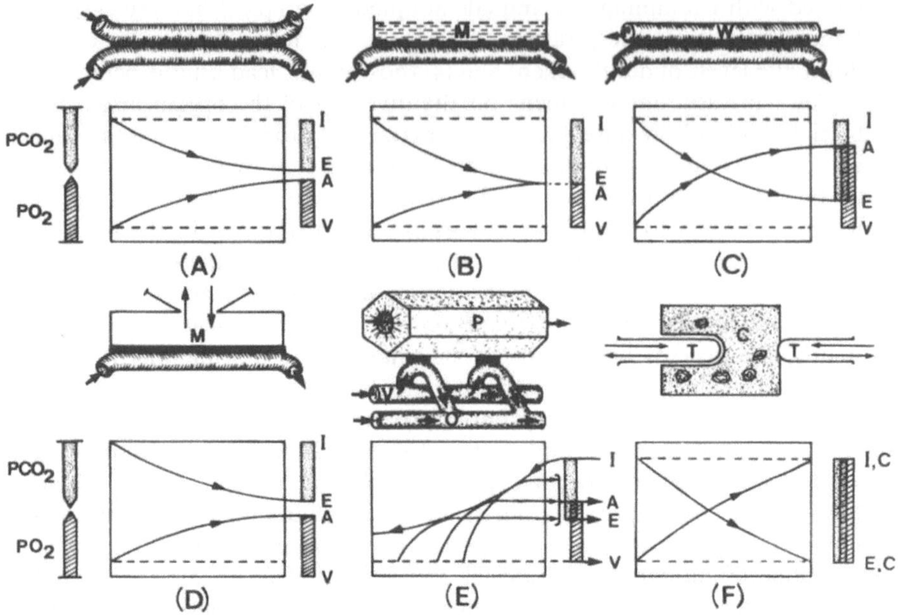

Fig. 2A–F. Physical models of gas exchangers. The efficiency of respiratory organs to a significant extent depends on the spatial arrangements of its structural components. In a co-current system (**A**) the gas-exchange media flow in the same general direction (*arrows*). The pressure gradients created by dissolved gases between the two media and hence the rate of gas exchange is low. Respiration across the skin (**B**) entails exposure of blood to an external respiratory medium (*M*) which could be either air or water and the level of arterialization depends on the partial pressure of O_2 in the external medium and the duration of exposure. In a counter-current system (**C**) the external gas-exchange medium (*W*) and the blood flow in opposite directions creating high gaseous partial pressure gradients, the arterial PO_2 (PAO_2) exceeding that of the end expired air (PEO_2). In a uniform pool design (**D**), the external gas exchange medium (*M*) is exposed to blood for a brief period of time when gases are exchanged. Overlap of PAO_2 and PEO_2 is not possible, the PO_2 differential depending on factors like ventilation-perfusion inequalities. The high efficiency of the cross-current system (**E**) is due to a serial arterialization of venous blood (*V*) and an additive effect of the oxygenated blood (*O*) along the lengths of the parabronchi (*P*). As a mark of its high efficiency, overlap between PAO_2 and PEO_2 occurs. In the tracheal system, (**F**) O_2 is delivered directly to the cells (*C*) by tracheoles (*T*) the intracellular PO_2 equating that of the inspired air. This efficiency is enhanced by intracellular tracheolar indentations. Partial pressures of O_2 and CO_2; *I* inspired air; *E* expired air; *A* arterial blood; *V* venous blood; *C* intracellular space

efficiency will partly depend on the manner and the rate the media are presented to each other at the respiratory site (Fig. 2). The three patterns which have received most attention are the counter-current one of fish gills (Fig. 2C), the uniform pool of the mammalian lung (Fig. 2D), and the cross-current of the avian lung (Fig. 2E) (Piiper and Scheid 1972, 1973). In many of the other animals the models are paradoxical as they variably present combined features of the primary designs. In fish (Sect. 10.2 and Fig. 5), water in the interlamellar space

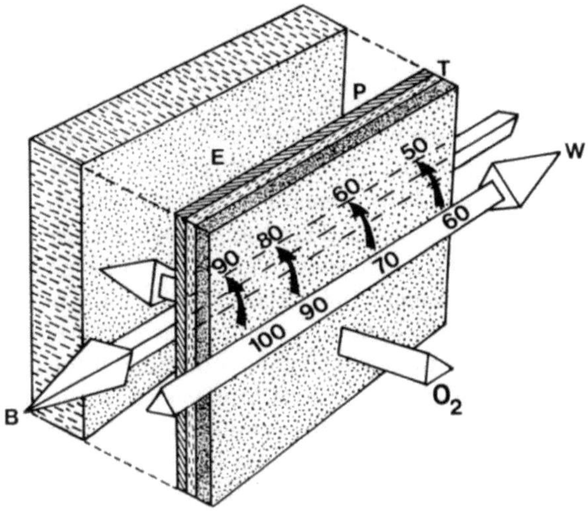

Fig. 3. Counter-current gas-exchange design. The ventilatory medium (*W*) and the perfusive one (*B*) flow in opposite directions. A concentration gradient of oxygen (O_2) is maintained as the two media interact across a tissue barrier (*T*) and the plasma layer (*P*), the erythrocytes (*E*) finally taking up the oxygen

and blood in the secondary lamellae flow in opposite directions (Fig. 2C). The water taken in with a large O_2 content meets and equilibrates with incoming venous blood at a low PO_2, a concentration gradient prevailing during the interaction (Fig. 3). Likewise, a high PCO_2 develops as the two media interact over the water-blood barrier. Typical PCO_2 values in blood from ventral and dorsal aorta are respectively 2.5 and 1.7 mm Hg in *Squalus suckleyi* (Lenfant and Johansen 1966), 3.3 and 1.9 mm Hg in *Scyliorhinus stellaris* (Piiper and Schumann 1967) and 5.7 and 2.3 mm Hg in *Salmo gairdneri* (Stevens and Randall 1967). In the cross-current design of the avian lung, the venous blood approaches the periparabronchial tissue essentially in a manner perpendicular to the bulk parabronchial air-flow (Fig. 2E, Fig. 22a, Sect. 13.4). The individual blood vessels contact a parabronchus at a single cross-sectional area (Maina 1988a) consequent to which the PO_2 of the end arterial blood results from an additive effect which occurs along the total length of parabronchi (Scheid and Piiper 1970). The mammalian uniform-pool model is typified by the capillary blood equilibrating with a pool of external medium, the alveolar air (Sect. 13.3; Fig. 15a, b, c) in which partial pressures are reasonably uniform (Fig. 2D). From theoretical analysis and experimental investigations, it has been observed that the counter-current design is potentially the most efficient respiratory system (Piiper and Scheid 1973) and may be viewed as an adaptation which was essential in overcoming the inherent disadvantages entailed in using water as the respiratory medium (Sect. 4). The arterial O_2 partial pressure (PAO_2) of 93 torr in the gills of the dogfish, *S. stellaris* (PIO_2, 149 torr and PEO_2 56 torr) is comparably very high (Baumgarten-Schumann and Piiper 1968), the corresponding values being 30 torr in man (Farhi and Rahn 1960) and 36 torr in birds (Bouverot et al. 1979). O_2 extraction in a normoxia-acclimated carp, *Cyprinus carpio* exceeds 80% (Lomholt and Johansen 1979) and the PAO_2 in the skipjack tunas, *Katsowonus pelamis* is as high as 90 torr (Muir and Kendall 1968), the diffusing capacity of 0.115 m lO_2 kg^{-1} min^{-1} torr being the highest value so far reported in fish (Brett 1964; Stevens 1972). The gills of the gastropod *Haliotis tuberculata*, in which the flow of blood manifests a counter-current disposition with that of the water, extract 56% of O_2 dissolved in seawater flowing through the mantle cavity (Yonge 1947). Such an arrangement is lacking in bivalve molluscs like *Mya arenaria* which extracts 3–10% (Van Dam 1935), *Cardium tuberculatum* 6–10% (Hazelhoff 1939), *Pecten irradians* 2.5–6.85% (Van Dam 1954), *Mytilus californianus* 3.6–10.4% (Bayne et al. 1976) and *Pholas dactylus* 6–10% O_2 (Knight 1984). Liem (1981) described in the larvae of an air-breathing teleost fish, *Monopterus*, a counter-current system (Fig. 3) between water currents produced by the fins and the blood flow in the fins and yolk sac: he opined that this efficient system enables *Monopterus* larvae to survive in hypoxic water. The cross-current organization is followed in efficiency by the uniform-pool system. In fish, PAO_2 is higher than that in the exhaled water (PEO_2) (Piiper and Scheid 1972). Whereas in the mammalian lung the alveolar and arterial PCO_2 ($PACO_2$) are about the same (at about 40 torr) (Piiper and Scheid 1980; Dejours 1988), in birds $PACO_2$ is lower than in the expired air (Scheid 1985). Such an overlap is theoretically and practically possible only in the counter-current and cross-current systems (Figs. 2C, E) and for somewhat different reasons in the insect tracheal system (Fig. 2F; Sect. 12.3). As pointed out by Scheid (1979), the implied levels of efficiency are only relative. Potentially, the more efficient designs like the counter-current and cross-current ones could operate at a lower level of efficiency than the uniform pool one in a resting animal. Furthermore, at high ventilation, all models become functionally identical (Piiper and Scheid 1975). Comparison of these models can only be made at identical component conductances (Piiper and Scheid 1975). Respectively, the medium to blood conductance ratio for O_2 in the counter-current, cross-current and uniform-pool designs were 0.61, 1.30, and 1.60 (Piiper and Scheid 1972). Birds and mammals are notably successful taxa with high metabolic rates. Divergently, the two groups of animals have evolved well-adapted lungs to support these adaptations.

9.2 Morphometric Models

The principal structural parameters of a gas exchanger which determine conductance or diffusing capacity of gases (DL) are surface area (S), diffusional distance (D), the volume of the capillary blood (VC) exposed to the external respiratory medium per unit time, and the materials property of the barrier(s) with respect to O_2 permeation (K). The DL of a gas exchanger is directly proportional to S and inversely proportional to D. Thus,

$$DL = S/D \times K .$$

The morphometric parameters are modelled (Weibel 1970/71) to give a more comprehensive measure of gas exchange capacity. With modifications, the

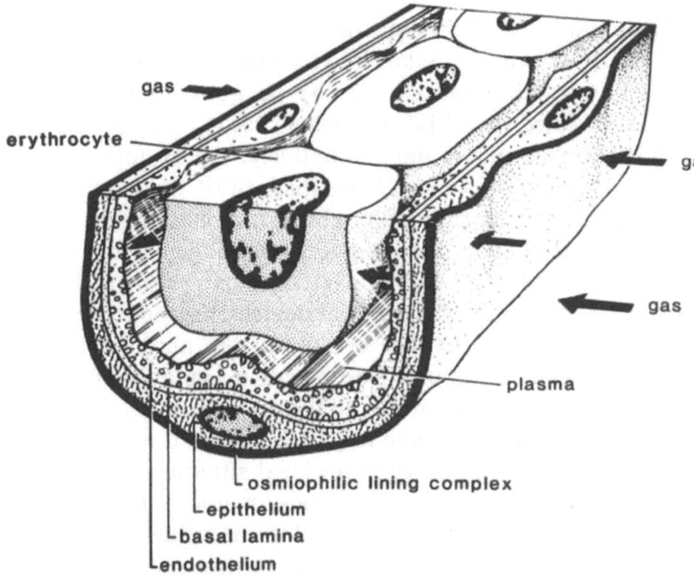

Fig. 4. A stereogram of a pulmonary blood capillary showing the general components of the air-haemoglobin pathway, namely the epithelium, basal lamina, endothelium plasma layer and erythrocytes. Oxygen diffusion is shown with *arrows*. The degree of diffusion of gases is directly proportional to the surface area and inversely to the diffusional distance. The parameters can be modelled to give the total morphometric pulmonary diffusing capacity which is the global indicator of a gas exchanger's structural refinement. (Maina 1989e, courtesy of Academic Press, UK)

model has been used to estimate the DL of O_2 of the gills (Hughes 1972b), the Dipnoan lung (Hughes and Weibel 1976; Maina and Maloiy 1985), the reptilian lung (Perry 1983), the mammalian lung (Gehr et al. 1981) and the avian lung (Maina 1989e). Integrating pulmonary parameters gives a better picture of a gas exchanger's total efficiency over that achieved by considering piecemeal data, as limitations or refinements can occur anywhere along the diffusional distance (i.e. the air-haemoglobin pathway, A-HP). The A-HP essentially consists of the blood-gas barrier (BGB), the plasma layer, and the capillary blood where O_2 is taken up by the haemoglobin (Fig. 4). Individual conductances, respectively Dto_2, Dpo_2, and Deo_2, are determined and the total DLo_2 derived from the three serial conductances. Thus,

$$1/DLo_2 = 1/Dto_2 + 1/Dpo_2 + 1/Deo_2 .$$

The basic assumption in such pulmonary morphometric analysis is that the measurements made on tissue preparations are representative of the functional state. Unfortunately, only few and notably variable O_2 permeation constants (Ko_2) and O_2 uptake coefficients (θ, o_2) are presently available. Such uncertainties have compelled workers like Hughes (1972b), Dubach (1981), and Perry

(1983) to restrict the estimates and comparisons to the anatomical diffusing factor (ADF) which only takes into account the surface area of the BGB and its thickness. Our observations on the penguin lung (Maina and King 1987), however, revealed that ADF in such a case (where modifications occur outside the BGB) can be misleading. Adaptively, the Humboldt penguin (*Spheniscus humboldti*), a champion diver, has a thick BGB and hence a low Dto_2 but on the other hand has an enormous Vc. This brings its DLo_2 to a value close to that of a terrestrial bird of comparable size. Despite the current inherent shortcomings in the model which should be overcome as pertinent data become available, DLo_2 is a useful index of the global degree of respiratory adaptation. It should, however, be pointed out that in spite of the uncertainties, in related animals (where pulmonary structural features are acceptably uniform) comparisons are tenable. Since in the estimation of DLo_2 it is assumed that the whole of the BGB and Vc maximally participate in gas exchange, the factor is a measure of a gas exchanger's optimal capacity in an ideal situation where ventilation-perfusion inequalities (V/Q) and other limitations are abolished. DLo_2 is thus consistently appreciably higher than the physiological value (Sect. 13.3). The difference between the two values is considered to express a safety margin which enables a lung to vary gas exchange commensurate with the extremes of the prevailing demands (Weibel 1973, 1984a). Powell and Scheid (1989) have, however, asserted that DLo_2 (should have been more accurately the ADF) is simply "a parameter to indicate the mean area-to-thickness ratio of the gas exchanging surface" and in their opinion has no relationship with the physiological diffusing capacity. While this view is bound to be debated in future, it seems to us rather imprudent to suggest that pulmonary structure and design do not influence function.

10 General Respiratory Sites

10.1 Integument

When well vascularized, the skin functions as a gas exchanger both in water and in air. Small animals like the larvae of most insects exchange O_2 and CO_2 exclusively through the skin (Fraenkel and Hertford 1938) and in molluscs substantial gas exchange occurs this way (Ghiretti 1966). In larvae of some invertebrates like the terebellid worms, localized areas of the body with reduced integumentally thickness serve a respiratory function (Weber 1978) while in larval lungfishes a transient ciliary apparatus promotes convective movement of water in contact with the skin (Whiting and Bone 1980). Although once imputed that the skin of fish is impermeable to gases, it has recently been found to posses an appreciable diffusion-limited conductance of O_2 much of which is utilized by the skin and the tissues immediately beneath it (Kirsch and Nonnotte 1977).

Scales do not curtail cutaneous respiration (Jakubowski 1963). The efficiency of the skin as a gas exchanger is, however, limited by the fact that the arterial PO_2 is dependent on the PO_2 in the external medium and optimally can only equilibrate with it (Figs. 1A and 2B). In hypoxic water (PO_2, 40 torr), a fish like plaice (*Pleuronectes platessa*) extracts 37% of its O_2 needs through the skin (Steffensen et al. 1981). Fish with a high capacity for transcutaneous respiration have a thin epidermis and the circulation is well developed. The thickness of the skin in flounder ranges from 31–38 μm and that of the eel is 263 μm (Jakubowski 1963). Comparison of the degree of vascularization of the skin of fish and Amphibia indicates that species like *Cyprinus carpio*, *Misgurnus fossilis* and *Nemachilus barbatulus* have as well vascularized skins as amphibians like *Rana terrestris* and *Xenopus laevis* (Satchell 1971): expressed as capillary length per gram of body weight, *C. carpio* has a value of 6 m, *M. fossils* 5.9 m, *N. barbulus* 5.8 m, *R. terrestris* 6.1 m and *X. laevis* 5.2 m. The European eel (*Anguilla anguilla*) acquires 28% of its O_2 needs across the skin (Kirsch and Nonnote 1977), a feature which may account for the relatively small branchial area in the species (Bennett 1988b). Remarkably high O_2 uptake values via the skin (60–70%) were reported in *Anguilla* when the fish was out of water (Berg and Steen 1965, 1966). In normoxic water ($WPO_2 = 150$ mm Hg), *Clarias batrachus* and *Saccobranchus* (= *Heteropneustes*) *fossilis* acquire 17% of their O_2 needs from water through the skin (Hughes and Singh 1971; Singh and Hughes 1971). Similar values were reported in *Anguilla vulgaris* (Berg and Steen 1965) and *Monopterus albus* (Liem 1967). Teal and Carey (1967) and Gordon et al. (1985) observed that 50–60% of the total O_2 needs of the mudskipper, *Periophthalmus sobrinus* is derived by cutaneous aerial respiration, while cutaneous respiration accounts for 40% of aquatic respiration. In *Mnierpes macrocephalus*, aquatic cutaneous respiration accounts for 40–48% of the O_2 needs (Graham 1973). Larger fish rely on gills more than the skin for gas exchange (Ojha and Munshi 1974). Among vertebrates, it is in amphibians where the skin best participates in gas exchange, the efficiency depending on factors like the degree of vascularity, magnitude of external surface area, diffusional distance and permeability characteristics of the skin tissue which may favour O_2 and CO_2 flux.

Body surface area in anuran amphibians scales moderately with mass ($Sa = 1.131\ M^{0.579}$) (Hutchison et al. 1968). Plethodontid salamanders are devoid of gills and lungs and rely entirely on the skin for gas exchange (Gatz et al. 1974) while aquatic urodeles like *Necturus* and *Siren* in which the lungs are very poorly developed rely on gills and the skin (Guimond and Hutchison 1973). The giant salamander (*Cryptobranchus alleganiensis*), which is lungless and is devoid of gills, to increase cutaneous respiration is dorsoventrally flattened with conspicuous folds of skin dangling from the margins of the body and legs, the skin being extensively vascularized (Guimond and Hutchison 1976). The lungs of amphibians living at high altitudes (O_2-poor air but cold freshwaters) have almost atrophied (Noble 1929; Vellard 1951) and the skin has become baggy (Parker 1940). In the common dusky salamander (*Desmognathus fuscus*), the mass-specific total surface area of cutaneous and buccopharyngeal capillaries is

3.922 $cm^2\ g^{-1}$ and the mean thickness of the skin 25.8 μm (Czopek 1961): about 50% of the cardiac output, estimated at 0.11–0.19 ml $min^{-1}g^{-1}$, is channelled to the skin where O_2 is absorbed, the diffusive capacity of the skin being about 0.3–0.5 nmol min^{-1} $torr^{-1}$ (Piiper et al. 1976). *Siren lacertina* up to 700 g in body mass and all sizes of *S. intermedia* and *Pseudobranchus striatus* can survive indefinitely in well-aerated water at 25 °C without access to air (Ultsch 1974): all three sirenids are metabolic regulators and at critical O_2 tension become O_2 conformers. Xerophilic amphibians like *Hyla arborea* and *Chiromantis petersi* are completely aerial relying virtually on the lung for gas exchange. At 25 °C, the skin accounts for nearly 80% of CO_2 and 30% of O_2 exchanged in the green frog, *Rana clamitans* (Vinegar and Hutchison 1965). The fraction of cutaneous gas exchange increases at lower temperatures and below 5 °C the lung may contribute very little in CO_2 elimination (Whitford 1973). To serve as a gas exchanger, the skin has to be perpetually kept moist. In some of the most terrestrial fish, the mudskippers (Gordon et al. 1969), a strong correlation between cutaneous respiration and gill development was observed (Tamura and Moriyama 1976; Tamura et al. 1976; Low et al. 1990). Of the three species studied, *Boleophthalmus boddaerti*, which has a greater affinity to water, had the longest and the greatest number of gill filaments and largest gill area to skin area ratio (GA/SA), indicating a greater role of its gills than its skin for respiration. In *Periophthalmodon schlosseri*, the GA/SA ratio increases as the fish grows, indicating that juvenile fish depend more on cutaneous gas exchange, while branchial respiration is more significant in mature individuals (Low et al. 1990). The species extracts 76% of its O_2 requirement across the skin and only 27% across the gills (Tamura et al. 1976). In the salamander (*D. fuscus*), the mean PO_2 between the environment and the capillary blood is about 100 torr for O_2 and 5–6 torr for CO_2 (Piiper et al. 1976). Reptilian species like the freshwater turtle (*Sternothaerus minor*) (Belkin 1968) and the sea-snake (*Pelamis platurus*) take up 22% of their O_2 needs through the skin (Graham 1974). The common iguana can stay submerged for more than 4 h when threatened (Morberley 1968). In the water turtle (*Chrysemys picta belii*), 70% of the total O_2 needs (during a dive) are met by the skin, and at a temperature near freezing (3 °C) the animal can be kept almost indefinitely in O_2-saturated water (Ultsch and Jackson 1982). At 10 °C, aquatic turtles examined by Ultsch et al. (1984) could survive submerged in anoxic water ($PO_2 < 2$ mm Hg) for 3–17 days and in normoxic water ($PO_2 = 150$–160 mm Hg) for over 100 days. The turtles *Trionyx spiniferus*, *Kinosternon subrubrum* and *Sternotherus odoratus* have highly exposed thi)n and well-vascularized integument (Dunson 1986) through which gas exchange occurs (Stone et al. 1992a); respectively, the three species derive 63, 86 and 74% of their O_2 needs from air. The stinkport (*S. odoratus*) can stay submerged in well-aerated water for over 50 min exchanging gases with water (Stone et al. 1992b). The aquatic elephant trunk snake (*Acrochordus javanicus*) (at 20–22 °C) derives 8% of its total O_2 needs across the skin while in the terrestrial boa constrictor (*Constrictor constrictor*), only 3% is exchanged this way (Staendaert and Johansen 1974). *Typhlonectes compressicaudus*, an aquatic caecilian can derive 28.3–55.2% of its VO_2 from water (Goncalves and Sawaya 1978). Correspondingly, *T. compressicaudus* has a very thin and well-vascularized skin (Fuhrmann 1912). Active recruitment of cutaneous blood capillaries has been observed in teleosts (Sacca and Burggren 1982), in amphibians (Burggren and Moalli 1984), and reptiles (Courtice 1981). Seasonal fluctuation in cutaneous vascularity has been observed in the hairly frog (*Astylosternus robustus*) by Noble (1925). The African swampworm, *Alma emini*, a glossoscolecid oligochaet which lives in swampy, waterlogged, strongly reducing soils of East and Central Africa, shapes and exposes (to air) its well-vascularized hind end which serves as a temporary lung (Beadle 1957). As the worm retreats in the substratum, it takes a bubble of air with it enclosed in the tube. The highly vascularized epithelium of the buccal cavity of *Amphipnous cuchia* is used for the dual function of breathing air and water (Singh et al. 1984). A close relationship between VO_2 and surface area was found in nine species of pulmonate and eight operculate snails (von Brand et al. 1948). When the lung of the land snail *Agriolimax agrestis* is filled with paraffin, 56% of the normal O_2 uptake takes place across the body wall (Schuurmans-Stekhoven 1920). Male lungfish, *Lepidosiren* has an enigmatic gill-like structure on the posterior appendage which has been presumed to serve as an accessory respiratory organ or as an oxygenation device during larval husbandry (Johansen 1982). Hughes (1966b) proposed that cutaneous respiration may have been crucial in the evolution of air-breathing by the early vertebrates as the skin augmented O_2 uptake and CO_2 elimination during the formative stages of the lungs.

10.2 Water Gills

Gills are undisputedly the archetype respiratory organs and take different sizes, shape and locations, and characterize all chordates at one stage of development (Mangum 1982). They range in complexity from simple tegmental evaginations like the tube-feet of some classes of echinoderms, the external gills of the annelid polychaetes, the larvae of teleosts, e.g. *Gymnarchus*, the book gills of the chelicerate arthropods, e.g. *Limulus polyphemus*, the horns of *Simulium* larvae, and the tadpoles of Amphibia (Fox 1921; Mill 1972; Mangum 1982) to the much elaborate internal ones of the crustaceans, molluscs and teleosts (Hughes 1984). Commonly, gills are evaginations of the ectoderm (Fig. 1B) and increase the respiratory surface area by hierarchical subdivision of the structural components (Figs. 5, 7 and 10). Lungfish larvae raised in well oxygenated water have diminished external gills, while in hypoxic water the gills are hypertrophied (Johansen 1982). Contraction of the gills, ciliary movements and mechanical ventilation (as in the decapods and fish), placing the body in a moving current of water or active body movement, maintain a PO_2 on the respiratory surface. Except in very few cases as in crustaceans, molluscs and fish, e.g. the mudskippers (Schlotte 1932; Saxena 1959; Graham 1976; Low et al. 1990), the gills of

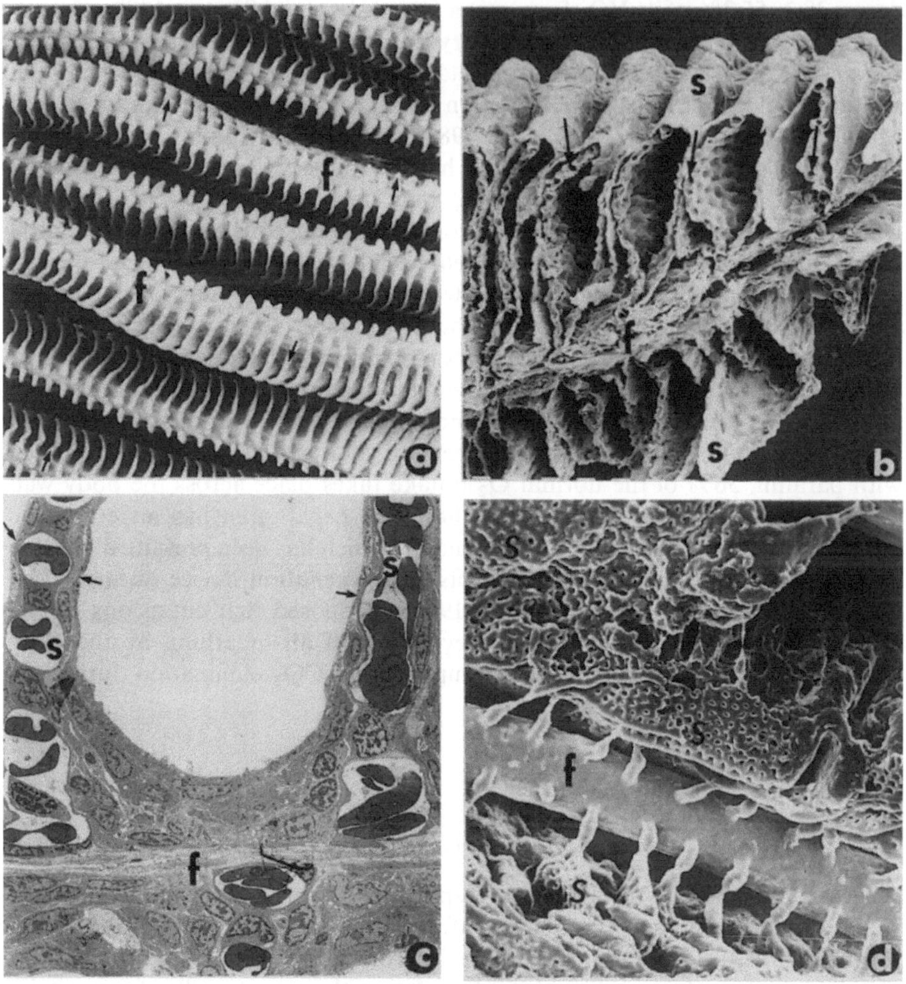

Fig. 5a–d. Structure of fish gills. The preparations are from *Oreochromis alcalicus grahami*. **a** Low-power view of gill filaments (*f*) and secondary lamellae (*arrows*), × 61; **b** longitudinal section of a gill filament (*f*) showing the blood channels (*arrows*) in the lamellae (*s*), × 1200. **c** view of a gill filament (*f*), secondary lamellae (*s*) and a thin water-blood barrier (*arrows*) overlying the lamellar vascular channels, × 360; **d** latex cast of gills showing a filamental blood vessel (*f*), secondary lamellae (*s*) the pits on the lamellar flaps marking the sites of the pillar cells, × 730. (**a** and **c** from Maina 1990a)

aquatic animals are rendered physiologically functionless in air due to the facts that: (1) surface tension makes the gill filaments and particularly the secondary lamellae adhere to each other, essentially creating large dead air-spaces in the gills, (2) there is a drastic reduction of the respiratory surface area and, (3) the gills dry out and thus become impermeable to gases. This explains why functional air-gills are extremely rare and evolutionarily are a dead end. However, several fish like *Hypopomus brevirostris*, *Symbranchus marmoratus* and *Hoplosternum litorale* to a notable extent use gills for air-breathing after definite modifications such as strengthening and spacing of secondary lamellae which prevent collapse (Carter and Beadle 1931; Johansen 1966). In *Chirolophis*, the lamellae are held apart by raised marginal ridges (Schmidt 1942). Gradual phasing out and finally extirpation of the gills and their subsequent replacement with air-breathing organs appear to have been the plausible stratagem for successful transition to air-breathing. Initially, the gills may have evolved as feeding organs (Gutman and Bonik 1981), a function which has hence been eclipsed by roles like respiration, osmoregulation, acid base balance and regulation of circulating hormones like catecholamines and angiotensin (Neckvasil and Olson 1986). Like in most multifunctional organs, over adaptation in one aspect engenders new problems and hence the best mutually acceptable functional compromise has to be sought. Gills of primitive actinopterygian fish, *Polypterus* (Hughes 1981) still bear cilia as a mark of their past trophic function. The trophic role (filter feeding) of the gills is still exhibited by primitive chordates in the subphyla Tunicata and Cephalochordata, e.g. *Amphioxus* and *Petromyzon* (Baskin and Detmers 1976; Youlson and Freeman 1976); the protochordata are presumed to occupy a transitional stage between invertebrates and vertebrates. Willmer (1970), however, suggested that the respiratory role of the gills may have evolved first. The definitive design of the gills is thus a compromise between all these functional attributes (McDonald et al. 1991). It has been proposed (Randall 1982; Butler and Metcalfe 1983) that fish are able to adjust the surface area of their gills as a means of balancing respiration with osmotic exchange. The change to pulmonary breathing may have been consequent to increased complexity of the organisms and impermeability of the skin to gases and ions (Gans and Northcutt 1983). The gills of lampreys are internal and their structural features are essentially similar to those of the more advanced teleosts (Lewis and Potter 1982). Larval caecilians use gills for respiration in the egg and are lost soon after hatching (Welsch 1981), while the neotenic larva of *Amblystoma* (Axolotl) has perennial external respiratory gills. The lugworm (*Arencola cristata*), has 11 pairs of branching filamentous gills with a water-blood barrier 13–16 μm thick (Mangum 1976). In the sabellid polychaete, *Eudistylia vancouveri* the gill divides to terminate in a double row of filaments called pinnules (Mangum 1982). Each branchial crown is composed of about 54 000 pinnules of which the surface accounts for more than 70% of the surface area of the body which is 30 cm^2 g wet wt (Vogel 1980): the diffusional distance across the pinnular surface is 10–12 μm (Mangum 1982). Though the parapodium of *Nereis succinia* does not strictly qualify as a gill, its respiratory role is irrefutable (Nicoll 1954; Mangum 1982): its water-blood diffusion distance is 1.5–2.3 μm.

Hughes and Morgan (1973), Laurent (1982), Hughes (1984) and Laurent and Perry (1991) have given good didactic accounts of the morphology and morphometry of teleost gills. Gas exchange takes place across the semicircular, bilaterally situated plates, the secondary lamellae (Fig. 5). The lamellae are divided into vascular channels by the pillar cells which send out cytoplasmic flanges overlaid by an interstitial layer and an epithelial cell, the three components forming the water-blood barrier (WBB). Among fish, the thickness of the WBB is highly variable (Table 4), ranging from 0.596 μm in *Euthynnus affinis* to 11.3 μm in *Scyliorhinus canicula* (Hughes and Wright 1970; Hughes and Morgan 1973). Active pelagic swimmers like tuna and mackerel have a thinner barrier and a more extensive gill surface area (Table 1) than moderately active ones (e.g. yellowfin-bream, sea-robin, and blueshark) and the sluggish benthic ones (e.g. toadfish, dogfish, icefish, flounders and caelacanth) (Gray 1954; Muir and Hughes 1969; Hughes and Morgan 1973; Hughes and Iwai 1978; Raubal 1987). The skipjack tuna which is warm-blooded (Carey and Teal 1969) has a WBB thickness of only 0.59 μm (Table 4) and a weight-specific respiratory surface area (RSA. W^{-1}) of 1840 $mm^2\ g^{-1}$ (Muir and Hughes 1969) while *Latimeria chalumnae*, a slow deep-water swimmer, has a WBB 6–8 μm thick and RSA W^{-1} of 18 $mm^2 g^{-1}$ (Hughes 1980). Generally, the morphometric features of fish gills commensurate with physiology and lifestyle. By varying features like filament length and size of secondary lamellae, the gills of the warm-blooded and more metabolically active white shark (*Carcharodon carcharias*), the Atlantic shortfin-mako (*Isurus oxyrinchus*) and common thresher shark (*Alopias vulpinus*) (Carey 1982) have a larger surface area than those of the pelagic ectothermic sharks like the sand-bar shark (*Carcharhinus plumbeus*), dusky-shark (*C. obscurus*), blue-shark (*Prionace glauca*), and scalloped hammerhead shark (*Sphyna lewini*) (Emery and Szczepanski 1986); the RSA W^{-1} of the gills of great white shark is 872 $mm^2\ g^{-1}$, while that of blue shark is 239 $mm^2\ g^{-1}$. The morphometric parameters of the gills of the bimodally breathing fish (Table 4) are lower than those of solely aquatic ones (Hughes and Morgan 1973). *Oreochromis alcalicus grahami*, a fish which lives in the hypoxic and highly alkaline Lake Magadi (Reite et al. 1974; Johansen et al. 1975) has a WBB 0.83 μm thick and gill filament length and lamellar numbers are relatively high (Maina 1990a, 1991). An energetically economic design of the gills calls for an increase in respiratory surface area with minimal resistance to water flow. In active fish, this has been achieved by increased filament length and lamellar number, while in the sluggish ones the filaments are greater in number, more spaced and shorter in size, while the lamellae are reduced in size (Hughes 1966a). The hierarchial subdivision of the gills from holobranchs, hemibranchs, gill filaments through to secondary lamellae and in some species like *Ophiocephalus* (Munshi 1960) development of tertiary lamellae increases the respiratory surface area within the restricted opercular space. The lamellae are covered by a thin squamous epithelium (Fig. 5c) which enhances gas flux while most of the tissue cells including the osmoregulatory chloride cells are relocated to the interlamellar space (Maina 1990a; Pisam and Rambourg 1991). Pillar cells subdivide the lamellar plates into channels (Fig. 5c, d) through which blood in a sheet-flow pattern (Fung and Sobin 1977) percolates. The gills are profusely perfused with blood by the branchial artery, a branch of the ventral aorta, and are well ventilated by a concerted action of the buccal pressure pump and opercular suction pump.

Table 1. Gill morphometric features of active, intermediate and sluggish fish[a]

Fish	Wt (g)	Total no. of filaments	Total filamental length (mm)	Sec. lamellae per mm (both sides)	Average filamental length (mm)	Gill area (mm^2)	
						Total per g	Single (lamella (both sides)
Sluggish							
Toad fish	316	688	4930	11	7.2	170	0.9
Fluke	404	1706	9747	19.3	5.71	247	0.53
Intermediate							
Sea Robin	365	1996	14022	20	7.04	432	0.56
Tautog	382	1564	11835	18	7.58	450	0.88
Active							
Mackerel	226	21814	19291	29.4	6.84	1040	0.41
Mullet	250	2190	16399	26.5	7.49	1010	0.58
Tuna	1667	6139	111296	64.57	17.93	1845	0.43

[a] Sources of data: Hughes (1966a) and Hughes and Morgan (1973).

Apart from a few air-breathing fish, the gills are the only organs which receive the whole of the cardiac output (Satchell 1971). The volume of the water which passes over the gills per unit time can be varied either by increasing the respiratory rate, the stroke volume, or both.

The spatial relationship of the flow of water and blood (in the secondary lamellae), which has been confirmed by blood-gas tensional measurements, is laminar and counter-current (Hughes 1966a, 1972a; Hughes and Hills 1971). The two gas exchange media run in opposite directions (Figs. 2c and 3). Such disposition maintains a greater tensional differential throughout the period of exposure of blood to water in the gills facilitating a higher degree of O_2 saturation of blood (Piiper and Scheid 1973; Sect. 9). It is probably due to this design and partly to factors like inactivity, low ambient temperature (usually below zero) and high O_2 content of water (11 ppm) that despite having a very low O_2-carrying capacity (0.67 vol%) (Rudd 1954, 1965) the oxygenation of blood in the haemoglobinless Antarctic icefish of the family Chaenichthyidae is very high (Holeton 1970; Hughes 1972a). Although laminar flow is advantageous as it reduces resistance, turbulence produces greater mixing and hence enhances gas exchange (Steen 1971a). Of the total resistance to O_2 diffusion, 50% is attributable to water, 37% to the tissue barrier and 7% to the erythrocyte (Hills et al. 1982). The remarkable efficiency of the gills is reflected in their high O_2 extraction which in *Scyliorhinus stellaris* is above 60%, and in teleosts as much as 80% (Saunders 1962; Baumgarten-Schumann and Piiper 1968). Hughes (1967) observed an O_2 utilization of 92% in the trigger fish (*Balistes capriscus*), a value far above that of mammals with only about 16% (Stahl 1967).

10.3 Water Lungs

For gas transfer to occur at a respiratory site, a concentration gradient must exist for the flux is physically diffusive. Experimentally, with effective mechanical ventilation, a rat can be kept in O_2-saturated water indefinitely. Aquatic animals like the sea-cucumber (holothurians) have a water lung which is connected to the anus and is ventilated by contractions and relaxations of the body muscles (Newell and Courtney 1965): in *Holothuria forksali*, cloacal O_2 uptake contributes about 60% of the total need and gas exchange is drastically reduced when cloacal respiration is arrested. The tracheal gills of the larvae of the aquatic dragon-fly nymph are in the rectum and are ventilated mechanically. In the freshwater pulmonate gastropods of the order Basommatophora, reversed evolution has taken place. After evolving air-breathing, the group has re-colonized water. This is evinced by the fact that the group lacks ctenidia and the mantle cavity has been converted to a water lung (Hunter 1953). In the freshwater pulmonates of the genera *Lymnaea* and *Planorbis*, the pulmonary cavity is converted into a water lung which is rhythmically filled with water (Precht 1939; Hunter 1953): if the O_2 drops below a critical level, the animal comes to the surface to breathe (Cheatum 1934). When exposed to a hypoxic environment or as the tide recedes, the bivalve *Pholas dactylus* collects water in the aperture on the inhalant siphon from which it draws 47% of its O_2 needs (Knight 1984; Knight and Knight 1986).

10.4 Buccal Membrane

During the day, at the surface of the ocean the water is usually saturated with O_2 while at a depth of about 100 m the concentration is as low as 0.013 mmol l^{-1}, which is equivalent to a partial pressure of 6.5 torr (Harvey 1960). Most aquatic animals when stressed skim the surface of the water, take in a layer of water with a high O_2 concentration and hold it in the buccal cavity before pumping it over the gills. The buccopharyngeal membrane (BP-M) is profusely vascularized and serves as an accessory respiratory organ. This behaviour was observed in *Oreochromis grahami* of Lake Magadi during the night and after the fish were stressed (Maina 1990a, 1991). In dividing turtles (Belkin 1968), the BP-M is utilized for gas exchange in water and during a dive 30% of the O_2 requirement is transferred across the membrane. In the large soft-shelled turtles (Agassiz 1857), the BP-M is modified by having papillae which increase the non-pulmonary respiratory surface area and enhance aquatic O_2 uptake (Girgis 1961; Zhao-Xian et al. 1989). The buccal cavity of the electric eel is well vascularized and functions as a gas exchanger both in and out of water (Johansen et al. 1968b; Farber and Rahn 1970). In the eel, as a mark of the efficiency of the cavity for gas exchange, only 10% of the inspired gas volume remains in the mouth after a breath-hold. The caudal part of gastrointestinal tract of the pond loach (*Misgurnus fossilis*) functions as a respiratory organ (Jeuken 1957): depending on the O_2 tension in water, the organism makes frequent excursions to the surface where it swallows air which after gas exchange is voided through the anus.

10.5 Gas Gills

The evolutionary success of insects is demonstrated by their remarkable adaptive radiation into every conceivable habitat (Callahan 1972). Strictly, all insects are air-breathers but a small number has atypically taken to water by evolving novel techniques of taking up and storing air under the water. The insects which are characterised by a compressible gas-gill like *Dytiscus* make occasional trips to the surface of the water and entrap a gas bubble with hairs. As the bubble is transported below the surface, CO_2 dissolves into the surrounding water faster than does O_2. With increasing depth, the partial pressure of N_2 in the gas bubble increases. As N_2 gradually dissolves into the surrounding water, the bubble collapses (Thorpe and Crisp 1947; Rahn and Paganelli 1968). In insects with an incompressible gas-gill like *Aphelocheirus*, the air-bubble (the plastron) into which the tracheae open is supported by groups of dense hydrofuge hairs. The firmly held plastron has a very small radius of curvature. Since it presents an extensive water – air interface (surface area per unit volume), the plastron has a constant volume (Hinton 1966). As the CO_2 dissolves in water, it is replaced by O_2 from the water and since the N_2 concentration in the bubble is at equilibrium with that in water, as long as the water is adequately aerated, the plastron can last indefinitely (Rahn and Paganelli 1968). Plastron respiration is restricted to well-aerated water as plastrons extract O_2 from tissues, should O_2 pressure of the environment fall below that of the issues (Hinton 1966). It is quite probable that the tracheal system (Sect. 12.3) evolved for terrestrial habitation and aquatic insects (counter to the process taken by other animals, i.e. from water to air) have secondarily acquired this capacity. The freshwater pulmonates which were essentially air-breathers have readapted very well to aquatic life (Sect. 12.1): the mantle cavity has no ctenidia and functions as a lung. In *Planorbarius corneus*, the mantle is regularly protruded into the water to serve as an accessory gill, while in *Lymnea peregra* air is taken in as the animal dives (gas-gill function) and in specimens living at considerable depths the lung functions as a water-lung (Zaaijer and Volvekamp 1958; Ghiretti and Ghiretti 1975).

10.6 Swim Bladder

Most modern fish have a hydrostatic organ, the swim bladder. The phylogenetic relationship between the swim bladder and the lung is controversial. The lung appears as a ventral diverticulum of the foregut while the swim bladder arises as a dorsal one (Romer 1970). Structurally, the layers of the wall of the swim bladder correspond with those of the gastrointestinal system (GIT) (Spengel 1904; Fange 1953). Advanced teleosts have physoclistic swim bladders i.e., without connection to the GIT, while swim bladders in the primitive ones have an external connection, i.e. are physostomus. Doubts have been raised as to whether the swim bladder and the lung are homologous and which arose first (Carter 1967; Schmalhausen 1968; Romer 1972). Intuitively, as a ventrally located air-space would cause an imbalance in a moving aquatic fish, the swim bladder may have evolved entirely for buoyancy control and in some cases (e.g. the Dipnoi, Sect. 11.2) acquired secondarily a respiratory role (Potter 1927): the two functions are by no means mutually exclusive. In *Hoplerythrinus unitaeniatus*, the swim bladder is very well vascularized with 60–80% of the cardiac output perfusing it (Smith and Gannon 1978). The regulation of efflux and influx of gases into and out of the swim bladder is under the control of the autonomic nervous system (Fange 1966, 1983; Steen 1971a, b; Nilsson 1983). Such control in euphysoclists like *Tetractenos* (= *Torquigener*) *glaber* (Green 1984), *Perca fluviatilisa* (Saupe 1940) and *Polachinus virens* (Ross 1978) takes place in the secretory gland/rete complex and the resorptive area. Gas gland region consists of a glandular epithelium and a parallel system of blood vessels, the rete mirabile (Green 1984). The two retia in the eel contain 88 000 venous and 116 000 arterial capillaries with a total length of 816 m and a total area of 211 cm^2 contained in a volume of 0.064 cm^3 of which 66.7% is occupied by blood (Krogh 1959). The surface-to-volume ratio of the rete mirabile in the eel was estimated at 1700 $cm^2\,cm^{-3}$ (Steen 1963) and the wall thickness at 1.5 μm (Scholander 1954). The resorptive area which lies anterodorsally is richly vascularized, the degree of exposure of the epithelium being controlled by relaxation and contraction of an oval sphincter. In *T. glaber* (Green 1984), the swim bladder epithelium at the resorptive area is 1 μm thick, while that in the gas-gland region is 20–30 μm. The current evidence of the functional mechanisms of gas production in the swim bladder suggests that lactic acid is produced by the gas-gland cells and then secreted into blood vessels where as a result of the counter-current multiplying function of the rete mirabile it causes the partial pressures of gases in the blood capillaries of the secretory area to rise (Fange 1966, 1983; Steen 1971a, b; Nilsson 1983). Gas production is regulated by the action of cholinergic enervation of the muscularis mucosa which regulates the exposure of the blood vessels, the gas-gland cells and the vasculature supplying the gas-gland region (Green 1984). Resorption of the swim bladder gases is enhanced by adrenergic nerves and occurs when the muscular sphincter opens to expose the resorptive vascular plexus, an event accompanied by dilation of the resorptive blood vessels (Fange 1953, 1983; Ross 1978).

The swim bladders of the air-breathing catfish *Pangasius sutchi* and *P. pangasius* (Browman and Kramer 1985; Zheng and Liu 1988) are a highly trabeculate gas-exchange organs. They are well vascularized and internally subdivided to form respiratory air-cells: the epithelium is made of cells which structurally and functionally resemble the type I and type II cells of the lungs of reptiles, mammals and birds. Further, the swim bladders are lined by a surpellic substance to reduce the surface tension at the air – liquid interface (Phlegar and Saunders 1978). These characteristics are suggestive of evolutionary or adaptive homology of such swim bladders with the lungs of the higher vertebrates. It is interesting to note that the dipnoan lungs (Sect. 11.2) which are thought to have given rise to the vertebrate lungs are in respect to the differentiation of the pneumocytes not as well developed as in the swim bladder of *Pangasius*. As has been alluded to earlier, the evolution of the respiratory organs has in most cases followed a parallel pattern and in some instances reversed evolution has occurred. It is hence unrealistic and futile to attempt to look for consistent and logical patterns in such a maze of concurrently occurring events.

10.7 Book Lungs

On extensively analyzing fossil scorpions, Kjellesvig-Waering (1986) concluded that the development of the book lungs enabled terrestrial existence and was the most important evolutionary change in scorpions. All the 48 species which had gills (belonging to suborder Branchioscorpionina) are now extinct while only nine families with book lungs (belonging to suborder Neoscorpionina) presently survive. The book lungs characterize the arachnids which constitute one of the largest taxon in the animal kingdom and were some of the first animals to conquer land (Foelix 1979; Paul et al. 1989). Scorpions are some of the most terrestrial creatures, most of them living in extremely xeric habitats. The book lungs probably derive from the book gills of *Xiphosura* (Anderson 1973). Morphologically, the scorpions have changed very little since the Devonian period (Paul et al. 1989). Like the Dipnoi (Sect. 11.2), scorpions may present the least adulterated aspects of the physiology and anatomy of the earliest animals. The structure of the book lungs in arachnids has been studied extensively (Pohunkova 1969; Moore 1976; Hill 1977; Anderson and Prestwich 1980, 1982; Hexter 1982; Strazny and Perry 1984, 1987). The lungs are formed from invagination of the surface ectoderm (Anderson 1973) and protrude into a sinus containing circulating haemolymph which moves in a cross-current manner (cf. air flow in the avian lung, Sect. 13.4) relative to the gradients in lamellar gas-tensions (Anderson and Prestwich 1980). The lamellae are covered by a cuticle of an average thickness of 0.5–1.0 μm in spiders (Anderson and Prestwich 1980). The cuticle provides support to the lamellae preventing collapse. In species like the desert scorpion, *Paruroctonus mesaensis* (Farley 1990) and spiders in general (Anderson and Prestwich 1980), further modifications in the form of cuticular spines which interconnect the lamellae on the atrial side and cross-bridges on the haemolyphatic side act as struts preventing lamellar collapse. In spiders (Moore 1976; Hill 1977; Foelix 1982), the cuticular spines are attached only to one lamellar wall, and on the haemolymphatic side hypodermal pillar cells, which are formed as bridges between lamellae (in the haemolymphatic passages), were observed (Hexter 1982; Foelix 1982). There are 100–130 lamellae in each of the four pairs of book lungs in *P. mesaensis* (Farley 1990). Zoond (1931) reported 130–150 lamellae (in each lung) of the scorpion *Opisthophthalmus capensis* and Awati and Tembe (1956) and Vyas and Laliwala (1976) observed 140–150 lamellae in *Mesobuthus tamulus*. Compared with spiders with paired lungs, those with one pair have larger lamellae (width 43% and length 23% greater) and wider interlamellar spaces (by 13%) (Anderson and Prestwich 1980). The lamellae are clearly surface amplifications to enhance the surface area available for gas exchange, a feature which has been attained in different ways in various animals. The interlamellar air-space width (IL-AS-W) in *P. mesaensis* is 2–10 μm (Farley 1990) and in the South American tarantula 5 μm (Steen 1971a). Generally, in spiders (Anderson and Prestwich 1980), IL-AS-W changes very little with increasing body mass (IL-AS-W = $0.90\ M^{0.21}$:W, μm; M, mg). This feature is typical of those highly important parameters in the body like the thickness of the blood-gas barrier (Fig. 20) which in the course of evolution appear to have become optimized.

Many spiders have book lungs and a tracheal system while others only have book lungs (Levi 1967; Foelix 1982; Strazny and Perry 1987). The book lungs appear to be the older system (Strazny 1982) and characterize the most primitive spiders, the orthognaths (Paul et al. 1987a, b): variably, they are replaced by a tracheal system. The water-spider (*Argyroneta aquatica*), has a well-developed tracheal system (Levi 1967). The tracheal system is the only respiratory system in insects (Sect. 12.3) where O_2 is delivered directly to the body cells. Lung volume in the house-spider (*Tegenaria domestica*) scales strongly with body mass ($VL = 2.805\ W^{1.157}$:W, g; VL, $mm^3\ 10^{-3}$) (Strazny and Perry 1984) and the haemolymph constitutes 53% of the lung volume. In spiders, the tracheal system appears to be crucial in the maintenance of a higher metabolic rate in the more active species (Anderson 1970). Scorpions and spiders with relatively low metabolic rates only have book lungs (Hardley and Hill 1969; Anderson 1970; Paul et al. 1989). The efficiency of the book lungs depends on a well-adapted cardiopulmonary coupling and the presence of the tracheal system where this has evolved. Scorpions have an intricate network of arteries and the ventilation and perfusion of the lung is well regulated (Farley 1990). The thickness of the chitinous layer on the surface of the lung of tarantula (*Eurypelma californicum*) is 0.2 μm (Paul and Fincke 1989), in the house-spider the value is 0.4 μm while in the African scorpion, *Pandinus imperator*, it is 0.3 μm (Rodl 1987). Though in *T. domestica* the chitinous layer is as thick as the hypodermis (Strazny and Perry 1984), it constitutes the highest resistance to O_2 diffusion due to the extremely low value of Krogh's diffusion constant (Kto_2) of chitin. A thin barrier coupled with passive diffusion (Paul et al. 1987a, b; Fincke and Paul 1989) together with heart-beats which agitate and cause gas-mixing in the lungs (Hill 1977; Herreid et al. 1981) enhance the efficiency of the book lungs. The anatomical diffusion capacity of O_2 by tissue barrier (Dto_2) in *T. domestica* is 0.035 $mm^3\ h^{-1}\ torr^{-1}$, a value which is similar to that of the turtle, *Pseudemys scripta* (Perry 1978). O_2 consumption in arachnids is generally very low (Anderson 1970; Paul et al. 1989). In some spiders such as *Dupiennius* and *Pholcus*, there is a general lack of mitochondria in the locomotory muscles which may largely work anaerobically (Linzen and Gallowitz 1975). The same state prevails in the prosoma and leg muscles of tarantula where there is a general lack of mitochondria (Paul and Fincke 1989).

The respiratory physiology of spiders and scorpions differ notably (Fincke and Paul 1989; Paul and Fincke 1989): fluctuations in CO_2 production occurs in the spiders but not in scorpions and the maximum CO_2 production is higher in scorpions and the recovery time shorter. For animals of similar body mass (15 g), the surface area of the four-lunged *Eurypelma* (50 cm^2) was smaller than that of the eight-lunged *Pandinus* with a value of 80 cm^2 (Paul and Fincke 1989). The higher value in the scorpion, however, is due to the greater number of lungs rather than to any morphological refinements. When normalized with the number of lungs, the surface area of the book lungs in *Eurypelma* is 12 cm^2 per lung and 10 cm^2 in *Pandinus*. With a lung volume of 0.03 cm^3 (Paul et al. 1987a, b), *Eurypelma* has a surface area per unit lung volume of 167 $mm^2\ mm^{-3}$, a value which is greater than that of 131 in a 16 g shrew, *Crocidura poensis* with a lung volume of 0.5 cm^3 (Gehr et al. 1980). From allometric extrapolation, a hypothetical 100 g specimen of *T. domestica* (Strazny and Perry 1984) has a lung volume of 0.578 mm^3 and a surface area of 69.8 mm^2. The surface-to-volume ratio for specimens of *T. domestica* (27–342 mg) ranged from

66–245 $mm^2 mm^{-3}$ values which fall between those of mammals and birds (Gehr et al. 1981; Maina et al. 1989a). It is evident from the above that increase in respiratory surface area by folding is notably more effective in increasing surface area than most of the other modes of compartimentalization found in aquatic and air-breathing animals. Comparison of the functional similarity between the book lungs of the spiders and the vertebrate lungs is justifiable on the basis of the empiric facts that linear dimensions of the alveoli (Tenney and Remmers 1963) and faveoli in amphibians and reptiles (Tenney and Tenney 1970) scale to the same power (0.2) of body mass as book lungs (Anderson and Prestwich 1980). Further, the mode of respiration, gas mixing and transfer in the book lung are essentially similar to those of mammals (Forster 1964). In the tarantula *E. carifornicum*, Paul et al. (1987a, b) observed that diffusion suffices in O_2 uptake and ventilation may only be necessary at high O_2 consumptions and in case there is a large intrapulmonary dead air-space. Like it is generally the case in most animals, morphometric features of the book lungs in many species of spiders and scorpions seem to have been optimized to interface with metabolic and physical needs (Paul and Fincke 1989). The only way to increase the respiratory surface area appears to have been to increase the number of lungs, but this in most cases is not possible due to lack of anatomical space.

11 Bimodal Breathers

11.1 Air-Breathing Fish

Air-breathing fish can be categorized into an amphibious group (those which resort to air-breathing during terrestrial excursions) and an aquatic group (those which are confined to water and only periodically surface to take in air) (Graham 1976). Both groups are largely found in ponds, swamps, marshes and sluggish or intermittently flowing rivers in the tropical regions of the world. Such habitats exhibit intensive putrefaction of organic plant materials, the derelict waters being hypercarbic and hypoxic (Dehadrai and Tripathi 1976). In the Nile swamps, the O_2 concentration is only 10% of the normal saturation level (Rzoska 1974). Upto 30 ppm of free CO_2 were recorded in some Indonesian swamps (Dehadrai and Tripathi 1976) and CO_2 tension as high as 60 mm Hg were measured in Floridan swamps (Ultsch 1976). In situations where dissolved O_2 is low and CO_2 concentration high, this poses serious respiratory physiological problems which include reduced affinity of blood to O_2, a decrease in O_2-carrying capacity of blood and an acid-base imbalance (Ultsch 1976). To compound the problem, the swamps generally have low pH, high concentration of gases like carbon monoxide, hydrogen sulphide, ammonia and methane, an environment apparently rather similar to the anoxic chemically reducing environment (Sect. 5) that characterized the earth during the formative stages of life (Tappan 1974). Fish in such habitats have developed a dual strategy of respiration: the accessory respiratory organs (AROs) supplement the limited amount of O_2 taken up by the gills from the O_2-deficient waters (Figs. 1D, E and 5). As an indication of the stressful environment such fish live in, air-breathing fish are generally small in size with the exception of *Ophicephalus marulius* and *O. striatus* of India (Dehadrai and Tripathi 1976). To varying extents, air-breathing is fairly frequent in marine fish which use an array of tissue surfaces for gas exchange (Graham 1976); none of these fish, however, is an obligate air-breather. Marine air-breathers live in well-aerated waters and while in air switch completely to aerial breathing (Graham 1976). Air-breathing may have occurred after geographical and ecological isolation of O_2-depleted waters (Das 1927, 1940; Carter 1931, 1957; Carter and Beadle 1931; Hora 1935; Fish 1956; Saxena 1963) or as a part of the general genetic evolutionary process of animals (Dubale 1951).

The morphophysiological pressures which lead to the evolution of air-breathing organs in diverse groups of fish which were confronted with similar hypoxic environment are conspicuous when their numbers and organs modified or adapted for aerial respiration are considered. About eight orders and 19 families of teleosts are known to use different organs, like the buccopharyngeal membrane, pharyngeal lung, opercular lung, stomach, intestine, air-bladder, and skin (Bertin 1958; Dehadrai and Tripathi 1976). The remarkable structural diversity accompanied by functional convergence is a clear indication of the varied stratagems adopted by remarkably different animals in response to comparable selective pressures, principally environmental hypoxia. The degree of reliance of air-breathing fish on atmospheric O_2 varies among species of fish (Table 2), some being facultative while others are obligatory air-breathers. The former resort to air-breathing only in extreme circumstances, while the latter group will in some cases even drown if prevented from surfacing. In normoxic waters, *Anabas testudineus*, *Clarias batrachus*, and *Saccombrancus* (= *Heteropneustes*) *fossilis* respectively obtain 54, 58, and 59% of their O_2 requirement from the air (Singh 1976). *Anabas* normally lives in waters with a PO_2 of 40–100 mm Hg and will only resort to air-breathing when the tension falls to below 100 mm Hg. The fish survives for a long time out of water in a non-desiccating atmosphere (Hughes and Singh 1970). *Clarias lazera* survives for 14–17 h even when kept in well-oxygenated water if prevented from surfacing (Moussa 1956, 1957). At an O_2 concentration of about 8.5 mg l^{-1} Abdel-Magid (1971), however, found that *C. lazera* could survive indefinitely in water. *C. batrachus*, which derives 16% of its O_2 need via the skin, can survive for a long time in water when refused access to free air (Singh and Hughes 1971). The piracucu (*Arapaima gigas*) relies on air-breathing for most of its O_2 supply and will die if refused access to the surface (Stevens and Holeton 1978), while the garfish (*Lepisosteus osseus*) is a facultative air-breather at low temperatures but an obligate air-breather when O_2 uptake increases at high temperatures (Rahn et al. 1971). *Piabucina festae* obtains 10% of O_2 from air if the water PO_2 (WPO_2) is high, but 70% of O_2 uptake from air when the WPO_2 is 35 mm Hg (Graham et al. 1977). *Ancistrus chagresi*, acclimated to hypoxia for 14–21 days,

Table 2. A comparison of the respiratory properties of blood and oxygen consumption in some air-breathing fish

Fish	Hb (g%)	O_2 capacity (vol%)	O_2 uptake (VO_2, ml kg^{-1} h^{-1})			Reference
			Aquatic	Aerial	Total	
Clarias mossambicus	13.3	17.82	64.6[a]	21.1[a]	85.7[a]	Fish (1956); Johnson et al. (1983)[a]
Clarias batrachus	12.59[a] (28.77)	16.87[a]	38.85[a]	54.4[a]	93.39[a]	Munchi and Singh (1968); Singh and Hughes (1971)[a]
Anabas testudineus	19.8	–	52.62[a]	60.80[a]	113.42[a]	Benerjee (1966); Hughes and Singh (1970)[a]
Saccobranchus fossilis	11.58[a] (30.20)	15.52[a]	50.10[a]	34.43	84.55[a]	Munshi and Singh (1968); Hughes and Singh (1971)[a]
Electrophorus electricus	11.2	13.90	6.92[a]	23.08[a]	30.00[a]	Johansen et al. (1968b); Farber and Rahn (1970)[a]
Amia calva	5.82	7.8	66.00	30.00	96.00	Johansen et al. (1970)
Lepisosteus osseus	–	–	14.40	39.99	53.40	Rahn et al. (1971)
Protopterus aethiopicus	6.2	6.80	1.26	10.14	11.40	Lenfant and Johnsen (1968)
Lepidosiren paradoxa	6.64	8.25	1.00	21.20	22.20	Lenfant et al. (1970)
Ancistrus chagresi	5.1	–	–	–	84.0	Graham (1983)

[a] Johnson et al. (1983) data based on fish acclimated to aerated water (PO_2 15.3 KPa) for 27 days.

exhibited definite adaptations like a higher blood O_2 affinity and higher Hb-concentration; the volume of the ARO increased by 25% and the fish could hold the breath longer and thus remove more O_2 (Graham 1983). Although in air-breathing fish most of CO_2 is voided through the gills and skin (into water), under stress the AROs of *Ophicephalus* (= *Channa*) *striatus* (Qasim et al. 1960) and *Notopterus chitala* (Dehedrai 1962) are important pathways for CO_2 excretion. These air-breathing fish will generally resort to air-breathing when the WPO_2 drops to 30 mm Hg (Gee 1976; Graham and Baird 1982). Physiological adaptations like higher blood O_2-carrying capacity and CO_2 less sensitive Hb (Willmer 1934), higher O_2-loading tension and higher Hb-concentration (Dubale 1959; Bernerjee 1966) and circulatory modifications (Johansen 1970) characterize air-breathing fish. Contrary to living in O_2-depleted water, some fish living in mountainous torrential streams (Hora 1922, 1933; Ojha and Singh 1986) and in waters with abundance of algae, phytoplankton and zooplankton (Rzoska 1974) are during the day exposed to water supersaturated with O_2 due to photosynthesis. In the latter, during the night the O_2 concentration may drop to zero and the fish may have to use AROs for gas exchange. In the gourami *Ostronemus*, to enhance O_2 uptake and probably CO_2 release, the air in the suprabranchial chamber cavity is replaced with water from the opercular cavities before the next inspiratory cycle (Peters 1978).

The AROs in fish such as *Clarias mossambicus* (Maina and Maloiy 1986), *C. batrachus* (Singh and Hughes 1971) and *A. testudineus* (Munshi 1968) are housed in concavities above the gills, the supraopercular chambers (Figs. 6 and 7a). Since these cavities are absent in the solely water-breathing teleosts, they should

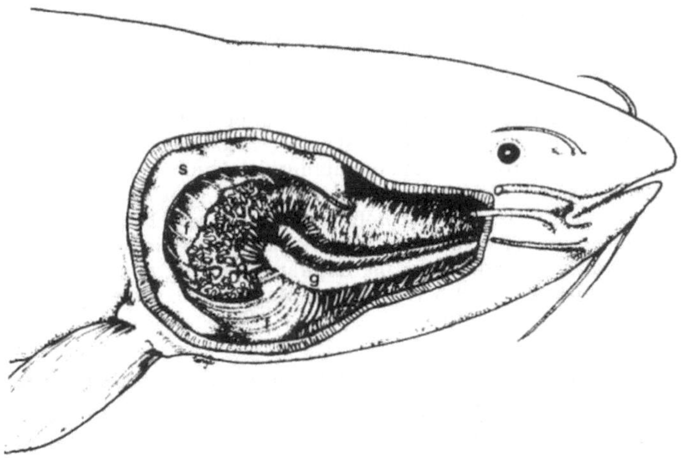

Fig. 6. A side view of the respiratory organs of a bimodal breathing catfish, *Clarias mossambicus*, with the operculum removed to show the topographic relationships between the gill arches (*g*) and the air-breathing organs, the labyrinthine organ (*o*) attached to the gill arches and the suprabranchial chamber membrane (*s*); *f* gill fan, approx × 1.6

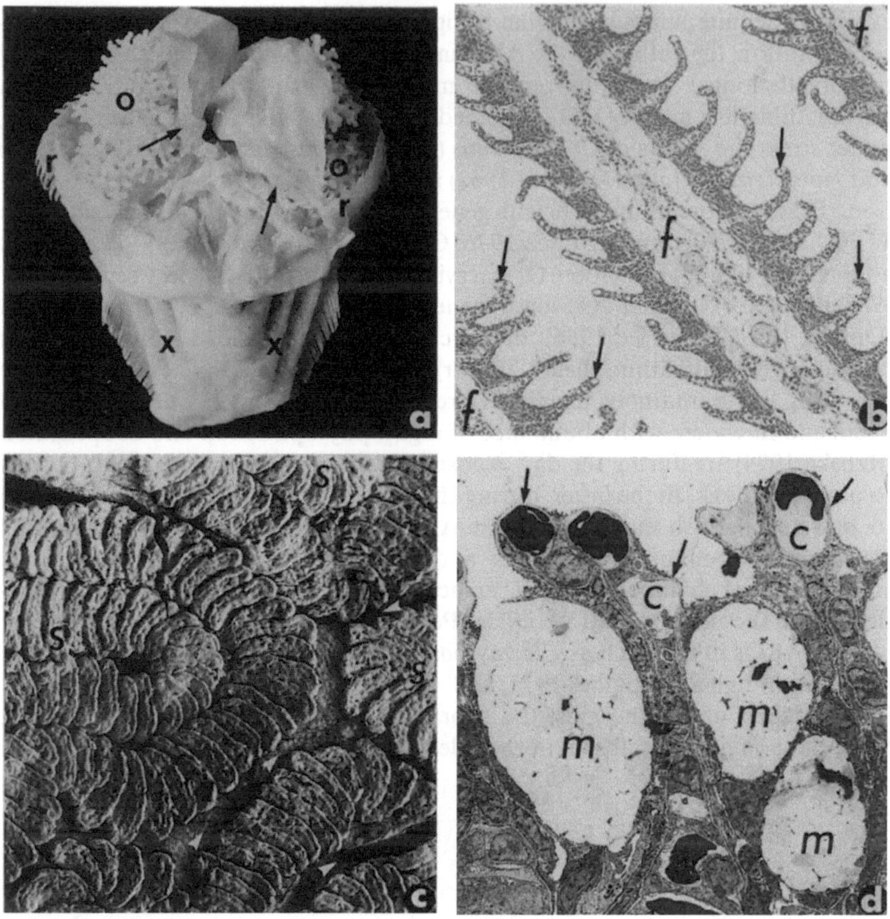

Fig. 7a–d. Respiratory organs of a bimodal breathing catfish *Clarias mossambicus*. **a** Gross specimen of the respiratory organs showing the gill arches (*x*) and the air-breathing organs, the labyrinthine organs (*o*) and the suprabranchial chamber membrane (*arrows*); *r* gill fans, × 110; **b** gill filaments (*f*) which give rise to secondary lamellae (*arrows*), × 398; **c** subrabranchial chamber membrane lined by profuse transverse capillaries (*s*) separated by narrow vascular lanes (*arrowheads*), × 389, **d** sectional view of the labyrinthine organs showing blood capillaries (*c*) which protrude above the respiratory surface; *m* mucous secretory cells; *arrows* the blood-gas barrier, × 1946. (Maina and Maloiy 1986, courtesy of Oxford University Press)

be a neomorphic development to house the AROs. In the bowfin *Amia calva* a facultative air-breather which estivates when its aquatic habitat dries up, the gills are well developed and gas exchange is supplemented by an air-bladder (Johansen et al. 1970). Aerial mode of ventilation notably differs from the aquatic one (Liem 1984): ventilation of the AROs is brought about by reversed flow of water caused by decreased pressure in the buccal cavity in which process air is drawn out of the suprabranchial chambers (SBC). AROs can be divided into monoaccessory type, those which consist of only a simple membrane lining the concavity (Fig. 1D), and diaccessory type (Fig. 1E) where the concavity contains an additional structure (Dehadrai and Tripathi 1976). In Anabantidae and Clariidae, the opercular lung consists of the SBC membrane (SBCM) lining the concavity and the tree-like labyrinthine organ (LO) supported on gill arches. The SBCM is a tough, well-vascularized, semitransparent membrane, while the LO is a dichotomously branching organ (Fig. 7a) which is supported by a fibrocartilaginous core made up of chondrocytes intermixed with collagen (Maina 1985a; Maina and Maloiy 1986). The respiratory surfaces of the SBCM and LO are lined by a dense arrangement of blood capillaries separated by narrow avascular tracts (Fig. 7c, d). Developmentally, it has been pointed out that the AROs are derived from the branchial arches through fusion of gill filaments and secondary lamellae (Munshi 1961a, b, 1968; Singh et al. 1984). The presence of pillar cells in the AROs, the morphological similarities among the respiratory organs, and the vasular interconnection between the gills and the AROs are taken to be compelling reasons for this proposition (Moussa 1956; Munshi 1961a, b, 1968, 1976; Singh and Thakur 1975; Munshi et al. 1986; Olson et al. 1990; Singh et al. 1991). Whereas the pillar cells which support the vascular lamellar channels in fish like *Heteropneustes* and *Clarias* are of gill tissue origin (Singh et al. 1981, 1982a), in fish like *Anabas*, *Channa* and *Amphipnous*, the pillar cells are formed by modified epithelial cells (Hughes and Munshi 1968, 1973a, b; Munshi and Singh 1968; Singh et al. 1982b): such cells are analogous to the pillar cells of other teleosts. The air-sac in the mud-eel (*Amphipnous cuchia*) develops from the gill mass tissue cells of arches 2–5 (Singh et al. 1984). Such a process, where similar tissue cells give rise to entirely different organs, illustrates the radical innovations and permutations which were necessary for air-breathing to have evolved in phylogeneticallly different animals which had different structural resources available to them. The sequence of evolution of the ROs in this taxon should have been in the order gills, the SBCM and LO. This contention is supported by the study on *A. testudineus* by Munshi and Hughes (1986) where at fingering stage, a time when the LO is not formed, the SBCM is used in aquatic respiration as the LO is not fully formed until about 8–12 months of life. In *Colisa fasciatus* (Prasad et al. 1982), air-breathing capacity is attained at 8–20 days of age.

Although osmiophilic lamellated bodies and surfactant characterize the swim bladders of some air-breathing animals (Browman and Kramer 1985; Zheng and Liu 1988) and practically all lungs of most advanced air-breathers (Pattle 1976), such features are absent in AROs like the SBCM and LO. This suggests that most AROs are not homologous to the lungs. It would, however, be of interest to investigate such organs in *Amphipnous* (= *Monopterus*) *cuchia* and *Channa* which do not appear to be derived directly from the gills. The circulatory system of the gills in species like *Anabas*, *Clarias*, *Saccobranchus*, *Macropodus* and *Channa* is connected in series to the AROs (Moussa 1956; Hughes and Munshi 1973b; Hughes et al. 1973; Ishimatsu et al. 1979; Ishimatsu and Itazawa 1983; Olson et al. 1986) and in parallel in *Saccobranchus*,

Amphipnous, and *Amia* (Olson 1981; Olson et al. 1990). The latter design is far form optimal in delivering O_2 to the tissues. In general, the gills of bimodal-breathing fish are relatively poorly developed and less well perfused with blood than those of solely water-breathing fish (Saxena 1962, 1963; Graham 1976; Munshi 1976; Harkim et al. 1978; Munshi et al. 1990; Olson et al. 1990): in *Amphipnous*, the gills are found only in the second arch, in *Channa striatus*, *Saccobranchus* and *C. batrachus* the fourth pair of gills is extremely reduced in size, while in *Anabas* all the gill arches are generally reduced with the 3rd and 4th arches being notably small and the gills having very few filaments and secondary lamellae. This feature may have occurred for the following reasons: (1) to avoid redundancy of the gills with the evolution of air-breathing organs, (2) to save on the energy entailed in ventilation of the gills, and (3) to avoid excessive loss of O_2 (taken up by the AROs) into the hypoxic water as the blood traverses the gills. *Anabas testudineus* retains the capacity of ventilating the SBCM with water even in adulthood (Hughes 1978; Peters 1978). The parts of gills which remain are only utilized in CO_2 efflux (Hughes 1966a; Singh 1976; Graham 1983) until an efficient active ventilatory mechanism is established to facilitate adequate CO_2 clearance for blood pH regulation (Randall et al. 1981). In *Anabas*, *Clarias* and *Saccobranchus*, 90% of CO_2 is eliminated through the gills (Singh 1976). Saxena (1963) suggested that high concentrations of free CO_2 in water may have played an equivalent if not a greater role in the evolution of air-breathing.

The water–blood barrier of gills of air-breathing fishes is structurally similar to that of solely water-breathing teleosts (Munshi 1976). The total surface area of the gills of the bimodal-breathing fish generally exceeds the combined one of the AROs (Hughes et al. 1973; Maina and Maloiy 1986) (Table 4). In *C. mossambicus* the weight-specific gill area is $17.3\ mm^2\ g^{-1}$, while the areas in SBCM and LO respectively are 7.8 and $4.7\ mm^2\ g^{-1}$. Further, the total respiratory area (TRA) in the bimodal-breathing fish is remarkably lower than in a water-breather (Hughes et al. 1973). In *Clarias* the TRA is about $30\ mm^2\ g^{-1}$ and in *Anabas* $89\ mm^2\ g^{-1}$, while in the salmon (*Salmo gairdneri*) and the tench, *Tinca tinca* (Hughes 1972b) the gill areas are respectively 240 and $250\ mm^2\ g^{-1}$. Despite the smaller TRA, due to the relatively thin air-blood barrier in the AROs which ranges from 0.188 μm in the intestine of the pond loach, *Misgurnus fossalis* (Jaskinski 1973) to 1.61 μm in *Saccobranchus* (Hughes et al. 1974a; Munshi 1976), the total anatomical diffusing capacity (DLo_2) of all the ROs (i.e. gills and AROs) in most species exceeds that of the gills of a solely water-breathing fish of similar size (Table 4). In *C. mossambicus* (Munshi 1976; Maina and Maloiy 1986), the weight-specific DLo_2 of the ROs is $0.1413\ ml\ O_2\ min^{-1}\ mm\ Hg^{-1}\ kg^{-1}$ while in the salmon and the tench the values are respectively 0.06 and 0.125 (Hughes 1972b). By varying the DLo_2s of the gills and the AROs, a bimodal-breathing fish adjusts its dependence on water or air. The bimodal-breathing fish tend to become more dependent on air with the development (Munshi 1976). The significant contribution the LO makes towards the overall gas exchange function is evident in that in fish like *Clarias* (Maina and Maloiy 1986) the organ contributes 50% of the DLo_2, the gills providing only 15% and the SBCM 35%. In *Anabas* the LO provides 80% of the O_2 influx and the gills only 2% (Hughes et al. 1973). The large contribution by the AROs may explain why species like *C. mossambius* and *A. testudineus* in particular are obligatory air-breathers. *Anabas* is known to survive out of water for a long-time as it "walks" from one pond to another "catching small arthropods on the way and even climbing trees" (Das 1927; Saxena 1963). Other air-breathing fish like *Clarias, Anguilla, Periophthalmus, Synbranchus, Saccobranchus*, and *Ophicephalus* are frequently found out of water (Hughes et al. 1974b; Dehadrai and Tripathi 1976). In *Anabas, Amphipnous* and *Saccobranchus*, dependence on air correlates well with morphometric dimensions of the aerial respiratory organs. Respectively, the DLo_2s in the three species (units: ml O_2 min^{-1} mm Hg^{-1} kg^{-1}) are 0.2896, 0.0165, and 0.0561 (Hughes et al. 1974a, b). Evidently, *Anabas* is best-suited for air-breathing in the three species. Correspondingly, at the same partial pressure of O_2 in water (WPO_2), *Anabas* extracts more O_2 from air (60.8 ml kg^{-1} h^{-1}) than *C. batrachus* (54.54) and *Saccobranchus* (34.45) (Singh 1976) (Table 2). The swamp eel (*Synbranchus marmoratus*) uniquely can ventilate its gills while it holds air in the ARO (Liem 1980; Heisler 1982; Graham et al. 1987). It is plausible that with gradual sophistication in the structure and function of the AROs, a facultative air-breather would become an obligate air-breather and the gills would progressively become redundant. Concomitant with requisite morphological adaptations like limbs and an impermeable skin, such an animal has much of the potential for adopting terrestrial habitation.

11.2 Lungfish (Dipnoi)

Air-breathing has independently evolved a number of times in the phylogenetic history of fishes in response (but not always) to aquatic hypoxia (Randall et al. 1981; Hughes 1984; Burggren and Johansen 1986). The oldest lungs are apparently those of the heavily armoured extinct placoderm fish *Bothriolepis* (Denison 1941). Some of the most primitive extant fishes, the bichirs (e.g. *Polypterus*) have a functional lung. The lungfish (Dipnoi) are of great interest mainly for reasons that: (1) they, together with the crossopterygians, are the closest ancestors to the terrestrial vertebrates, (2) they represent some of the earliest air-breathers, and (3) they illustrate how changes in the respiratoy system have influenced the design and physiology of other related organ systems, particularly the circulatory system. The Dipnoi are monophyletic and have a history of about 400 m years having first appeared in early Devonian period (Rosen et al. 1981; Gans 1986). They, together with scorpions (Sect. 10.7) are probably some of the oldest morphologically unadulterated extant animals. The intense selective pressures this group has endured are illustrated by the fact that there are 50 extinct dipnoan genera which represented about 120 species and arguably only 3 species exist today (Gans 1986). Though dipnoan fossils are found in all continents, the contemporary genera are restricted to entirely different continents,

Lepidosiren to South America, *Neoceratodus* to Australia and *Protopterus* to Africa (Greenwood 1986). The similarities between the physiology of the lungfish and the tetrapods particularly amphibians in respect to gas exchange, excretory and muscle physiology and cardiopulmonary design and function, are remarkable (Johansen and Hanson 1968; Dunn et al. 1981; Burggren and Johansen 1986). These and other morphological similarities have prompted the thinking that the Dipnoi should taxonomically be grouped together with the tetrapods. Having developed the capacity to breathe air, the Dipnoi had solved one of the primary requirements for successful terrestrial habitation. As has been proposed by some palaeontologists like Rosen et al. (1981), it is possible that the fins, as in lobe-finned fish like *Eusthenopteron* and the mudskippers, became strengthened into lateral appendages which could support the body on land. The gross morphology of the gills of the highly elusive benthic coelacanth, *Latimeria chalumnae* closely resembles those of its nearest living relative *Neoceratodus forsteri* (Hughes 1980). Air-breathing was crucial for the survival of these fish particularly through the arid Devonian period, though why they retained lungs across the wetter Carboniferous is difficult to explain. It is debatable whether the dipnoan lung initially evolved as a hydrostatic organ (a swim bladder) and subsequently acquired a respiratory role or vice versa (Sect. 10.6). The air-breathing organs of the lungfish are, however, true lungs for the reasons that (1) their pneumatic duct originates as a ventral diverticulum of the foregut (Grigg 1965a; Klika and Lelek 1967; Romer 1970; Hughes and Weibel 1976a; Maina 1987a), (2) their blood supply comes from the sixth branchial arch (Grigg 1965a; Johansen et al. 1968a; Burggren and Johansen 1986), and (3) there are structural features present, like surfactant (Hughes 1978) and immunoreactive cells (Zaccone et al. 1989), which are common to the lungs of the primitive and advanced animals. In some cases, e.g. *Amia*, *Lepisosteus* and *Erythrinus*, the swim bladder has acquired an accessory respiratory role (Carter and Beadle 1931).

Of the three genera of lungfishes, *Neoceratodus*, which has well-developed gills, is least adapted for aerial respiration and is an obligate water-breather (Burggren and Johansen 1986; Fishman et al. 1986; Kemp 1986). *Lepidosiren* and *Protopterus* are obligate air-breathers and will drown if they are refused access to air for an extended period of time. Aerial respiration is induced at a water PO_2 (WPO_2) of 85 mm Hg in *Neoceratodus*, while in *Protopterus* WPO_2 does not influence gill or aerial breathing (Johansen and Lenfant 1968). In *Protopterus* (Lenfant and Johansen 1968), at 25 °C, about 90% of O_2 uptake is effected through the lungs. In adult *Lepidosiren*, at 20 °C, 96% of the O_2 needs are acquired through the lung (Sawaya 1946), while in *Neoceratodus* nearly 100% of both O_2 and CO_2 exchange is aquatic (i.e. via the gills and skin) (Lenfant et al. 1966; Burggren and Johansen 1986). Corresponding to these observations, the frequency of the buccal/opercular movements in *Neoceratodus* (at room temperature) are about 30 times per minute (Johansen et al. 1967), in *Protopterus* the movements are 0.5–10 times min^{-1} at 20–24 °C (Lenfant and Johansen 1968) and in *Lepidosiren* the movements are very weak and range from

1–20 breaths per minute (Johansen and Lenfant 1967). At a temperature of 25 °C, freely moving and respiring *Neoceratodus* had a VO_2 of 0.7 ml g^{-1} h^{-1} acquired from branchial respiration at a rate of 67 beats min^{-1} supplemented by the lung (Grigg 1965a, b). In *Neoceratodus*, 0.05 ml O_2 g^{-1} h^{-1} are derived from the gills alone at a rate of 80 beats min^{-1}. Combined O_2 uptake from water and air in *Protopterus amphibius* ranges from 30–60 ml O_2 kg^{-1} h^{-1} (STPD) with the mature fish acquiring only 10–15% of their O_2 needs from water (Johansen et al. 1976a). In contrast to the aquatic life, during aestivation, gas exchange in lungfish is effected entirely by lungs (Lahiri et al. 1970) as the circulatory pattern changes such that the gills are by-passed (Johansen and Reite 1968). An increase in O_2-Hb affinity is the principal adaptational feature which characterizes aestivation in lungfish (Johansen et al. 1976b) and enables the fish to take up the required O_2 needs in a subterranean environment with a reduced ambient O_2 tension and altered mechanics of pulmonary ventilation (Lomholt et al. 1975). While *Neoceratodus* cannot live for long out of water, an aestivating *Protopterus* encased in a mud cocoon, has been known to live for 4 years (predictably 7) in a dormant state (Coates 1937; Smith 1961). In such a state, VO_2 falls to 10% of its non-aestivating state and the respiratory rate, which is characterized by very small tidal volumes and a large residual volume, drops at one to two times per hour (Lomholt et al. 1975) as the heart rate drops to only three times per minute, and excretion and elimination of metabolic end products except CO_2 are suspended (Smith 1930; Swan 1969; Fishman et al. 1986).

Neoceratodus (Grigg 1965a; Kemp 1986) has a single lung while *Protopterus* (Maina and Maloiy 1985; Maina 1987a) and *Lepidosiren* (Hughes and Weibel 1976, 1978) have a paired lung. Dipnoan lungs are ventilated by action of a positive pressure buccal pump (Grigg 1965a, b; Burggren and Johansen 1986) which is followed by a period of apnea of variable length. Exhalation is effected by contraction of the smooth muscle and the elastic tissue fibres which are an integral part of the structural components of the lung and may further prompt internal gas distribution (Grigg 1965a; Johansen and Reite 1967). Inhaled air O_2 concentration is reduced from the normal atmospheric concentration of 20.9 to 14% in the resting *Neoceratodus*, and to as low as 5% in a struggling animal (Grigg 1965a). Aestivation appears to cause changes in the pattern and mechanics of air-breathing. *Protopterus amphibius* has been said to use suctional breathing when aestivating (Lomholt et al. 1975). This yet to be fully substantiated observation would place the Dipnoi above all other air-breathing animals, below the reptiles as aspiration pulmonary ventilation replaced buccal force pump ventilation as the ribs became an integral part of the tetrapod terrestrial locomotory skeletal framework. In *Protopterus aethiopicus* (Maina and Maloiy 1985; Maina 1987a) and *Lepidosiren* (Hughes and Weibel 1978), the lung has a central air-duct supported by strong trabeculae which open into peripherally situated gas exchange cells (about 1 mm wide) which are supported and delineated by strong internal septa (Fig. 8). Blood capillaries cover the septa and budge out into the air-spaces (Johansen et al. 1968b; Hughes and Weibel 1976, 1978; Maina and Maloiy 1985), increasing the respiratory surface area by

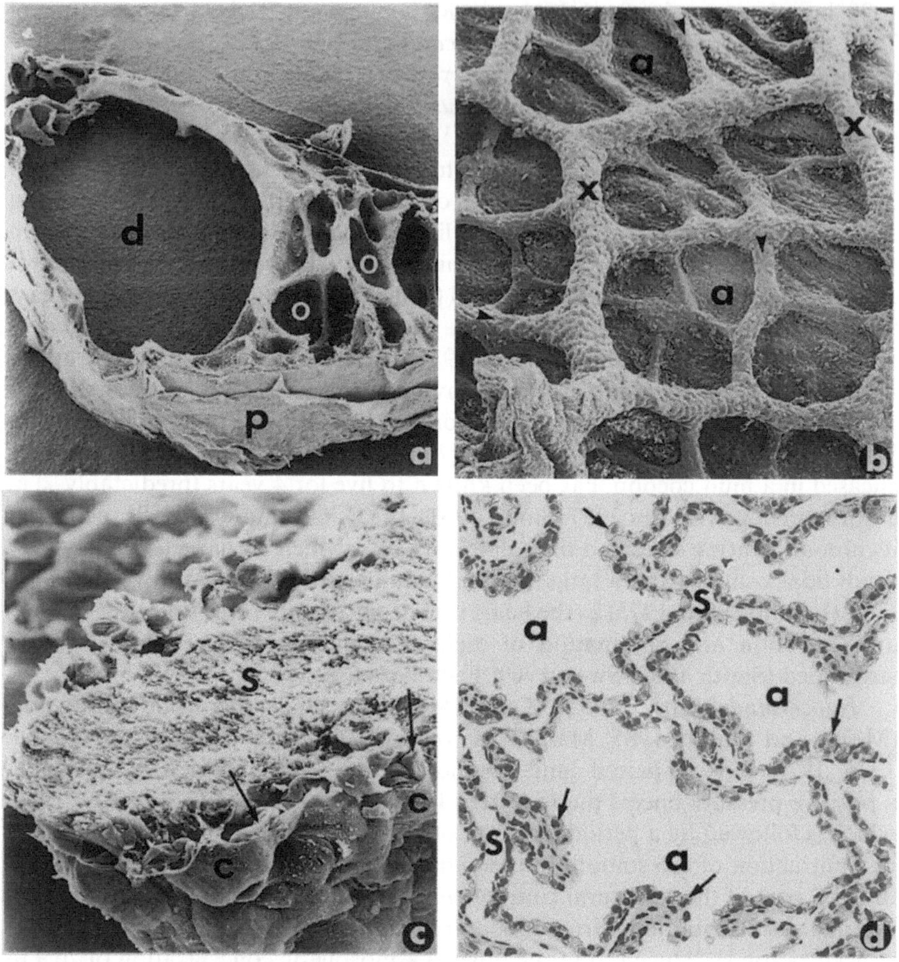

Fig. 8a–d. Structure of the lung of the lungfish *Protopterus aethiopicus*. **a** Cross-section of the lung which is covered by a pleural membrane (*p*) and made up of an eccentrically located air-duct (*d*) and respiratory air-cells (*o*), × 25; **b** internal surface of the lung showing the subdivision of the lung by primary septa (*x*) and secondary septa (arrowheads) into terminal gas-exchange spaces (*a*), × 18; **c** the double capillary disposition of the blood capillaries (*c*) on the surface of the septum (*s*); *arrows* red-blood cells, × 355; **d** sectional view of the air-spaces (*a*) and the double capillary disposition of blood capillaries (*arrows*) along the septa (*s*), × 296. (**a**, **b**, and **c** from Maina 1987a, **d** from Maina and Maloiy 1985, courtesy of the Royal Society of London)

a factor of as much as 25% (Perry 1989). During expiration, virtually all the air is removed from the lung of *Lepidosiren* (Bishop and Foxon 1968), While in *Protopterus* about 80% of the tidal volume is voided (Jesse et al. 1967). This is facilitated in part by the well-developed smooth muscle and elastic tissues which form a part of the trabeculae and septae (Maina and Maloiy 1985; Maina

1987a). Blood flow to the lung in *Protopterus* increases fourfold following inspiration (Johansen et al. 1968a) and constitutes 3.5% of the lung volume in *Lepidosiren* (Hughes 1978). In both *Neoceratodus* (Johansen et al. 1967) and *Protopterus* (Wood and Lenfant 1976), intrapulmonary pressure is positive throughout the respiratory cycle. The compression of the intrapulmonary air raises the alveolar PO_2 enhancing the Hb-loading and may simultaneously be important in counteracting the high perfusion pressure of the lung (consequent to the incomplete separation of the pulmonary and systemic circuits) thus preventing oedema (Johansen and Reite 1967).

The lungs of the lower vertebrates show a trabecular type of architecture (Hughes 1970, 1978), a common feature which shows a design solution of compromising dynamic support and flexibility in a tubular organ. Commensurate with its facultative air-breathing characteristics, the lung of *Neoceratodus forsteri* shows the least degree of septation and compartmentalization (Grigg 1965a, b). In *Lepidosiren* (Hughes and Weibel 1978) and *Protopterus* (Klika and Lelek 1967; Maina and Maloiy 1985; Maina 1987a), pulmonary pneumocytes consist of an undifferentiated population of cells characterized by somewhat blunt microvilli on the surface. Comparatively, the dipnoan lung in general exhibits a greater degree of internal complexity than that of the amphibian lung (Sect. 13.1). On the other hand, morphologically, the dipnoan lung shows a closer similarity to the tetrapod lung than to the respiratory organs of the air-breathing fish. Inasmuch as these features may present convergent evolution, they may as well illustrate an important phylogenetic stage of the progression in the development of the respiratory system. Swim bladder cum lungs (Sect. 10.6) are widely distributed in fish (Smith 1982), some of the notable groups being *Polypterus, Amia, Lepisosteus, Umbra* and uniquely, the recently described South Asian catfish *Pangasius sutchi* (Browman and Kramer 1985; Zheng and Liu 1988). In general, these organs are morphologically simpler than the dipnoan ones. From comparative morphological studies, it is becoming clearly evident that the lung may be a remarkable case of a non-recapitulative ontogeny.

The mass-specific respiratory surface area of *Lepidosiren* is $0.85\ cm^2\ g^{-1}$ (Hughes and Weibel 1976), while that of *P. aethiopicus* was estimated at $14\ cm^2\ g^{-1}$ (Maina and Maloiy 1985). Despite this large difference between the two species, due to the fact that *Lepidosiren* has a relatively thick blood-gas barrier (BGB) of 0.86 μm and *Protopterus* a thin one of 0.37 μm, on modelling the lung (Maina and Maloiy 1985), the diffusing capacities respectively 0.003 and 0.002 ml O_2 $mbar^{-1}kg^{-1}$ are comparable. The BGB in *Protopterus annectens* is about 1 μm thick (Klika and Lelek 1967). Nearly 96% of the lung of *Lepidosiren* is covered by respiratory epithelium (Hughes and Weibel 1978). When compared with other air-breathing fish, *Lepidosiren* has a respiratory surface area twice that of the air-breathing organs of *Anabas* (Hughes et al. 1973). The BGB consists of an epithelial cell lining, an interstitial space, and an endothelial cell (Maina and Maloiy 1985). Like in many other air-breathing organs of the vertebrates, the respiratory surface is lined by a thin film of surface-active material, the surfactant (Hughes and Weibel 1978; Maina and Maloiy 1985), which is presumed to reduce surface tension and hence ascertain patency of the gas exchange components. The surface film is, however, not present on the air-breathing organs of fish (Sect. 11.1), which are not homologous to the air-bladders and lungs of higher vertebrates (Hughes 1973). Except for *P. aethiopicus*, the thickness of the BGB in Dipnoi ranges from 0.5–1.0 μm (Klika and Lelek 1967; Hughes 1970) and hence falls within the range of the terrestrial mammals (Gehr et al. 1981), values which are thinner than those of the lungs of amphibians and reptiles (Meban 1980). Though much has yet to be known on the ontogeny and phylogeny of the lungfishes (Gans 1986), it is undoubtedly clear that the taxon provides an important area in the study of the evolution of air-breathing, terrestriality, and the advent of the tetrapods. *Neoceratodus* is of particular interest due to its atypical non-dipnoan and rather piscine cardiorespiratory characteristics and to its being maybe somewhat more primitive than the other living genera (Johansen and Hanson 1968). In *Protopterus* and *Lepidosiren*, the circulatory pattern in the ROs and in the heart closely resembles that of amphibians (Satchell 1976).

12 Invertebrate Air-Breathers

12.1 Pulmonate Gastropods

Air-breathing is not a domain of vertebrates. Some invertebrates have evolved internalized vascularized cavities for air-breathing which functionally serve as lungs (Yonge 1947, 1952; Ghiretti 1966; Ghiretti and Ghiretti 1975). Pulmonate snails and slugs have modified mantle cavities (Hyman 1967; Runham and Hunter 1970), spiders have book lungs (Strazny and Perry 1987) (Sect. 10.7), terrestrial crabs have gas-chambers (Greenway 1984) (Sect. 12.2) the insects (Sect. 12.3) have a tracheal-air-sac system (Maina 1989b), and the gills of the limpet *Siphonaria capensis* serve both as water- and air-lungs (deVilliers and Hodgson 1987) (Sect. 10.2). The phyllum Mollusca has undergone remarkable adaptive ecological radiation, as reflected in its remarkable specific diversity, the group being numerically exceeded only by the taxon Arthropoda (Ghiretti 1966). Although the majority of molluscs live in water (marine- and freshwater), a few species lead a successful terrestrial life. The molluscs are, however, generally poorly adapted for life on land and it is only by adopting specific behavioural patterns, like residing in crypts, damp environment and nocturnality, that most of them have been able to inhabit land (Macchin 1974). The pneumonate gastropods in particular exhibit an extensive ecological radiation from marine- and freshwater and land. This in part has been facilitated by radical respiratory morphological changes in the mantle cavity. The early molluscs were probably flat-bodied animals close to the Turbellaria that crawled on a flat sole. They probably relied on the skin for respiration. At present, some prosobranchs like Lepetidae and some nudibranchs and pulmonates like Onchidiidae rely entirely on cutaneous respiration. Air-breathing in the pulmonate gastropods probably arose in O_2-deficient water (cf. air-breathing teleosts, Sect. 11.1) such as in estuaries, swamps and muddy rivers (Seed 1983). The freshwater pulmonates of the genera *Lymnaea* and *Planorbis* use their lungs for gas exchange in both air and water (Cheatum 1934; Precht 1939).

The molluscs were faced essentially with the same challenges which the other animals faced in transition from water- to air-breathing. This called for total overhaul of the aquatic respiratory organs (ROs). Problems of terrestrial locomotion, reproduction, excretion, thermal regulation and desiccation followed. The development of the lung in the pulmonate gastropods as well as the actual origin of this taxon are unclear (Morton 1958). Doubts have been raised as to whether the lung is homologous to the ctenidia of the mantle cavity of the other gastropods (Seed 1983): it has been proposed that the lung is an entirely new structure unique to the gastropod evolution. This view is supported by the fact that in most pneumonate gastropods the lung develops independently of the mantle cavity and particularly in the genera *Onchidella* a mantle cavity and a lung coexist with separate apertures (Ghiretti 1966). A direct conversion of the gills (ctenidia) to a lung would entail a transitional stage where the prolung would be an organ capable of functioning both in water and air as evinced by the bimodal-breathing freshwater pulmonates of the genera *Lymnaea* and *Planorbis* (Precht 1939). Due particularly to the notably different rheological and physical properties of water and air (Sect. 4), it is difficult to envisage such efficient organ, though some terrestrial crabs (Greenaway et al. 1988) (Sect. 12.2) and some fish (Graham 1973) (Sect. 11.1) significantly utilize gills on land but after requisite anatomical modifications.

Functionally, the lungs of the pulmonate gastropods are basically a diffusive type where the respiratory site is a passive region through which respiratory gases diffuse of their own concentration gradients (Ghiretti 1966). However, small pressure changes which could only result from intrapulmonary volume changes have been reported (Ghiretti 1966; Ghiretti and Ghiretti 1975). A PO_2 of 15 torr at 15 °C [PIO_2 152 torr and PEO_2 137 torr] has been reported in the lung of the land snail, *Otala lactea* (Barnhart 1986). The lungs of the pneumonate gastropods like *Trichotoxon copleyi* (Maina 1989a) and *Helix pomatia* (Pohunkova 1967) are sac-like cavities which are well vascularized (Fig. 9a, b). The blood capillaries bulge into the air space (Fig. 9b, d) increasing the respiratory surface area by a factor of 3–4 over that which could be provided by a smooth lung of equivalent size (Ghiretti 1966). In most pneumonate gastropods, no features of gill-like ctenidia are evident, suggesting a total extirpation of the ctenidia or independent evolution and development of the two organs. The pulmonary-sac is connected to the outside through a small opening, the pneumostome, which opens rhythmically to let out vitiated air and allow fresh air in, while preventing excessive water loss across the pulmonary sufaces (Wit 1932; Prior et al. 1983). A muscular layer (Fig. 9a) constitutes 52% of the wall of the lung in *T. copleyi* (Maina 1989a) and may be used in mechanically ventilating

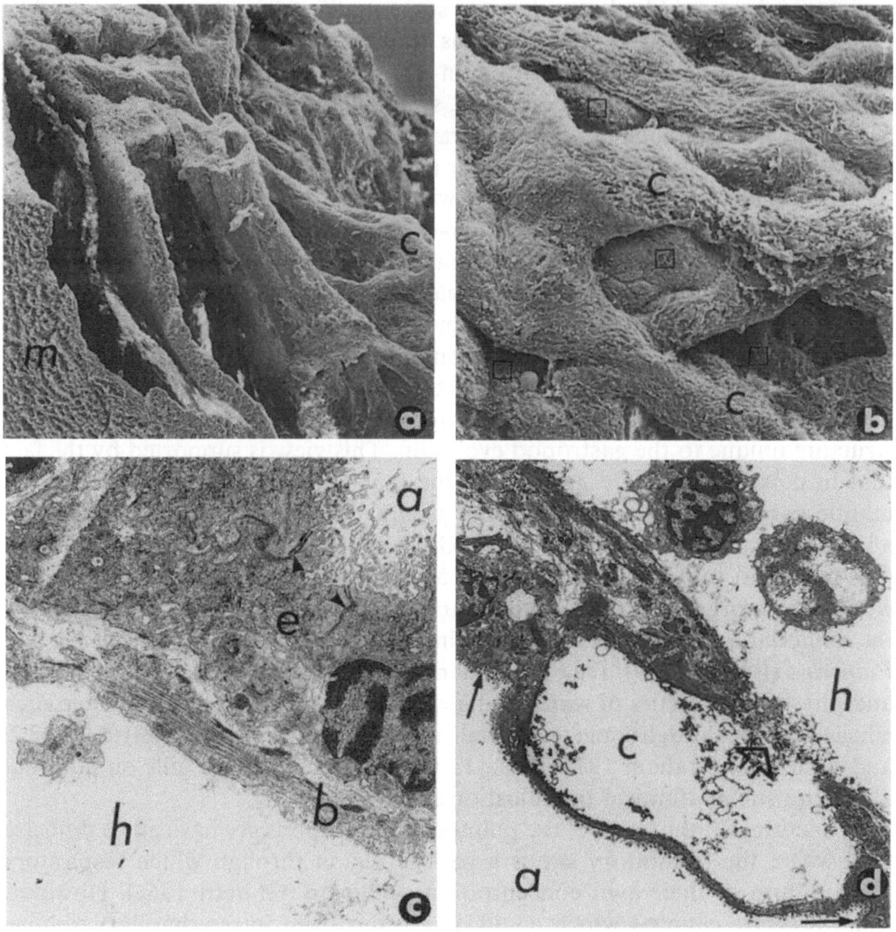

Fig. 9a–d. The lung of a pulmonate gastropod *Trichotoxon copleyi*. **a** Side-view of the lung showing the haemolympatic capillaries (*c*) with the lung being surrounded by a well-developed muscle layer (*m*) which mechanically ventilates the arguably diffusive lung, × 34; **b** close-up of the surface of the lung showing blood capillaries (*c*) separated by subepithelial crypts (*squares*) which increase the respiratory area, × 170. **c** pulmonary pneumocytes (*e*) separating air (*a*) from haemolymph (*h*), the pneumocytes being attached to an interrupted basement membrane (*b*) and attaching to each other through distinct tight cell junctions (*arrowheads*), × 5460; **d** close-up of a blood capillary (*c*) protruding into the air-space (*a*) with the epithelial cell nuclei (*arrows*) displaced to the intercapillary crypts making the haemolymphatic-air barrier particularly thin. The haemolympatic capillaries are connected to the subepithelial haemolymphatic channel (*open arrow*), × 4265. (Maina 1989a, courtesy of Oxford University Press)

the lung. With a well-vascularized lung, diffusive respiration and possible mechanical ventilation (Ysseling 1930; Mass 1939) are adequate in effecting optimal gas exchange in the lung of *Helix pomatia* (Ysseling 1930). Pressures as low as 2 mm Hg are enough for gas exchange to take place (Dahr 1927). The closure of the pneumostome followed by contraction of muscles is said to compress the air in the lung enhancing its diffusion into the blood (Dahr 1927; Ysseling 1930; Mass 1939). The *Arion*, the pneumostome closes at a PO_2 of 15 mm Hg and the intrapulmonary pressure increases to 20–30 mm Hg due to muscular contraction (Dahr 1927). Krogh (1941), however, refuted that these movements have any respiratory significance. Low O_2 tension stimulates the opening of the pneumostome which may remain permanently open at a concentration below 10% (Mass 1939) and remains open longer at a CO_2 concentration of 3–5% (Dahr 1927). Ghiretti and Ghiretti (1975), however, observed that CO_2 has very little influence on the frequency of ventilation. In *Helix*, the volume of the lung is 5–7 cm^3 when the animal is out of the shell but decreases to 0.5 cm^3 when the animal retracts into the shell (Krogh 1941): the slugs in general have small lung volumes. In *Arion* (10 g) the volume is about 0.3 cm^3 (Ghiretti 1966). The volume capacity of the pulmonary chamber is not proportional to body mass as small snails have a greater chamber than larger ones (Cheatum 1934). A respiratory surface area of 107.5 cm^2 has been determined in *Helix pomatia* (Pelseneer 1935), the mass-specific respiratory area ($RA\,W^{-1}$) of 8.3 $cm^2\,g^{-1}$ in this species being close to that of the ctenidia of the marine forms. In most pulmonate gastropods, the $RA\,W^{-1}$ ranges from 7–9 $cm^2\,g^{-1}$ (Pelseneer 1935). These values are higher than those of 2.5 $cm^2\,g^{-1}$ in the teleost *Tinca tinca* (Hughes 1972b), 3.2 $cm^2\,g^{-1}$ in the lizard *Tupinambis nigroupunctus* (Perry 1981) and compare with that of 10 $cm^2\,g^{-1}$ in *Gallus gallus* (Abdalla et al. 1982). The similarity in the magnitudes of the pulmonary morphometric parameters in slugs with those of apparently more advanced vertebrates was remarked on by Ghiretti (1966). As pointed out by Maina (1989a), comparison of such structural features in animals so phylogenetically different is riddled with onerous assumptions and shortcomings and inferences should be made cautiously. Further, structural parameters are but some of the contributive factors to the global functional efficiency of a respiratory system (Sect. 10.7). The thickness of the blood-gas barrier (BGB) in slug lungs ranges from 6–10 μm (Runham and Hunter 1970) but a thinner barrier (4 μm) was reported in *T. copleyi* by Maina (1989a), some areas being as thin as 0.26 μm. In the lung of *T. copleyi*, the basement membrane in most areas lies directly on the haemolymph as the endothelial cell is lacking (Fig. 9c). The open type of circulation (Vosswinkel 1982) greatly reduces the thickness of the BGB, enhancing gas exchange. The lungs of the pulmonate gastropods are capable of exchanging gases at six times the rate of the resting values (Runham and Hunter 1970), a feature which indicates notable functional reserves. This potential facilitates the remarkable physiological lability of molluscs (Ghiretti 1966; Jones 1983; Gregor and Prior 1986). Even under constant external conditions, the VO_2 of a given specimen is highly variable (Ghiretti 1966; Ghiretti and Ghiretti 1975). In *Myra arenaria*, at 14 °C, extreme levels ranging from 14–57.8 $\mu l\,g^{-1}h^{-1}$ have been measured (Collip 1921; van Dam 1935). When active or dehydrated, *Limax maximus* can increase its VO_2 by 400–500% (Runham and Hunter 1970; Prior et al. 1983), as heart rate increases by 15–38% (Mackay and Gelperin 1972; Gregor and Prior

1986). During such times, the lung contributes 50% of the VO_2 (Prior et al. 1983). After attaining air-breathing and terrestriality, some pneumonate gastropods in a classic case of evolutionary regression have reverted to aquatic life (Hunter 1953). Neomorphic [pseudobranchs = adaptive = secondary] gills (Hyman 1967) have developed within the former lungs as in the intertidal pulmonate limpets constituting the taxon Siphonariidae (Yonge 1952), but in the related *Trimusculus* (*Gadinia*) the lung is unchanged as the animal has continued to breathe air (Yonge 1958). Astoundingly, the lung in the family Athoracophoridae has numerous blind-ending tubules which extend deeply into the surrounding tissue (Hyman 1967). From resemblance to the arrangement of the tracheal system of insects (Sect. 12.3), such snails have been termed tracheopulmonates.

12.2 Crustacea

Probably no other animal group exhibits clearer gradations towards terrestrial habitation and transition to air-breathing than the phylum Crustacea and in particular crabs (McMahon and Wilkens 1984). At one extreme, the non-malacostracan crustaceans depend largely on cutaneous respiration, whereas most crabs belonging to the infraorders Brachyura and Anomura have colonized land very effectively. Air-breathing in crustaceans has independently been achieved in different ways either directly from seawater (Anomura and most Brachyura) – these are restricted to narrow coastal ridges – and those of freshwater origin, mainly the crayfish of the families Astacidae and Parastacidae and the crabs of the superfamilies Potamoidea and Parathelphusoidea (Edney 1960; Bliss 1968). A few species in the families Parathelphusidae and Grapsidae have evolved a capacity for direct development from egg to an immature crab (Hartnoll 1988), some species are capable of producing isosmotic urine by external urine processing (Wolcott and Wolcott 1984; 1985; Greenaway 1988; Morris et al. 1991) and a good number live far away from the sea (Dunham and Gilchrist 1988; Wolcott 1988). Paradoxically, in spite of these innovations which portend adaptations to success on land, the decapods have not quite completely dissociated themselves from water (Powers and Bliss 1983; Wolcott 1988). In a perpetually dry environment, even the most highly land-adapted crab soon succumbs. Terrestrial crabs are relatively of recent origin. Their fossils do not appear before the Cretaceous [5–60 million years] (Warner 1977). This may explain why only 10% of the decapods are presently appreciably terrestrial and also the rather tenuous hold on land the group exhibits. The taxon may illustrate an early stage of an ongoing (though latent) land invasion by the molluscs (Bliss 1968). This makes their study even more interesting and challenging. It may provide an ideal practical and real time study model for population ecological dynamics and transition from water to dry land.

The gills of crabs (Fig. 10) lie protected in the branchial chambers which are extensions of branchiostegites and in some cases thoracic walls. The chambers

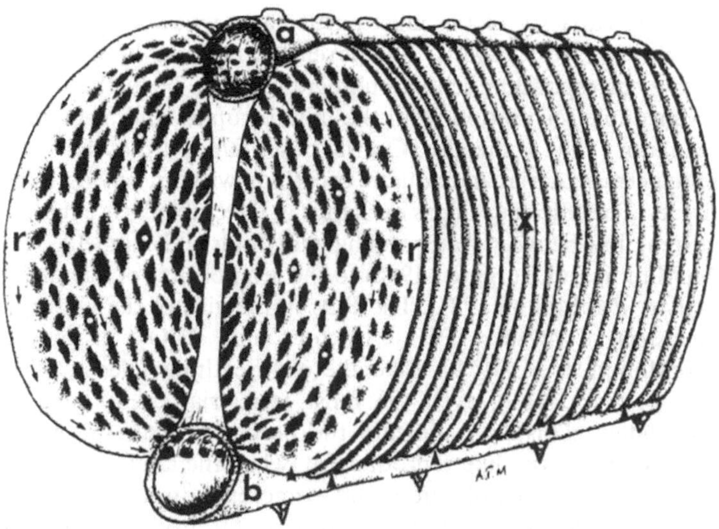

Fig. 10. A transverse view of a gill filament of a semiterrestrial freshwater crab *Potamon niloticus* showing the afferent blood vessel (*a*) and the efferent vessel (*b*). the haemolymph flows from the afferent vessel (*arrows*) through the lamellae (*x*) in an acriform manner percolating through channels delineated by pillar cells (*white dots*). Much of the haemolymph passes through the much wider marginal channel (*r*) interacting with the water (*arrowheads*) in a counter-current manner; *t* interlamellar septum; approx × 91

are well vascularized to form a "lung" in most terrestrial crabs (Diaz and Rodriquez 1977; Wood and Randall 1981; Greenaway and Farrelly 1990). Air-breathing crabs have two lungs and some, the Scopimerinae, have sternal and podal gas exchange sites (Maitland 1986). Some land based crabs with shells like *Hemigrapsus nudus* and *Pachygrapsus crassipes* carry water during land excursions (Burnett and McMahon 1987), while yet others like *Cyclograpsus lavauxi* (Innes et al. 1986) recycle a small amount of water over the gills. Such water may serve as a source of O_2, ions for acid-base balance and other haemolympatic ionic adjustments. It may also prevent desiccation of gills, and CO_2 and NH_3 may be excreted into it. Though fragile, this ability enables land habitation during intertidal air exposure by essentially aquatic crabs. In both brachyuran and anomuran species, with increased terrestriality and evolution of alternative respiratory sites, there is a reduction in the number and surface area of the gills and a wider spacing of the gill lamellae (Gray 1957; Bliss 1968; Newell 1976; Farrelly and Greenaway 1992). The gill surface area in the sublittoral *Callinectes sapidus* is 1367 $mm^2 g^{-1}$, in the supralittoral *Cardisoma carnifex* 80 $mm^2 g^{-1}$, and in the terrestrial *Birgus latro* only 12 $mm^2 g^{-1}$ (Gray 1957). In *B. latro*, the gills are remarkably ineffective in O_2 uptake but play a major role in CO_2 elimination (Greenaway et al. 1988). These progressive changes are consequent mainly to the need for water conservation and the higher arterial O_2 tension. Large surface area may not be necessary to maintain the required diffusing capacity of the gills, particularly when supplemented by secondary aerial respiratory organs. The development for a lung in the terrestrial crabs with a corresponding reduction in gill size may be further viewed as an attempt to uncouple respiration from the initial trophic and at present primary osmoregulatory role of the gills. On the other hand, it may be a structural compromise between aerial and aquatic breathing without having to retain superfluous tissues. The efficiency of gas exchange in the crab gills is enhanced by the counter current flow between water and haemolymph in the lamellae (Hughes et al. 1969; McMahon and Burggren 1988) to the level of that of fish under similar state of activity (McMahon 1981). The PO_2 in the gills of the Dungeness crab (*Cancer magister*) of 44 torr (PIO_2 141 torr and PEO_2 97 torr) is notably high (McMahon et al. 1979). Like the air-breathing teleosts (Sect. 11.1) and the Dipnoi (Sect. 11.2), even the most terrestrial crabs have not lost the gills completely. This may indicate the need to retain the gills to enable those essential processes like reproduction (when the land crustaceans have to revert to water), CO_2 elimination and osmoregulation. It may also suggest that absolute land colonization by crabs is far from complete and we may be witnessing an ongoing aggressive though inconspicuous process. As pointed out elsewhere (Sect. 5), a critical step in the evolution of air-breathing and terrestriality was that of relocation of CO_2 removal with the extirpation of the gills.

Notwithstanding a few exceptions, the vascular arrangements of the gills of land and aquatic crabs are fairly similar (Taylor and Greenaway 1979; Burggren and McMahon 1988). Secondary lamellae receive deoxygenated haemolymph from an afferent branchial vessel (Figs. 10 and 11a, b). After entering the marginal channel, the haemolymph percolates between the pillar cells that support the two walls of the lamella into the efferent channel (Maina 1990b). There is a striking external resemblance between the general organization of the gills of crabs and those of fish (cf. Figs. 5 and 11). Generally, the terrestrial crabs have very well vascularized branchial chambers (Farrelly and Greenaway 1987; Greenaway and Farrelly 1990). In the coconut crab (*B. latro*), the chamber is lined by labyrinthine channels which greatly increase the respiratory surface area, while the soldier crab (*Mictyris longicarpus*) has an invaginated lung where the air-blood barrier is 1.2–3.6 μm thick (Farrelly and Greenaway 1987; Maitland 1987). *M. longicarpus* has an aerobic capacity comparable to that of most vertebrates and can increase it tenfold and sustain it for as long as 2 h (Maitland 1987). The exercising VO_2 in the crab is 406 μmg h^{-1} and 50–70% of its O_2 needs are transferred across the lung. As a mark of the high efficiency of the respiratory organs (ROs) of the soldier crab, the PAO_2 is 95.5 torr (Farrelly and Greenaway 1987). The ghost crab (*Ocypode gaudichaudii*) has a comparable aerobic capacity (Full 1987) while *O. saratan* has an efficient and highly adapted lung through which much of O_2 and CO_2 exchange occurs (Al-Wassa et al. 1989). Ghost crabs are the fastest running crustaceans and can attain a speed of 3.4 m s^{-1} (Herrnkind 1983; Herreid and Full 1988). On submersion, VO_2 in *O. saratan* declines drastically (Eshky et al. 1988). These metabolic capacities are facilitated by a very well developed lung. *O. saratan* is one of the fastest moving

Table 3. Gas diffusional distances and respiratory parameters in some terrestrial and aquatic crabs[a]

	Diffusional distance		VO_2	Pao_2	Pvo_2
	Gills (μm)	Lung (μm)	(μmol g^{-1} h^{-1})	(mmol l^{-1})	
Terrestrial					
Holthuisana transversa	5–8	0.22–0.30	0.80	–	–
Birgus latro	–	0.5–1.2	1.41	27	13
Pseudothelphusa garhami	5	0.4–1.0	1.51	124–140	–
Mictyris longicarpus	3.5–7.0	1.2–3.6	–	79–95	20–27
Ocypode ceratophthalmus	2.9–3.5	0.25–0.30	–	–	–
Cardisoma hirtipes	5–12	2.5–4.5	2.29	–	–
Aquatic					
Procambarus clarkii	3.15–8.7	–	–	–	–
Astacus pallipes	1.4–5.7	–	–	–	–

[a] Sources of data on thicknesses summarized in Burggren and McMahon (1988) and respiratory parameters in McMahon and Burggren (1988).

terrestrial crabs (Hartnoll 1988) and has a VO_2 of 5 ml O_2 h^{-1} (Al-Wassa et al. 1989). In *M. longicarpus* the PO_2 in the lung is nearly 100 mm Hg, while in *Pseudothelphusa garhami*, another species with a highly elaborate invaginated lung, the PO_2 ranges from 120–140 mm Hg (Innes and Taylor 1986), and $PACO_2$ levels are at the modest level of 8–10 mm Hg (Table 3). Gas exchange in the crab lung is of the uniform pool type (Piiper and Scheid 1975) (Sect. 9). McMahon and Burggren (1981) have, however, suggested that a different spatial arrangement may occur between the air and the haemolymph. The haemolymph perfuses the lung in an anteroposterior direction (Harms 1932), whereas air-flow is in the reverse direction (Cameron and Mecklenburg 1973), the two media providing the highly efficient counter current design (Sect. 9.1) which may contribute to increased O_2 and decreased CO_2 levels in arterial blood. When in air, about 90% of the haemolymph in the arid-zone crab *Holthuisana transversa* perfuses the lungs (Taylor and Greenaway 1984). The sand-bubbler crabs, *Scopimera inflata* and *Dotilla myctiroides*, breathe air through membranous disk-like modifications on their legs (Maitland 1986). The two species have very small gills and lack a lung. At high tide, the crabs are found in burrows that contain trapped pockets of air which the crabs are able to breathe from until low tide. The total surface area of the gas-windows in *S. inflata* exceeds that of the

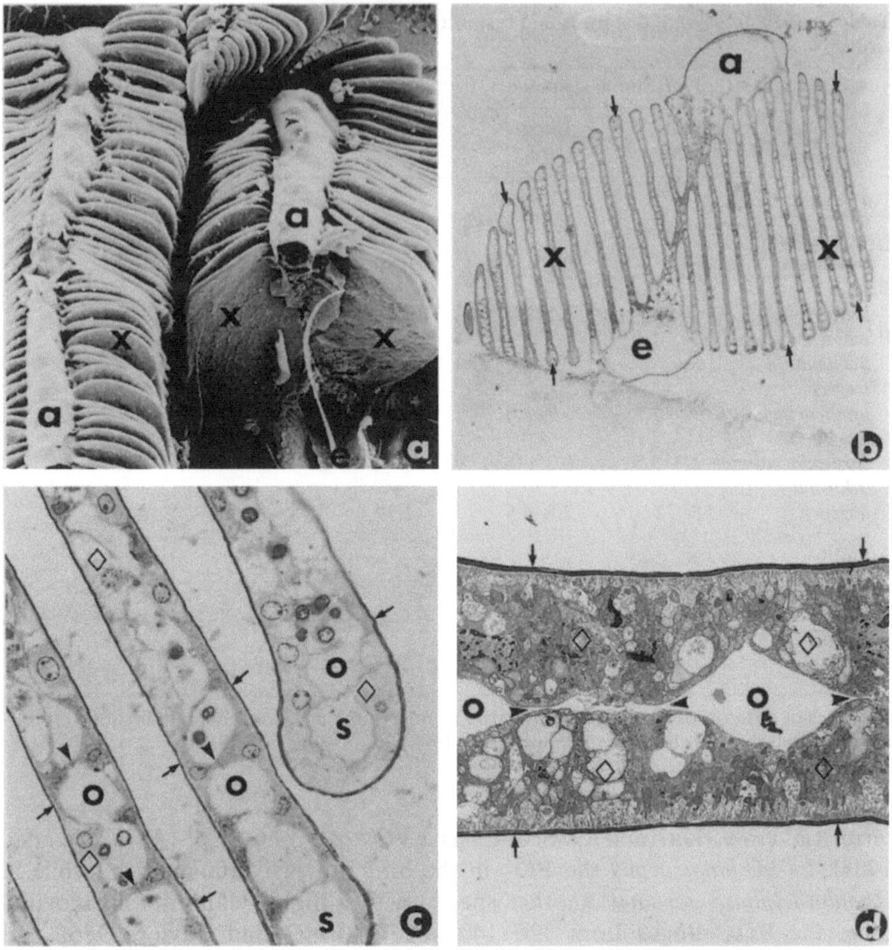

Fig. 11a–d. Gill filments of a semiterrestrial freshwater crab *Potamon niloticus*. **a** Gills consist of filaments which bear lamellae (*x*) on both sides and are supplied with blood through a filamental afferent vessel (*a*) and the haemolymph is drained by an efferent one (*e*), × 51; **b** transverse section of a gill filament showing an afferent vessel (*a*) and an efferent one (*e*) interconnected by secondary lamellae (*x*) through which the haemolymph percolates, much of it along the much wider marginal channel (*arrows*). × 34; **c** close-up secondary lamellae showing the epithelial cells (*diamonds*) lined by a cuticular layer (*arrows*) and haemolymphatic channels (*o*), the most peripheral ones (*s*) constituting the marginal channels which are formed by fusion of epithelial cells (*arrowheads*), × 367; **d** secondary lamella which is lined by a cuticular layer (*arrows*) and epithelial cells (*diamonds*) with abundant mitochondria with the cells fusing to separate the lamellae (*arrowheads*) into vascular channels (*o*), × 1620. (Maina 1990b, courtesy of Oxford University Press)

gills of the highly xeric species *Holthuisana* (= *Austrothelphusa*) *transversa* (Greenaway 1984). When the gas windows were painted over, VO_2 dropped by about 60%, and when sealed completely the crab died (Maitland 1986). The semaphore crab, *Heloecius cordiformis* (Hartnoll 1988; Maitland 1990a, b) has modified branchial chambers which are used for aerial and aquatic breathing; the crabs (7 g, W) carry between 0.35 and 0.5 ml of water in the branchial chambers.

The malacostracan gills are basically vascularized outgrowths of the coxae of the thoracopods. In crabs, the number of gills differ between species, ranging from 5–9 pairs. The gill lamellae are lined by epithelial cells that are covered by a thin cuticular layer (Taylor and Greenaway 1979; Maina 1990b). These cells are thinner where they overlie a vascular channel and contain numerous mitochondria with well-developed cristae (Maina 1990b) (Fig. 11a, d). Knob-like spacing protrusions from the marginal channels have been observed in gills of land crabs, like the soldier crab (Farrelly and Greenaway 1987, 1992), whereas other groups of crabs, like gecarcinids, have evolved the capacity of utilizing gills on land by having rigid highly sclerotized gills and widely spaced lamellae (von Raben 1934; Copeland 1968; Taylor and Butler 1978; Farrelly and Greenaway 1992). In *Ocypode saratan* and *B. latro*, the gill area is increased by extensions of the gills to form a structure similar to the labyrinthine organ of some air-breathing fish (Harms 1932) (Sect. 11.1). Gas exchange in *Callinectes sapidus* drops by a factor of 3 on exposure to air (O'Mahoney 1984) and by nearly a factor of 5 in *Cancer productus* (de Fur and McMahon 1978). The thickness of the epithelial cell and the cuticular cover, the diffusive barrier, on average ranges from 3–12 μm in the gills of most crabs (McMahon and Burggren 1979, 1988; Burggren and McMahon 1988; Farrelly and Greenaway 1992). The thickness of the water-blood barrier (WBB) in the soldier crab (*Mictyris longicarpus*) was 3–6 μm (Maitland 1987) (Table 3): on average, the chitinous cuticular layer which forms only about 30% of the thickness of the WBB (Farrelly and Greenaway 1992) constitutes as much as 80% of the gas-diffusive resistance (Taylor and Butler 1978). O_2 diffusion through chitin is ten times slower than in other tissues (Krogh 1919) and on drying up diffusion decreases 30 times (Harms 1932). The critical cuticular thickness appears to range from 1–2 μm (Taylor and Greenaway 1979). Probably to enhance conductance, the lamellar epicuticle is lacking in the gills of the ghost crab, *Ocypode platytarsis* (Diaz and Rodriquez 1977. The air-blood barrier in the branchial chambers of the land crabs is notably thin (0.2–0.5 μm), the values falling within the range of the lungs of the air-breathing vertebrates (Storch and Welsch 1975; Burggren and McMahon 1988) (Table 3). Generally, in crustaceans the WBB in the gills is 20 times thicker than the blood-gas barrier of the lungs (Taylor and Greenaway 1979). While the gills may have a larger surface area, the main factor limiting gas exchange in the crustacean gills is their high diffusive resistance mainly resulting from their much thicker WBB (Redmond 1955; Hughes et al. 1969; Burggren et al. 1974). Air-breathing gills are designed to harmonize mechanical support and gas exchange, features which are structurally diametrically opposed. Further, gills carry out other important functions like osmotic regulation, acid base balance and nitrogenous excretion, features which have to be accommodated in the overall design of the gills. The retention of the gills and their lack of absolute quantitative refinement may be consequent to inability to relocate these functions. The weight-specific respiratory surface area of the gills in crabs (on average $6\,cm^2\,g^{-1}$) is comparable to that of some indolent mammals and is generally greater than in most fish. The conversion of the branchial chamber into a lung to supplement gill function in crabs is strikingly similar to that of the ctenidia of the molluscan pneumonate gastropod lung (Sect. 12.1) and the accessory respiratory organs of some fish (Sect. 11.1). This is a classic case of developmental and morphological convergence in phylogenetically different and temporarily separated animals in response to similar selective pressures. Corresponding with most other animals, the organization and adaptations of the respiratory organs of the crabs, be they aquatic or aerial, correlates with O_2 demands which in turn depend on factors like activity, size and environment. In the notably active ghost crab where a species like *Ocypode ceratophthalmus* has been clocked at an amazing speed of $3.4\,m\,s^{-1}$ and may cover a distance of $300\,m\,day^{-1}$ (Herreid and Full 1988), the branchial chambers are very well developed for gas exchange. While in air, the chambers are efficiently ventilated by the scaphognathites at a rate of nearly $1\,l\,kg^{-1}\,min^{-1}$ (Cameron and Mecklenburg 1973), a value similar to the water-flow produced by a 1 kg *Cancer magister* (McMahon et al. 1979) and the crayfish, *Orconectes virilis* (Burggren and McMahon 1983). Where water and air are available, crabs like *Cardisoma* can ventilate gills and lungs simultaneously, a feature invaluable to a species dwelling in areas susceptible to flooding (McMahon and Burggren 1988). Similarly well-developed and highly inflated are the branchial chambers of species like *Gecarcinus lateralis* and *Cardisoma guanhumi.* These species frequently cover several kilometers in a day (Herreid and Full 1988). The most advanced degree of aerial breathing among the terrestrial crabs has probably evolved in the anomuran coconut crab (*Birgus latro*) which can survive in air even when the gills are removed and will drown when refused access to air (Harms 1932): the lung in this species is characterized by folds and corrugations that increase the respiratory surface area. Unique mechanical ventilation is exhibited by the extremely xeric Australian crab, *H. transversa* where branchial chambers are tidally ventilated in a similar manner and probably to the same level of efficiency as in the vertebrates (Taylor and Greenaway 1979); in the species, the air-blood barrier of the branchial chambers is remarkably thin, the distance being generally less than 1 μm, while the **WBB** (of the gills) ranges from 5–8 μm. In species like *Gecarcinus lateralis, C. guanhumi* and *B. latro*, the scaphognathites ventilate the branchial chambers, while gas exchange in *Trichodactylus* depends on diffusion (Valente 1948) and in *Uca* movements of the carapace ventilate the chambers (Carter 1931). The restrictive demands for terrestrial colonization is probably nowhere else better illustrated than in the decapods where of the about 10 000 species 89% are marine, 10% are freshwater and only 1% are terrestrial (Hartnoll 1988). This, though limited, success has provided important materials for comparison with the more successful arthropods and tetrapods. The land crabs in particular provide fundamental modification for terrestrial habitation. Unfortunately, paucity of ecological, behavioural, morphological, functional and palaeontological details make extensive extrapolation of the present data impossible and at times risky.

12.3 Insects

The insects are probably the first animals to successfully colonize dry land and subsequently take to the air. They appeared on land at the middle of the Paleozoic era (about 400 million years ago) and had effectively established themselves by the Carboniferous period (Callahan 1972). Their long existence has produced extremely aggressive and remarkably successful organisms through calculated miniaturization of body size and ingenious perfection of their physiological processes. The energetic demands of insect flight are enormous. When resting, the insect flight muscle has a metabolism similar to that of other passive muscles of the higher vertebrates, but when active its rate of transformation of chemical to mechanical energy is ten times greater and in terms of power-to-weight ratio it compares with a small aero-engine (Pringle 1983). The mechanical efficiency of flight muscles of insects like the hoverfly is as high as 58%, a value well beyond empirical thermodynamic expectations (Weis-Fogh 1973). A 0.1 g free-hovering euglossine bee (*Eulaema* and *Eufreisea* sp.) has a VO_2 of 154 ml $O_2\,g^{-1}\,h^{-1}$ (Casey et al. 1985). Hovering, a mode of flight which is the most highly energetically demanding form of locomotion and is close to the upper theoretical limit for power generation by striated muscle (Casey et al. 1985), has been perfected by insects. The metabolism of flight muscles of the most active insects like the honey-bee and the blowfly is as high as 2500 kcal g muscle^{-1} h^{-1} (Weis-Fogh 1961; Pringle 1983). This level of metabolic activity is only equalled by pure microbial cultures. The flight muscles of insects are ultrastructurally unique by having an abundance of myofibrils (Hodge 1955; Smith 1961); efficiency of the final conversion of ATP to work is about 50% in fibrillar muscle. As they lack the essential enzyme lactic dehydrogenase, the fibrillar muscles of the insects are unable to function without a continuous and sufficient supply of O_2. Mitochondrial volume density in the insect flight muscle has been observed to be as high as 50% (Ready 1983). The O_2 uptake by the flight muscle mitochondria at normal room temperature and pressure is 0.1 ml $O_2\,ml^{-1}\,s^{-1}$, a value which is one-tenth the mitochondrial volume (Pringle 1983). The asynchronous muscles of some insects produce as many as 100 contractions per second and wing beats have been reported to be as high as 1100 times per second in the midge *Forcipomyia* (Pringle 1983). The highly energetic species of insects like dragon-flies (Odonata), most bees and wasps (Hymenoptera), some flies (Diptera) and some beetles (Coleoptera) are endothermic and regulate their body temperature through internally generated heat (Heinrich 1977, 1982; Heath and Heath 1982). These exceptional physiological feats could only have been achieved if commensurate with an efficient

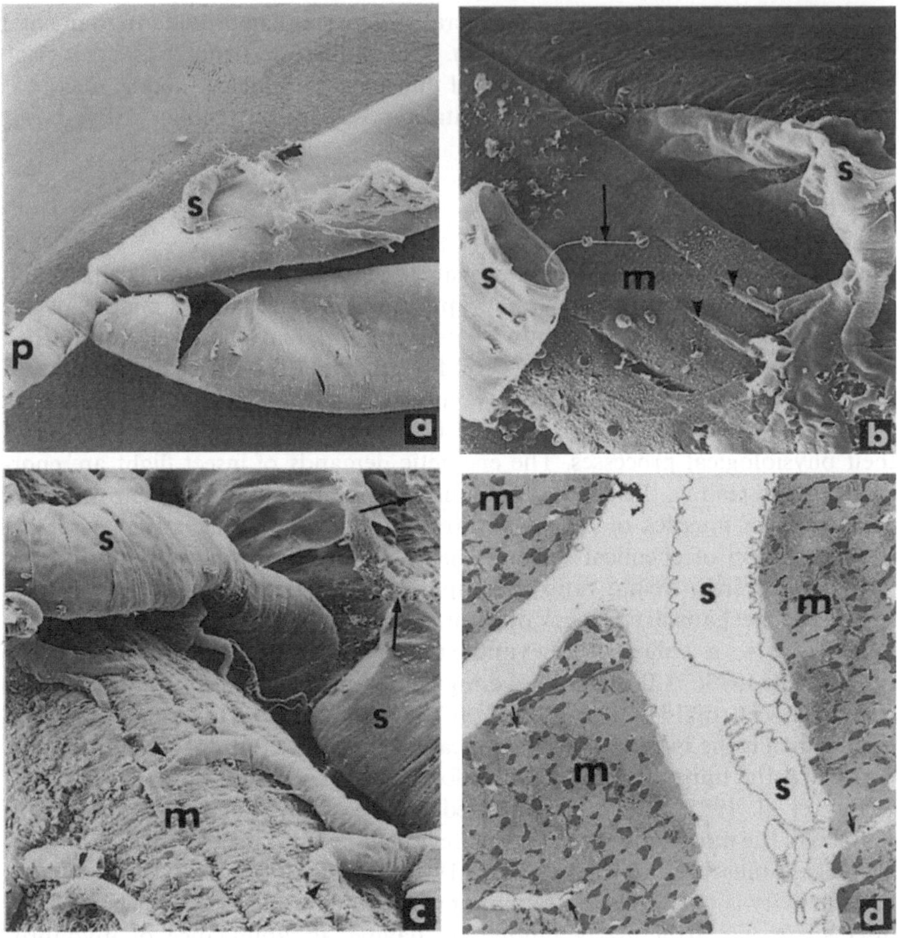

Fig. 12a–d. The air-sac tracheal system of a grasshopper *Chrotogonous senegalensis*. **a** Paired lateral air-sacs showing the connection to the spiracle (*p*) and to the tracheal system (*s*), × 17; **b** tracheal supply (*s*) to the flight muscles (*m*), the terminal trachea (*arrowheads*) elaborately indenting the muscle and arrow showing taenidium × 230; **c** tracheal supply (*s*) to the abdominal muscles (*m*) with the *arrowheads* showing indentations by the terminal trachea, and the *arrows* Malphigian tubules × 34; **d** sectional view of the trachea (*s*) lying on a flight muscle (*m*) sending finer branches (*arrows*) into the muscle fibres, × 2268. (**a**, **b**, and **c** from Maina 1989b, courtesy of Wiley/Alan R. Liss Inc.)

respiratory system. By uncoupling the respiratory and the circulatory systems, insects have evolved a unique and probably the most efficient mode of gas exchange in living animals, the tracheal system. It has apparently taken eons to mold.

Like all fully terrestrial animals, the insects faced the problem of desiccation. A chitinous cuticular surface developed to curtail excessive water loss, in the process rendering the skin non-functional in gas exchange. The insect respiratory system (Figs. 1F and 12) formed as invagination of the integumentary ectoderm into the body, the spiracles constituting the portals of influx and efflux of respiratory gases (Richards and Anderson 1942; Locke 1958a, b; Whitten 1972). In mechanical terms, the tracheal system functions both as a compressor and an exhaust of an internal combustion engine. The primary and secondary tracheae (Fig. 12b, c, d) range in size from 0.2–3 μm and the terminal tracheoles are about 0.08 μm in diameter (Richards and Korda 1950; Edwards et al. 1958). Although most terminal tracheae lie in the intercellular spaces, in some cases and particularly in the more metabolically active tissues they indent cells in a manner of a finger poked into the surface of a balloon (Fig. 2F). The terminal tracheoles end blindly (Richards and Korda 1950) though possible anastomoses has been reported (Buck 1948). In flight muscles of the desert locust (*Schistocerca gregaria*), shunts between the primary tracheae and the air-sacs (Fig. 12a) ensure continuous O_2 supply to the flight muscles (Weis-Fogh 1964c). Theoretically, the tracheoles supply O_2 to each individual body cell. In the silkworm, there are 1.5 million tracheoles (Buck 1962) and the tracheal volume which for a 5.7-g worm is 48.4 $\mu l\,g^{-1}$ makes about 5% of the body volume (Bridges et al. 1980). Demoll (1927) reported a tracheal volume of 585 $\mu l\,g^{-1}$ in the adult cockchafer, *Melolontha*. In the wing muscle of the locust, between 10^{-1} and 10^{-3} (volume of tracheole per volume of muscle) is taken up by the tracheolar system (Weis-Fogh 1967). Tracheal O_2 intake and CO_2 removal is entirely by diffusion across a concentration gradient. In small insects, passive diffusion suffices in gas exchange (Wigglesworth 1935), whereas in larger insects like the grasshopper, locust and dragon-fly, diffusion is aided by auxiliary mechanisms like abdominal and the less well understood thoracic pumping (Weis-Fogh 1964a). During steady flight, in the desert locust (Weis-Fogh 1967) about 320 $l\,kg^{-1}\,h^{-1}$ of air can be ventilated into the tracheal system by abdominal and thoracic pumping at an average tidal volume of 167 mm^3 at a frequency of 30–60 times per minute. The giant beetle *Petrognatha gigas* has a ventilatory rate of close to 2000 $l\,kg^{-1}\,h^{-1}$ (Miller 1966). Intratracheal pressure increases to 10–25 mm Hg at the peak of abdominal contraction (Weis-Fogh 1967). It is probably limitations by factors like diffusive pathway and ventilatory mechanical adjustments that have restricted the size of insects. Average tracheolar length for optimal diffusion appears to be 10 mm and the minimum diameter 0.2 μm (Krogh 1922a, b). O_2 concentration drops only by 1% from the spiracle to the tissues (Buck 1962) and the PO_2 between the tracheoles and the metabolizing tissue has been estimated at 40 torr (Weis-Fogh 1967). As an indication of the efficiency of the tracheal system in conducting O_2 to the tissues, in adult *Aphelocheirus* (Thorpe and Crisp 1947) the drop in PO_2 between the spiracles and the tracheoles is only 0.3 kPa. Much of the resistance to tracheolar air-flow comes from the spiracles (Buck and Keister 1958; Scheid et al. 1981; Tenney 1985). Like in the mammalian lung (Sect. 13.3), developmentally the tracheae branch in such a way that the cross-sectional area (CSA) of the diffusive path increases with distance (Krogh 1922a; Locke 1958b, c). Resistance thus remains fairly constant until the mean free path of diffusing O_2 molecules is reached when further reduction limits diffusion (Weis-Fogh 1964b; Pickard 1974). At the critical diffusive diameter, it is plausible that tracheolar density would have become adequate to meet the metabolic demands of a particular tissue. Buck (1962) observed that the maximum tracheolar CSA may be reached half-way the tracheolar length and subsequently decreases to 1/60th of the initial value. The development of the tracheal system appears to be partly determined by intrinsic factors in the target tissues (Locke 1958b, c) and the distribution to specific organs by local aerobic capacities (Edwards et al. 1958; Locke 1958b). In larval mealworms (*Tenebrio molitor*), hypoxia was observed to influence tracheal growth and development at an ambient PO_2 of about 76 mm Hg, leading to wider tracheae (Loudon 1989).

The primary and secondary tracheae, the air-sacs and the tracheoles (Fig. 12) are supported by well-developed taenidia which are annular or helical continuations of the epicuticle (Richards and Anderson 1942; Edwards et al. 1958; Shafiq 1963). The taenidia enable stretching while resisting compression (Locke 1958a). As the critical pressure of collapsing thin-walled tube (with constant wall dimensions) varies inversely with the cube of the radius (Boresi 1955), it is physically more difficult to buckle smaller tubes. Hence, the critical tracheolar diameter may partly be determined by the nature of the air-flow rather than by mechanical attributes. Nevertheless, the structure of the taenidia is a design that compromises the necessity for support and the need to maintain a thin wall that will not impede gas diffusion. Air-sacs (Fig. 12a) are absent in the subclass Apterygota and are well developed in Diptera and Hymenoptera. They play an important part in respiration by providing a large tidal volume and in ventilating the tracheal system (Bursell 1970). In cicada *Fidicina monnifera*, the air-sacs together with the tracheal system constitute 45% of the body volume (Bartholomew and Barnhart 1984). A good correlation between the development of the air-sacs, body size and activity occurs in Hymenoptera (Tonapi 1958 cited from Miller 1974). Such insects inspire a volume of air equal to two-thirds of the total volume of the tracheal system. As much as 70% of the total tracheal volume may be ventilated in a respiratory cycle in some insects (Bursell 1970). The pattern of air-flow in the tracheal system has been debated for a long time (Weis-Fogh 1964c). While it has been asserted that there is no law governing inhalation and exhalation and that the respiratory currents may alternate even in the same individual, some studies have shown what appears like a definite and consistent pattern of air-flow which may change with prevailing demands (Weis-Fogh 1967; Miller 1974). In *S. gregaria*, during flight the air flows unidirectionally through the thoracic spiracles supplying the head and central nervous system, and out through the abdominal spiracles, and at rest 50% of the total volume of air ventilated by the abdomen passes unidirectionally. The tracheal air-flow in the honey-bee has a strong unidirectional component during flight (Bailey 1954). In *Sphodromantis*, 95% of the inhaled air passes unidirectionally, while only 5% passes tidally (Miller 1974). Synchronized spiracular movements that characterize a large number of orders of insects suggests that unidirectional air-flow may be a common if not a universal feature, particularly in the larger insects. It is notable that unidirectional mode of ventilation characterizes diverse groups of animals such as molluscs (Sect. 9), fish (Sect. 10.2), and crustaceans (Sect. 12.2), but among the fully terrestrial species this feature is only encountered in birds (Sect. 13.4) and insects. Corresponding with observations made by Weis-Fogh (1964c), the similarities in functional (mode of ventilation) and structural (presence of air-sacs and gas conduits like tracheae and parabronchi), common features in insects and birds, animals which are phylogenetically remarkably different, are intriguing. It is one of the many classic paradigms of convergent evolution culminating in animals arriving at a common solution to the problem of undertaking to mold an efficient respiratory system. The plasticity of biological systems in the face of selective pressures is clearly evident. Even after mastering the terrestrial niche, some insects, though strictly air-breathers, have reverted to living in water after evolving ingenious methods of carrying and storing air underwater as they exchange gases by using physical gills (Sect. 10.2).

Insects, in a refinement of engineering almost past belief, have evolved a respiratory system comparable to the vertebrate circulatory system, in the process uncoupling circulatory and respiratory systems. As O_2 is delivered directly to the tissue cells with minimal drop in O_2 tension, the PO_2 in the body cells should theoretically equal that in the inspired air (Fig. 2F). Only insects living in O_2-deficient waters have a respiratory pigment in form of a simpler form of haemoglobin. In most insects, the pigment plays no significant role in gas transport but may be useful in O_2 storage in the tissues (Florkin and Jeuniaux 1974; Loudon 1989). Many spiders (Sect. 10.7) exhibit a system somewhat between the insects and the vertebrates where a tracheal system coexists with a book lung (Sect. 10.7). It has been suggested that in insects O_2 supply would be inadequate unless the tracheoles indented the cells. In general, terminal tracheoles are never more than 0.2–0.5 μm from a mitochondrion, in some cases being as close as 0.005 μm (Hodge 1955; Wigglesworth and Lee 1982; Maina 1989b); a cluster of mitochondria surrounding a terminal tracheole has been termed a mitochondrial continuum by Edwards et al. (1958). The terminal tracheoles are fluid filled (Wigglesworth and Lee 1982). During exercise, the fluid is osmotically removed from the terminal tracheoles, the process enhancing interaction between air and body cells. The optimization of the insect tracheal system may be attributed to their long evolutionary history and may partly account for their exceptional ecological successes as evinced by their remarkable miniaturization and adaptive radiation culminating in exceptional specific diversity and numerical density.

13 Vertebrate Air-Breathers

13.1 Amphibians

Although the actual origin of the extant amphibians is not clear (Szarski 1962; Lovtrup 1977; Milner 1988), the taxon occupies a central place in the process of attainment of air-breathing, terrestrial colonization and consequently evolution of tetrapods (Szarslo 1977). The larval forms are fully adapted to aquatic respiration by having external gills which are shed at maturity (on terrestrial habitation) after acquiring a lung. Romer (1946, 1972) opined that transition from aquatic to terrestrial life by the vertebrates was initiated by amphibians in the Devonian period and completed by the reptiles at the end of the Paleozoic era, a time interval during which some bony fish of the group sarcopterygii had developed primitive lungs. It is possible that any of these fish or the lungfish (Dipnoi) may have branched to give rise to the amphibians (Romer 1972). Developmentally, the lungs of the tetrapods seem to be homologous to the dipnoan lung (Wassnetzov 1932). Indeed, the early stages of the development of the lungfish show clear morphological similarity to the amphibians in having features like suckers and external gills. The extant amphibians, however, did not evolve until the Jurassic period (150 million years ago) and their continued reliance on water is still evident for activities like breeding and cutaneous gas exchange. While the anurans and some urodeles have generally well-developed lungs, in apoda the degree of pulmonary development is highly variable with some species having only a single tubular lung (Pattle et al. 1977; Maina and Maloiy 1988), a feature which resembles that in snakes (Ophidia) (George and Shah 1956).

The versatility of the methods and the routes for gas exchange in amphibians are unique among vertebrates (Szarski 1964; Rahn and Howell 1976). Aquatic salamanders have trimodal gas exchange in using gills, lung and skin (Guimond and Hutchison 1972). Four main features have dependently influenced the design and development of the amphibian lung. These are (1) ectothermy, (2) bimodal breathing, (3) relatively less energetic life-style and (4) their rather transitional phylogenetic position between the lower and higher vertebrates. VO_2 in a mammal is well over 100 times that of an amphibian of the same mass (Sect. 7). Ectothermy, nevertheless, enables the amphibians to attain small sizes (e.g. tadpoles) which due to their high surface-to-volume ratio are beyond reach of the endotherms. Bimodal breathing has greatly influenced pulmonary structure and function in amphibians. In the small and entirely aquatic amphibians like newts and some salamanders (Plethodontidae), the skin and the buccal cavity are the sole gas-exchange organs (Gatz et al. 1974; Gratz et al. 1974). Such animals have long and cylindrical bodies to maximize their surface areas and normally live in cool, damp and well-aerated environments. *Amphiuma* and the *Siren* have the lowest metabolic rates among the amphibians when measured under similar conditions (Guimond and Hutchison 1976). The lungs of *Necturus* and *Cryptobranchus* are thin-walled, transparent, poorly vascularized and are non-septed, while those of *Amphiuma* and *Siren* are more elaborate (Guimond and Hutchison 1976) and have a respiratory surface area four times greater (Czopek 1965). The lung of the European salamander (*Salamandra salamandra*) has two orders of septa which are covered by a ciliated epithelium (Goniakowska-Witalinska 1978) while those of the tree frog, *Hyla arborea* (Goniakowska-Witalinska 1986), terrestrial toad, *Bufo marinus* and the aquatic

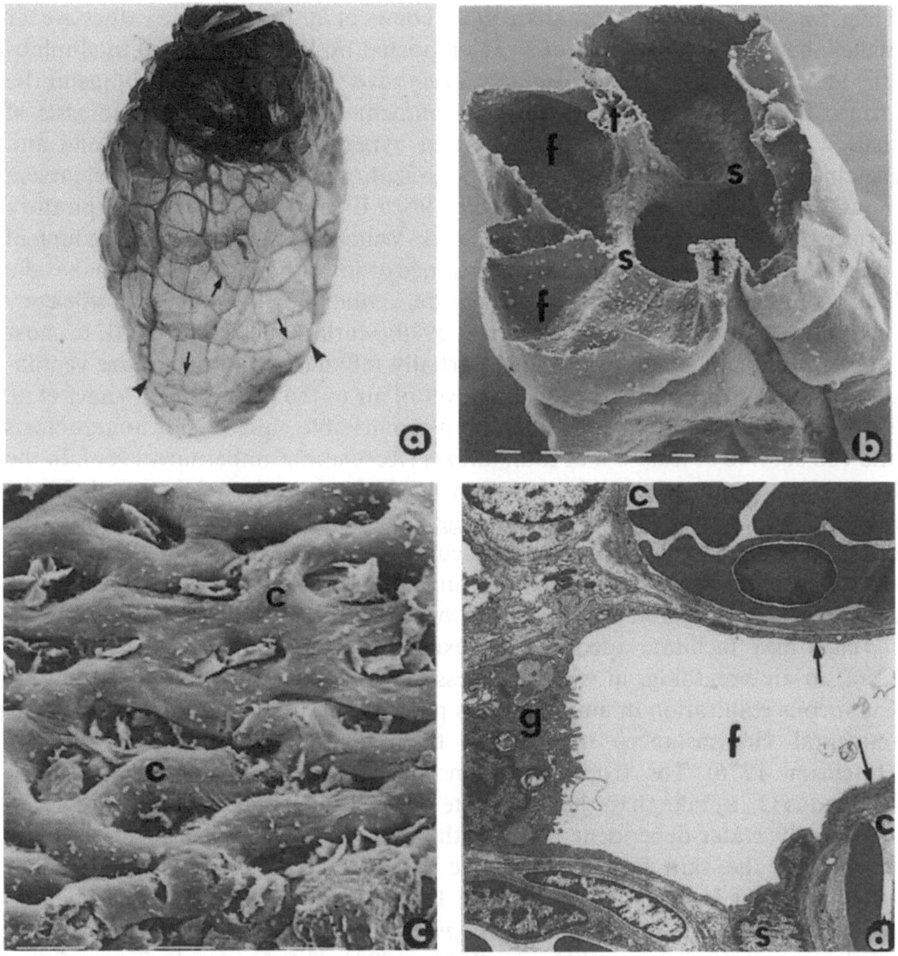

Fig. 13a–d. The amphibian lung. **a** Simple lung of the tree frog *Chiromantis petersi* showing the internal subdivision with trabeculae (*arrowheads*) and septa (*arrows*), × 6; **b** tubular lung of a caecilian (*Boulengerula taitanus*) supported by diametrically located trabeculae (*t*) and internally subdivided into air-cells (*f*) *by septa* (*s*), × 160; **c** surface of the lung of *C. petersi* showing abundant anastomosing blood capillaries (*c*), × 714; **d** sectional view of the surface of the lung of *C. petersi* covered by granular pneumocytes (*g*) and smooth pneumocytes (*s*) with the blood capillaries (*c*) protruding into the air-spaces (*f*), the air and blood being separated by a thin blood-gas barrier (*arrows*), × 4536. (**b** from Maina and Maloiy 1988, courtesy of Oxford University Press, **c** and **d** from Maina 1989c, courtesy of Cambridge University Press)

one, *Xenopus laevis* (Smith and Rapson 1977), and the tree frog, *Chiromantis petersi* (Maina 1989c) are more elaborate in having three hierarchical septa (Fig. 13a, b). In keeping with their terrestrial habits, the toads have a more extensive pulmonary surface area than the frogs (Tenney and Tenney 1970). *Siren* utilizes 16–26% of the inhaled air between breaths (Guimond and Hutchison 1973) while *Amphiuma* utilizes more than 50% (Toews et al. 1971). These values are far higher than the average ones of 1–3% reported by Whitford (1973) in amphibians and suggest that *Amphiuma* and *Siren* have the most efficient lungs in the group. This corresponds with the environment they live in which consists of shallow, warm, occasionally hypoxic and hypercarbic waters (Guimond and Hutchison 1976). Further to other specializations like long well-vascularized lungs with greater septation (Czopek 1962a, b) and presence of respiratory surface throughout the lung (Bell and Stark-Vancs 1983), the septa of the lung of *Amphiuma tridactylum* have a preponderance of smooth muscle which makes the lung completely collapse on exhalation, enhancing ventilatory efficiency (Czopek 1962b; Martin and Hutchison 1979; Stark-Vancs et al. 1984). In most other amphibians, the lungs remain partially inflated during the entire ventilatory cycle resulting in incomplete turnover of air in the lungs (Stark-Vancs et al. 1984). *Amphiuma* and *Necturus* exhibit a remarkable capacity for anaerobiosis and can survive at least 6 h of total anoxia (Rose and Zambernard 1966). In the larval tiger salamanda (*Ambystoma tigrinum*), O_2 capacity of the whole blood increases with that of body weight (Burggren et al. 1987). This may be due to the heavy dependence of amphibians on cutaneous respiration and the fact that with increasing size, surface area per unit mass decreases (surface law). It was suggested by Burggren et al. (1987) that haematological adaptations in *A. tigrinum* may facilitate adequate gas exchange without the animal having to result to air-breathing, in which process it would be vulnerable to predators. Cutaneous respiration in amphibians is relatively important and in some environmental circumstances may equal the pulmonary one (Guimond and Hutchison 1976). The tropical anuran species have a higher pulmonary to cutaneous O_2 uptake than the temperate ones (Hutchison et al. 1968). However, the largely water-dependent anuran, the clawed toad (*Xenopus laevis*) relies heavily on the skin for gas exchange as reflected in its physiological and morphological features (Czopek 1955; Emilio and Shelton 1974). Despite its reliance on water, the lung of *Xenopus laevis* is well developed for aerial respiration (Pohunkova and Hughes 1985a): 80% of its O_2 uptake is transferred through the lung (Emilio and Shelton 1974). The lung of *C. petersi* is very well vascularized (Fig. 13c, d). During aestivation, O_2 uptake is drastically reduced and may be entirely cutaneous. The epidermis in *Siren* is twice as thick as that of *Amphiuma* (Czopek 1965). In *Triturus cristatus*, the length of skin capillaries when expressed as a percentage of total respiratory capillaries is 73.7 as compared to 23.3% for the pulmonary capillaries (Czopek 1965). Like the African lungfish *Protopterus aethiopicus* (Sect. 11.2), in response to periodic droughts, *Siren* aestivates in a cocoon as the metabolic rate and heart rate are drastically reduced (Gehlbachet al. 1973).

From quantitation of the blood capillaries in the lung and the skin, Czopek (1965) observed their relative distribution to be 62% in the lung and 38% in the skin, values which correlate closely with the distribution in O_2 uptake in the two sites (Rahn and Howell 1976). In *S. salamandra*, the blood capillaries have a diameter ranging from 6.8–20.7 μm and bulge out into the pulmonary air-space, one-half to two-thirds of their surface being exposed to air (Goniakowska-Witalinska 1978); the arithmetic mean thickness (τt) of the blood-gas barrier (BGB) of the lung in the species is 2.05 μm (range 1.12–3.80 μm), while the harmonic mean thickness (τht) is 1.98 μm. Barrier thickness in *Triturus alpestris* is 1.23 μm and in *T. cristatus carnifex* 0.849 μm (Goniakowska-Witalinska 1986). Increased pulmonary vascularization in amphibians correlates with the degree of terrestriality, behaviour, distribution, tolerance to dehydration and decrease in cutaneous respiration (Czopek 1962a, b; Guimond and Hutchison 1976). Two-thirds of the lung surface in amphibians is covered by blood capillaries (Krogh 1904). Development of the external gills in amphibians closely correlates with functional needs (Noble 1931): *Siren*, which has a well-developed lung and a high capacity for cutaneous gas exchange, has very poorly developed gills, while *Necturus*, which has to rely heavily on pulmonary gas exchange, has well-developed gills which supplement pulmonary function. *Necturus* kept in cool, well-aerated water has less well-developed gills than those of animals kept in warmer poorly aerated water (Guimond and Hutchison 1976). The lung of *C. petersi*, which is probably one of the most xeric of all anuran amphibians, and of some aquatic frogs such as *Xenopus laevis* and *Pipa pipa*, which live in hypoxic waters, are profusely subdivided, a clear indication of their higher gas-exchange efficiency. Pulmonary vascularization in the toad *Bufo* is more profuse than that of the aquatic *Xenopus* (Czopek 1965). The length of the lung blood capillaries per gram body weight in *Rana esculenta* is 11.57 mg^{-1}, in *Hyla arborea* 34.36 mg^{-1}, while in the European salamander the value is only 5.2 mg^{-1} (Czopek 1965). As respiratory organs, lungs are by far more efficient than the skin for reasons that the lungs have a thinner BGB and denser capillary network with individual capillary diameters some 20–30% larger than those of the skin (Czopek 1965). The skin of amphibians, especially that of the European salamander is more important for CO_2 removal than O_2 uptake (Whitford and Hutchison 1965) due to its considerable thickness (59 μm) and relatively poor capillary network (Czopek 1965).

Even though amphibians have numerous options for respiration available to them, in any instance, only one surface is best developed, while the others are only supplemental (Guimond and Hutchison 1976). Whereas the perfusion of the skin can be regulated by vasodilation and vasoconstriction (Poczopko 1959), ventilation which is largely beyond the control of the animal is usually inadequate for efficient gas exchange. In *Bufo marinus* and *Rana catesbeiana* (Hillman 1987a, b), dehydration results in increased peripheral resistance caused by haemoconcentration influences on blood viscosity, while in *Rana pipiens* and *Bufo americanus* (Gatten 1987) dehydration induces no change in resting VO_2 but significantly reduces the rate of O_2 uptake during exercise to a higher extent in *Rana* than in *Bufo*. Similarly, dehydration in the terrestrial frog, *Eleutherodactylus coqui*, reduces the capacity to utilize aerobic metabolism during activity (Pough et al. 1983). In the xeric anurans, *Phyllomedusa sauvagei* and *Chiromantis xerampelina*, at the resting evaporative water loss (EWL) cutaneous respiration is insignificant but with increased water loss fractional cutaneous gas exchange correlates with cutaneous EWL (Stinner and Shoemaker 1987). Terrestrial anurans like *C. xerampelina* have evolved notable adaptations to withstand dehydration in having an impermeable skin (Stinner and Shoemaker 1987), uricotelism (Shoemaker et al. 1972) and can tolerate water loss amounting to more than 60% of their body mass (Loveridge 1970). As consistent with other amphibian species (Hillman 1978b; Pough et al. 1983), a decline in aerobic capacity correlates with the decline in blood flow rates during dehydration. Hillman et al. (1985) contended that systemic O_2 transport may be the major limiting factor to aerobic capacity in aerial breathing in amphibians. At elevated temperatures, the participation of the gills and the skin decreases in *Amphiuma* and *Siren* (Guimond and Hutchison 1968, 1976). Further, the significance of pulmonary and aquatic respiration correlates with adaptive changes in the respiratory properties of blood in amphibians, an increased dependence on pulmonary gas exchange being accompanied by lowered O_2 affinity and an increased Bohr shift (McCutcheon and Hall 1937; Lenfant and Johansen 1967; Johansen and Lenfant 1972). The simplicity of the amphibian lung (Fig. 13a, b) should be viewed in a paedomorphic context rather than in a primitive one. A compromise between desiccation and gas exchange has led to the fact that there are no large amphibians today. However, the xeric *C. petersi* (Loveridge 1970, 1976) has acquired adaptations like tolerance to high ambient temperature (prefers direct solar exposure and temperatures of 40–42 °C), low O_2 affinity and temperature-insensitive O_2-Hb binding (Johansen et al. 1980): at 25 °C, O_2 uptake was 81 $\mu l\, g^{-1} h^{-1}$ and rose to 253.5 $\mu l\, g^{-1} h^{-1}$ at 40 °C, the skin contributing 38.5% of the total O_2 needs. At 25 °C, O_2 uptake in the caecilian, *Boulengerula taitanus* is similar to that of other amphibians (Hutchison 1968; Wood et al. 1975). However, caecilian blood has a high O_2 capacity (14%) and the RBCs are smaller and greater in number than in other amphibians. These features may have adaptive significance in a fossorial amphibian which lives in a hypoxic and hypercarbic environment. Cutaneous respiration (CR) offers very few possibilities for innovative designs and thus, in itself evolutionary, appears to be a dead end. Gans (1970), however, observed that CR could be a unique feature of the modern amphibians, an inference which may correspond with a suggestion made by Romer (1972) that Carboniferous amphibians were well scaled and thus the water-permeable skin of modern amphibians represents a specialized secondary condition.

Structurally, the lungs of Amphibia exhibit remarkable variations which correlate with habitat and metabolic activities (Szarski 1964; Goniakowska-Witalinska 1985). These differences, while principally phylogenetic, may in part illustrate the multifunctional nature of the amphibian lung in features like vocalization, buoyancy and defence. With few exceptions, the lungs are paired thin-walled rather smooth air-sacs, which in some species are internally divided to varying degrees into peripheral air-cells (Fig. 13) which range in diameter from 1.45 mm in *Rana pipiens* to 2.3 mm in *Bufo marinus* and *Rana catesbeiana* (Tenney and Tenney 1970). In *S. salamandra*, the thickness of the lung varies from 22.2–142 μm, in the tree-frog *Hyla arborea* 11.3–36.8 μm, in *Triturus alpestris* 14.4–80.6 μm, and in *Triturus cristatus carnifex* 132.0–137.6 μm (Goniakowska-Witalinska 1978, 1986). Ontogenetically, the thickness of the lung wall in *S. salamandra* does not change substantially with age (Goniakowska-Witalinska 1982) but does so appreciably in the newt, the values increasing from 16.2 μm in the larval stage to 137.6 μm in the adult (Goniakowska-Witalinska 1980a). Septa are absent in the newt lungs (Goniakowska-Witalinska 1980a, b) and are on average 25 μm thick in the tree-frog *Hyla arborea* (Goniakowska-Witalinska 1986). They provide mechanical support to the lung, anchorage for the blood capillaries and increase respiratory surface area. Abundance of smooth muscle and presence of elastic tissue in the septa (Smith and Rapson 1977; Pohunkova and Hughes 1985a; Goniakowska-Witalinska 1986; Maina and Maloiy 1988; Maina 1989c) may account for the greater compliance of the amphibian lung, which reduces the work of breathing (Hughes and Vergara 1978). Exhalation may also be effected by the elastic recoil of the contractile elements of the lung. In contrast to Anura, the Urodela in general have structurally less complex lungs (Goniakowska-Witalinska 1986) and the Apoda (caecilians) have a tubular poorly compartmented lung (Welsch 1981; Maina and Maloiy 1988), which was considered by Marcus (1937) to be more primitive compared to that of the lungfish (Dipnoi). The internal aspect of the lung consists of low first-order septa which delineate the rather flat air-spaces, the faveoli. Like all the other gas-exchange organs derived from organs like swim bladder and gas-glands (Hughes 1970), the amphibian lung is lined by a surfactant (Welsch and Storch 1973; Pattle et al. 1977; Maina and Maloiy 1988). Studies of the surfactant of *Triturus vulgaris* by Pattle et al. (1977), however, showed that it only partly reduces surface tension to 2 $mN\,m^{-1}$ compared to the more efficient mammalian one which reduces it to 0.1 $mN\,m^{-1}$ or less. An alternative antioxidant role of the surfactant has been suggested by Brooks (1970). The amphibian lungs are ventilated by a buccal force pump which works against the elastic forces of the lung, the efficiency of the pump being about 8% and the O_2 cost of breathing 5% of the total energy budget (West and Jones 1975); comparatively, the energetic cost of breathing in amphibians is twice that of the human and is generally comparable to that of fish. The respiratory surface of the amphibian lung is lined by epithelial cells which, except in *Triturus pyrrogaster* (Okada et al. 1962) and *C. petersi* (Maina 1989c) (Fig. 13d), are apparently not differentiated into smooth (type I) and granular (type II) cells of the reptilian (Hughes 1970; Solomon and Purton 1984; Maina 1989d), mammalian (Weibel 1984a, b; Maina 1987b, 1988b, 1990c) and avian (Maina et al. 1989a) lungs. Phylogenetically, the two populations of pneumocytes are derived from differentiation of the single type of cell present in the lungs of amphibians and lungfishes (Goniakowska-Witalinska 1982). Pulmonary macrophages are very rare in the amphibian lung (Meban 1973; Goniakowska-Witalinska 1986). In the lung of the tree-frog, a large number of these cells were observed (Maina 1989c) and a macrophagic response was experimentally induced through carbon particle aspiration in anurans *Bufo fowleri* and *Kaloula pulchra* (Welsch 1983). Septal and free macrophages were observed in the apodan lungs of *Chthonerpoton indistinctum* and *Ichthyophis paucesulcus* (Welsch 1981). The scarcity of free (monocyte derived) pulmonary macrophages may be due to the fact that the epithelial cells of the lung can serve as phagocytes (Welsch 1983) and a thick blood-gas barrier (BGB) keeps away invasive agents. In general, the thickness of the BGB in amphibians ranges from 2.59 μm in the urodeles, 2.35 μm in apodans to 1.89 μm in anurans (Meban 1980); these values are relatively higher than the average ones of 2.04 μm in reptiles (Meban 1980), 1.61 μm in mammals (Gehr et al. 1981), and 0.86 μm in birds (Maina 1989e; Maina et al. 1989a). The thinnest parts of the BGB in the caecilian lungs of *C. indistinctum* and *I. paucisulcus* are only 1 μm thick (Welsch 1981), while the tree-frog *H. arborea* (Goniakowska-Witalinska 1986) has one of the thinnest barriers (0.6 μm) reported in amphibians. *H. arborea* is characterized by a high cellular metabolism (Goniakowska-Witalinska 1973) and its skin and the lungs are richly vascularized (Czopek 1965). The BGB consists of a pulmonary epithelium, an interstitial space and an epithelium (Fig. 13d). The interstitium constitutes the thickest part (about 42%) of the barrier (Meban 1980). With development, the thickness of the BGB of the lungs of *S. salamandra* and *T. cristatus* respectively increases from 1.61–2.05 μm and 0.304–0.849 μm (Goniakowska-Witalinska 1980a, 1982). This may correspond with decrease in metabolic rate with increasing size (Czopek 1959, 1965). In morphological and morphometric features, the lungs of anurans and apodans are evidently more advanced than those of urodeles (Meban 1980). The extant amphibians can be looked at (from a respiratory perspective) as representing an evolutionary stage between air-breathing fish on the one hand, and reptiles on the other. After giving rise to successful amniotes, the amphibians have literally lingered on essentially with one foot in water and the other on land. Though the first animals to conquer land, due to their inflexibility as a result of the firm link to water, amphibians are a defeated group and constitute an insignificant taxon among the extant tetrapods.

13.2 Reptiles

The reptiles were apparently the first vertebrates to be adequately adapted for terrestrial habitation by dissociating themselves from dependence on water for functions like reproduction, in this respect by evolving a cleidoic egg and an impermeable cover to prevent desiccation. The skin, except for a few species like the soft-shelled freshwater turtles (Robin et al. 1964; Stone et al. 1992a, b), was rendered non-functional as the lungs became the singular gas exchangers. To enhance respiratory efficiency, the reptiles developed and co-opted the ribs for suctional pump ventilation of lungs (Glass and Wood 1983). Contemporary reptiles range from entirely aquatic to extreme terrestrials. The mainstream transition from aquatic to land habitation by the vertebrates was started by the amphibians during the Devonian period, through the stem reptiles (cotylosaurs) and completed by the true reptiles by the end of the Paleozoic era (Romer 1967). The Mezozoic era, which lasted nearly 200 million years is often referred to as "the age of the reptiles" as this taxon was the dominant animal group.

Reptilian lungs remarkably differ in complexity, to the extent that there is no simple model for lungs in the group. Snakes (Ophidia), particularly exhibit remarkable variation in lung development and morphology (Cope 1894; Verde 1951; Pohunkova and Hughes 1985b). Serpentine animals (Klauber 1956) as well as vermiform amphibians, the caecilians (Maina and Maloiy 1988) show a reduction to single lungs (Sect. 13.1). In one of the many classifications which have been attempted (Milani 1894; Marcus 1937; Baudrimont 1955; Duncker 1978a; Perry 1981, 1983), the reptilian lungs have on the basis of structural complexity been grouped into three categories. The simplest lung which characterizes lizards like the tegu *Tupinambis nigropunctatus* (Klemm et al. 1979) is termed unicameral. Such lungs have a central air-pathway and peripherally appressed shallow air-spaces giving a low surface-to-volume ratio. The paucicameral lung typifies most chameleons and iguanids and consists of a subdivided central lumen leading to air-cells, while the most elaborate of the reptilian lungs is the multicameral one of the turtles, varanid lizards and crocodiles (Perry 1978; Perry and Duncker 1978, 1980). Such lungs consist of a short intrapulmonary bronchus which gives way to an elaborately subdivided gas exchange tissue. This classification is clearly too simplistic as transitional forms and gradations exist. The multicameral lung is thought to have played an important role in the evolution of the amniotes (Duncker 1978a; Perry 1983). Such lungs could easily have given rise to the bronchoalveolar lung of mammals and the avian lung-air sac system (George and Shah 1965; Perry 1989). A major difference between the reptilian and the mammalian lung is that of development of the airways. The reptilian lung fills from the peripheral walls into the axial air-space, the air-ducts constituting the unfilled spaces while in the mammalian lung the alveoli form as the terminal bifurcations of the bronchial system (i.e. filling from inside).

The structure of the reptilian lung is similar to that of the diving mammals (Perry et al. 1989a, b). In the tegu, the thickness of the parenchyma (the gas exchange tissue) is 1.3 mm while in the monitor lizard (*Varanus exanthematicus*) it is 2.0 mm (Perry 1981). Differences in tissue thicknesses were observed in the tegu lung (Hlastala et al. 1985) with the cranial region 2.8 μm, the middle 2.1 μm and the caudal 1.4 μm thick. Gas exchange in the reptilian lung occurs in the honeycomb modifications of the pulmonary tissue, the faveoli (Fig. 14a, b, c) (Maina et al. 1989b). In the tegu, the volume density of the parenchyma is 25.2%, in the monitor lizard 32.1% (Perry 1981) and in the Nile crocodile about 25% (Perry 1988). Respectively, the parenchyma in the mammalian and avian lungs constitute 85–90% (Gehr et al. 1981) and about 50% (Maina et al. 1989a) of the lung volume. Parenchymal air-spaces constitute 40% of the intrapulmonary air in the crocodile lung (Perry 1988), a value notably lower than in mammals (about 56%) (Maina and King 1984) and 53% in the avian lung (Maina 1989e). The septa consist of a core of smooth muscle, collagen and elastic tissue and bear

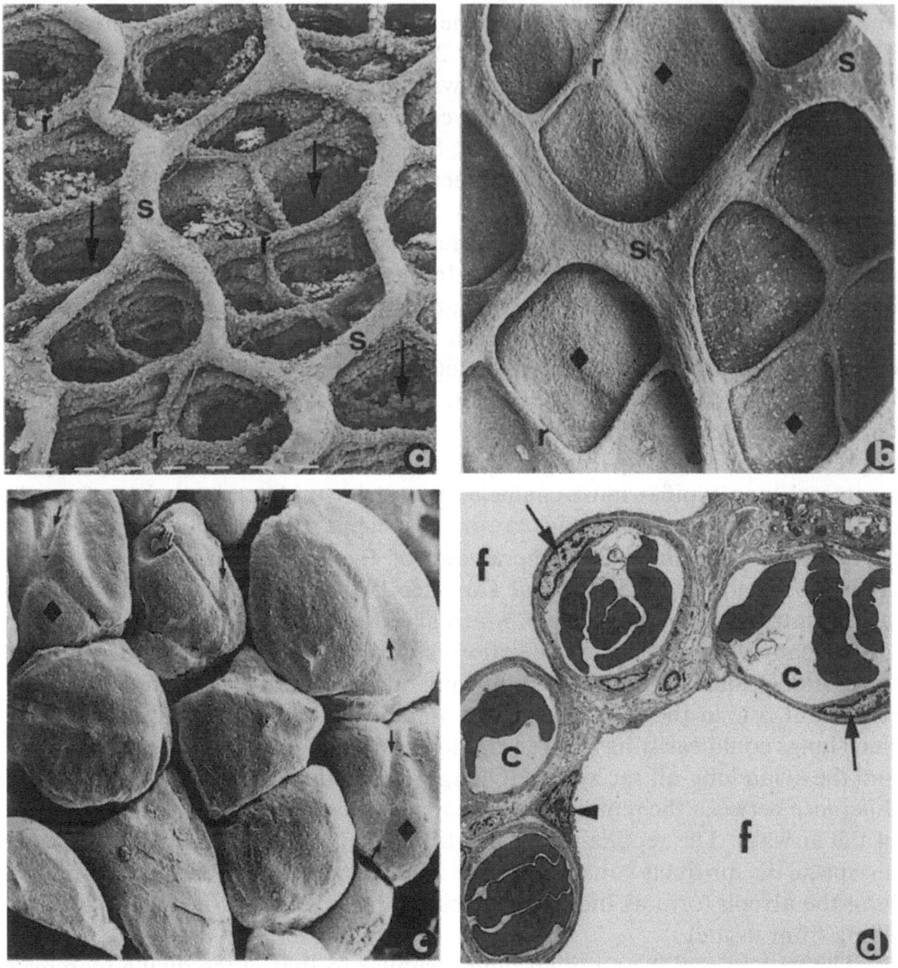

Fig. 14a–d. The reptilian lung. **a** Internal view of the lung of a snake, the black mamba (*Dendroaspis polylepis*), showing primary septa (*s*) and the secondary ones (*r*) which form openings into the faveoli (*arrows*), the gas-exchange components, × 30; **b** faveoli (*diamonds*) of the lung of the monitor lizard (*Varanus exanthematicus*) formed by hierarchically arranged septa, the primary (*s*) and secondary septa (*r*), × 51; **c** latex cast impressions of the gas-exchange components of the lung of the pancake tortoise (*Malacochersus tornieri*), the *arrows* showing septa and *diamonds* faveoli, × 51. **d** a sectional veiw of the blood capillaries (*c*) lining the surface fo the lung of *D. polylepis* exposed to the faveolar air-spaces (*f*) with the *arrowhead* showing the nucleus of a pulmonary pneumocyte and *arrows* the endothelial cell nuclei, × 973. (**a** and **d** from Maina 1989d, courtesy of Cambridge University Press, **c** and **b** from Maina et al. 1989b, courtesy of Wiley/Alan R. Liss Inc.)

blood capillaries on their surface (Fig. 14d). Smooth muscles have been implicated in the convective movements of intrapulmonary air (Klemm et al. 1979). In the tegu, smooth muscles constitute 7.4% of the non-trabecular tissue, while the value in the monitor is 1.3% (Perry 1981). Non-vascular smooth muscle in the trabeculae of the lung of the Nile crocodile (*Crocodylus niloticus*) forms 64 and 36% of the interedicular walls (Perry 1988). The compliance of the lung of the garter snake (*Thamnophis sirtalis*) of 0.042 ml cm H_2O^{-1} g^{-1} (Bartlett et al. 1986) is 50 times that reported for a mouse, a mammal of about equal body weight (Bennett and Tenney 1982). In the crocodile lung (Perry 1988), the residual lung volume (18 cm^3 kg^{-1}) is only 13% of the maximal lung volume and is 1.8% of the body volume. The compliance of the lung of the crocodile (0.7 ml cm H_2O^{-1} cm^{-3}) is over four times that of the body wall and may partly depend on the contractile elements and the saccular nature of the lung (Crag 1975; Perry 1988). The high compliance may constitute an energy saving design. More elaborate reptilian lungs have a lower compliance but not as low as that of a mammal of similar size (Perry and Duncker 1978; Tenney et al. 1984). The volume-specific lung compliance of the multicameral crocodilian lung was observed to be similar to that of the much simpler one of the gekko (Perry 1988), a feature which suggests that lung compliance may be a function of parenchymal structure and not lung type. Pulmonary blood capillaries bulge out from the epithelial surface into the faveolar lumen constituting a double capillary system whereby only one face of the capillary wall participates in gas exchange (Fig. 14d). Besides supporting the capillaries, the septa impart some degree of rigidity to the somewhat delicate lungs (Fig. 14a, b), while the smooth muscles and elastic fibres may contribute to the elasticity of the lung.

The distribution of the parenchyma in lungs of most reptiles, particularly the snakes and lizards, is remarkably inhomogeneous (Perry 1981, 1990; Maina et al. 1989b). Much of it is located at the anterior part of the lung, while the posterior part is saccular and smooth (Kardong 1972; Klemm et al. 1979; Stinner 1982; Maina 1989d). In the crocodile lung, most of the parenchyma is located in the caudal two-thirds of the lung where the blood makes 38–50% of the total volume (Perry 1988). The parenchyma in such lungs is ventilated both during inhalation and exhalation, a feature which may enhance pulmonary gas exchange and may partly explain the exceptionally long breath holds that characterize the reptiles (Zhao-Xian et al. 1991). In the chameleon and most snakes, the posterior region is highly enlarged and extends into the spaces of the coelomic cavity, a feature which emphasizes the evolutionary closeness between reptilian and avian respiratory systems (Sect. 13.4). Morphologically, the chelonian lung exhibits reptilian, mammalian and avian characteristics (Solomon and Purton 1984). The putative similarities in design and function of reptilian and avian lungs are not surprising due to the close evolutionary relationship between birds and reptiles (de Beer 1954). Pneumocytes are distinctly differentiated into type I and type II cells in the reptilian lung (Luchtel and Kardong 1981; Pohunkova and Hughes 1985b; Perry 1988; Maina 1989d). A mitochondrial-rich pneumocyte to which function has not yet been ascribed has been described by Bartels and Welsch (1984) in the lung of the freshwater turtle *Pseudemys* (= *Chrysemys*) *scripta*. Probably as a feature of primitivity, the type II cells in the lung of the gecko *Rhacodactylus leachianus* (Perry et al. 1989a) lack lamellar bodies. To minimize the thickness of the blood-gas barrier (BGB), the attenuated type I cells line most of the faveolar surface, while the rather cuboidal (and thicker) type II cells are located in the epithelial crypts formed by adjoining bulging blood capillaries. Type II cells are metabolically more active as evinced by the numerous and diverse organelles which they contain (Maina 1988b, 1989d, 1990c). They are known to produce the osmiophilic factor (the surfactant) which lines the pulmonary surface reducing surface tension at the air – tissue interface and hence increasing the ease of inflation of the lung and preventing transudation of plasma. Analysis of pressure volume curves (Perry and Duncker 1978) suggest that reptilian lungs possess very effective surfactant. The BGB of the reptilian lung essentially consists of an epithelial cell, an interstitial space (which contains supportive elements such as collagen, fibrocytes and elastic tissue) and an endothelial cell. The interstitium constitutes about 45% of the thickness of the BGB, the endothelium 25% and the epithelium 30% (Meban 1980). In eight reptilian species examined by Meban (1980), the harmonic mean thickness (τht) of the BGB ranged from 0.73 μm in the American alligator (*Alligator mississippiensis*) to 1.38 μm in the European tortoise (*Testudo graeca*). Perry (1981) reported lower τht values of 0.36 μm and 0.70 μm in the tegu and monitor lizards, respectively. Hlastala et al. (1985) reported the thickness of the BGB in the tegu to range from 0.35–0.90 μm (mean value 0.50 μm). In the snake, Stinner (1982) found a value of 0.46 μm and Perry (1978) reported a comparable value of 0.55 μm in the red-eared turtle (*Pseudemys scripta elegans*).

The reptilian lung constitutes 5% of the body weight (Tenney and Tenney 1970) and the respiratory surface area strongly scales with body mass ($Sa = aW^{0.74}$). The volume of the lung appears to increase proportionately with metabolic rate rather than body size, as in amphibians. Bennett and Dawson (1976) established that at 20 °C, the resting VO_2 in reptiles scaled to a power of 0.80 ($VO_2 = ml\ O_2\,h^{-1} = 25.6\ W^{0.80}$), a relationship which indicates that reptiles resemble other ectothermic vertebrates in their general metabolic levels. When stressed, reptiles like *P. scripta* can, however, increase their VO_2 20 times from rest (Gatten 1974). In reptiles, lung volume is comparable to that of the lung-air sac system of birds but larger than in mammals (Duncker and Guntert 1985a; Perry 1990). In animals of comparable size, a reptile has a lung volume two to seven times greater than a mammal (Crawford et al. 1976; Glass and Johansen 1981; Perry 1989). Alligators have an O_2 store 85% of the total lung volume (Andersen 1961). The semiaquatic turtle *P. scripta* has a lung volume of 160 $ml\,kg^{-1}$ and an O_2 delivery to the tissues of 0.9 $ml\,kg^{-1}\,min^{-1}$ at 20–23 °C (Crawford et al. 1976). The tegu and the monitor lizards have mass-specific lung volumes of 87.4 and 294 $cm^3\,kg^{-1}$ (Perry 1981), the snake *Pituophis melanoleucus* a value of 97.5 $ml\,kg^{-1}$ (Stinner 1982) and the Nile crocodile 109 $cm^3\,kg^{-1}$ (Perry 1988). These values are remarkably higher (Table 4) than the average one in birds of 26 $cm^3\,kg^{-1}$ (Maina et al. 1989a) and mammals 20 $cm^3\,kg^{-1}$ (Gehr et al. 1981). The large lungs provide greater storage of air and probably have dispensed with the need for the reptiles to breathe continuously (Glass and Wood 1983-review) and enable diving reptiles to stay under water for an appreciably long time (Burggren et al. 1978; Ultsch et al. 1984). Though the actual control mechanism is not well known, arrhythmic breathing appears to be an energy-saving adaptation in animals with a low aerobic capacity (Milsom 1984) where continuous respiration is not necessary. In marine turtles, there is a strong correlation between breath-hold length and resting lung volume (Milsom and Johansen 1975). Rhythmic breathing has further done away with the need for a sophisticated and energetically expensive ventilatory mechanism (Wood and Lenfant 1976). Dive duration in *Pseudemys concinna* may reach 141 min (Belkin 1964). Further to the periodic pattern of breathing, the group has an exceptional tolerance to hypoxia (Belkin 1963). Alligators take up to 12 h to repay an O_2 debt and turtles can tolerate total anoxia for a long time, in the process accumulating lactic acid levels up to 2200 $m\,mol\,l^{-1}$ and can survive a decrease in brain pH to 6.4 (Coulson and Hernandez 1980). The general tolerance to hypoxia in the ectotherms is largely due to their low metabolic needs. Allometric comparison of pulmonary variables in reptiles with those of terrestrial mammals and birds (Duncker and Guntert 1985a, b) shows that reptiles in general have morphometrically inferior lungs compared with the other two taxa. However, when normalized with body mass, the respiratory surface area in reptiles is generally greater than in amphibians (Table 4). The tegu has a mass-specific respiratory surface area of 4.2 $cm^2\,g^{-1}$ and the monitor lizard 6 $cm^2\,g^{-1}$ (Perry 1981), values which are remarkably lower than those of mammals and birds (Gehr et al. 1981; Maina 1989e; Maina et al. 1989a). The surface density of the respiratory area (surface area of the blood-gas barrier per unit volume of the parenchyma, S_VBGB, p) is a measure of the degree of subdivision of the lung. This factor differs remarkably among reptiles, the values in this group being remarkably lower than in mammals and birds (Table 4). In the red-eared turtle, Perry (1978) reported a value of 1.8 $mm^2\,mm^{-3}$, in the snake Stinner (1982) reported a value of 10.1 $mm^2\,mm^{-3}$ and in the tegu and monitor lizard respectively Perry (1981) found values of 19 and 7.1 $mm^2\,mm^{-3}$. These parameters and those outlined below illustrate that pulmonary structural and functional characteristics in reptiles correlate with individual aerobic capacities. The ventilatory volume in a reptile is about 20% of a mammal of comparable size (Glass and Wood 1983). At body temperatures of 37–40 °C, the mass-specific VO_2 of a reptile is much less than that of an endothermic vertebrate (Glass and Wood 1983). Bennett and Dawson (1976) calculated that a 1-kg lizard (at 37 °C) consumes 122 ml $O_2\,h^{-1}$ which is 18% of the VO_2 of an equivalent-sized mammal. These differences are reflected in the tissue structure of the two groups of animals. Bennett (1972) and Else and Hubert (1985a) observed that mitochondrial volume density is higher in mammals than reptiles. The muscle capillary surface area per unit muscle mass of a reptile is about 20% the value of a mammal of similar size (Pough 1980). Reptiles themselves exhibit

Table 4. Comparative morphometry of the lungs of lungfishes, amphibians, reptiles, mammals and birds. *VL* Volume of the lung; V_{vp} volume density of the exchange tissue in the lung; *Sa* respiratory surface area; *W* body mass; V_p volume of the parenchyma; τht and $\bar{\tau} t$ harmonic and arithmetic mean thicknesses of the blood-gas tissue barrier; *ADF* anatomical diffusion factor; DLo_2 total anatomical diffusing capacity*

	VL W^{-1} ($cm^3\ g^{-1}$)	V_{va} (%)	Sa W^{-1} ($cm^2\ g^{-1}$)	Sa Vp^{-1} ($mm^2\ mm^{-3}$)	τht (μm)	$\bar{\tau} t$ (μm)	ADF W^{-1} ($cm^2\ \mu m^{-1}\ kg^{-1}) \times 10^3$	$Dto_2\ W^{-1}$ ($ml O_2\ min^{-1}\ mmHg^{-1}\ kg^{-1}$)	$DLo_2\ W^{-1}$ ($ml O_2\ min^{-1}\ mmHg^{-1}\ kg^{-1}$)
A. Water-breathing fish									
Tuna sp.[a]	–	–	23.80	–	–	0.58	–	6.00	–
Salmo gairdneri[a]	–	–	2.6	–	–	6.37	–	0.118	–
Tinca tinca[a]	–	–	1.83	–	–	2.47	–	0.149	–
Opsanus tau[a]	–	–	2.10	–	–	5.00	–	0.063	–
Latimeria chalumnae[a]	–	–	0.19	–	–	5.00	–	0.006	–
Geotria australis[b]	–	–	23.64	–	2.01	4.14	–	2.093	–
Chaenocephalus aceratus[a]	–	–	1.17	–	–	–	–	–	–
Anguilla anguilla[c]	–	–	1.94	–	–	6.00	–	–	–
Scyliorhinus canicula[a]	–	–	–	–	–	11.27	–	–	–
Solea solea[a]	–	–	–	–	–	2.80	–	–	–
B. Air-breathing fish									
Channa punctata[d]	–	–	0.72[v]	–	–	2.03[v]	–	0.0530[v]	–
	–	–	0.39[w]	–	–	0.78[w]	–	0.0753[w]	–
Channa striatus[e]	–	–	–	–	–	1.36[v]	–	–	–
Anabas testudineus[f]	–	–	0.47[v]	–	–	10.00[v]	–	0.0071[v]	–
			0.08[w]			0.21[x]		0.0539[w]	
			0.32[x]			0.21[x]		0.2281[x]	
Saccobranchus (= *Heteropneustes*) *fossilis*[g]	–	–	0.58[v]	–	–	3.58[v]	–	0.0242[v]	–
			0.31[w]			1.60[w]		0.0288[w]	
			2.00[x]			98.00[x]		0.0031[x]	
Clarias mossambicus[h]	–	–	0.17[v]			1.970[w]	–	0.0213[v]	–
			0.08[w]			0.313[w]		0.0500[w]	

			0.05[x]			0.287[x]		0.0700[x]	
Clarias batrachus[i]	–	–	–	–	–	0.550	–	–	–
Amphipnous cuchia[j]	–	–	0.05[y]	–	–	0.435[y]	–	0.0165[y]	–
			2.28[z]			0.440[y]		–	
C. Lungfish									
Protopterus aethiopicus[k]	0.23	51	14.3	217	0.370	–	4.35	14.34	0.192
Lepidosiren paradoxa[l]	0.05	44.1	0.85	–	0.850	–	3.69	3.33	0.271
D. Amphibians									
Hyla arborea[m]	–	–	–	–	0.565	0.600	–	–	–
Triturus alpestris[m]	–	–	–	–	–	1.230	–	–	–
Salamandra salamandra[m]	–	–	–	–	1.780	2.050	–	–	–
						(2.300)[n]	–	–	–
Triturus cristatus carnifex[m]	–	–	–	–	–	0.849	–	–	–
Triturus vulgaris[m]	–	–	–	–	2.340	2.630	–	–	–
Xenopus laevis[n]	–	–	–	–	1.210	1.710	–	–	–
Xenopus tropicalis[n]	–	–	–	–	1.230	1.750	–	–	–
Rana temporaria[n]	–	–	–	–	1.500	2.060	–	–	–
Rana catesbeiana[n]	–	–	–	–	1.530	2.040	–	–	–
Bufo bufo[n]	–	–	–	–	1.670	1.850	–	–	–
Paramesotriton hongkongensis[n]	–	–	–	–	1.810	2.600	–	–	–
Triturus cristatus[n]	–	–	–	–	2.200	2.810	–	–	–
E. Reptiles									
Tupinambis nigropunctus[o]	0.844	26.0	2.94	13.4	0.46	–	6.39	0.002	–
Varanus exanthematicus[o]	0.307	28.7	5.43	6.5	0.65	–	8.35	0.003	–
Pseudemys scripta[o]	0.262	44.9	2.06	1.8	0.50	–	4.12	0.001	–
					1.19[n]	2.10[n]			
Lacerta sp.[p]	0.948	34.6	4.25	13.0	0.53	–	8.01	0.003	–
Crocodylus niloticus[o]	0.109	42.2	1.09	2.4	1.00	–	1.09	0.0003	–

Table 4. (*Contd.*)

	VL W^{-1}	V_{va}	Sa W^{-1}	Sa Vp^{-1}	τht	$\bar{\tau}t$	ADF W^{-1}	Dto_2 W^{-1}	DLo_2 W^{-1}
	(cm^3 g^{-1})	(%)	(cm^2 g^{-1})	(mm^2 mm^{-3})	(μm)	(μm)	(cm^2 μm^{-1} kg^{-1}) × 10^3	($ml0_2$ min^{-1} $mmHg^{-1}$ kg^{-1})	
Lacerta viridis[n]	–	–	–	–	0.90	1.96	–	–	–
Lacerta muralis[n]	–	–	–	–	0.840	1.690	–	–	–
Calotes nemoricola[n]	–	–	–	–	1.030	1.990	–	–	–
Anguis fragilis[n]	–	–	–	–	0.900	2.060	–	–	–
Chamaeleo chamaeleon[n]	–	–	–	–	0.730	1.680	–	–	–
Natrix natrix[n]	–	–	–	–	1.340	2.420	–	–	–
Coronella austriaca[n]	–	–	–	–	1.020	2.310	–	–	–
Alligator mississipiensis[n]	–	–	–	–	0.730	1.650	–	–	–
Testudo graeca[n]	–	–	–	–	1.380	2.490	–	–	–
F. Mammals									
Shrew sp.[q]	30	80–90	32.55	146.84	0.334	0.756	97.66	32.23	3.73
Mus musculus[q]	32	80–90	27.87	96.44	0.322 (0.440)[n]	– (1.27)[n]	86.55	28.56	7.47
Rattus rattus[q]	66	80–90	20.08	68.04	0.385	1.400	52.17	17.21	3.36
Cavia porcellus[q]	30	80–90	19.23	70.30	0.416	1.586	45.79	15.11	3.34
Oryctolagus cuniculus[q]	22	80–90	14.83	74.07	0.500 (0.650)[n]	– (1.400)[n]	29.66	9.79	2.06
Helgale pervula[q]	58	80–90	29.12	56.12	0.394	–	75.31	24.85	4.24
Mungos mungo[q]	56	80–90	28.73	57.49	0.409	–	70.24	26.42	2.67
Genetta tigrina[q]	72	80–90	35.93	56.77	0.506	–	71.02	32.17	3.99
Nesotragus moschatus[q]	63	80–90	27.00	47.33	0.562	–	48.04	52.32	3.78
Homo sapiens[r]	58	80–90	19.32	36.95	0.620	2.200	31.17	10.29	2.62
Papio anubis[s]	46	90	36.96	92.20	0.478	1.12	74.95	31.70	4.01

G. Birds									
Anas platyrhynchos[t]	29.5	41	28.7	239.8	0.133	0.903	215.13	72.0	3.83
Anser anser [t]	24.8	40	23.1	253.0	0.118	0.887	195.86	68.3	3.59
Falco tinnunculus[t]	18.7	52	32.4	323.8	0.210	1.663	150.60	16.6	1.01
Alca torda[t]	37.2	33	29.9	236.5	0.230	0.803	125.26	44.5	4.91
Cephus carbo[t]	32.7	33	27.0	240.0	0.193	0.850	135.69	46.9	4.83
Ambryospiza albifrons[t]	25.9	49	40.7	314.0	0.121	0.584	304.95	106.0	6.40
Chrysococcyx klaas[t]	24.4	53	31.9	274.2	0.157	0.679	203.25	75.8	4.75
Hirundo fuligula[t]	24.1	51	86.5	352.9	0.090	0.613	973.24	318.3	15.3
Colibri coruscans[u]	42.9	46	87.1	389.0	0.099	0.183	879.80	282.0	–

Sources of data and definition of some structures:
[a] Hughes and Morgan (1973), Hughes (1984). [b] Lewis and Potter (1982). [c] Bennett (1988b). [d] Hakim et al. (1978). [e] Munshi (1976). [f] Hughes and Munshi (1973a). [g] Hughes et al. (1974a). [h] Maina and Maloiy (1986). [i] Munshi (1961b). [j] Hughes et al. (1974). [k] Maina and Maloiy (1985). [l] Hughes and Weibel (1976). [m] Goniakowska-Witalinska (1986). [n] Meban (1980). [o] Perry (1983). [p] Cragg (1975). [q] Gehr et al. (1981). [r] Gehr et al. (1978). [s] Maina (1987c). [t] Maina (1989e). [u] Dubach (1981). [v] Gills. [w] Suprabranchial chamber. [x] Labyrinthine organ. [y] Air-sac. [z] Skin.
* Some of the parameters like ADF and Dto_2 were calculated from the available data where these values were not directly reported.

remarkable differences in metabolic and structural characteristics. The capillary density per fibre in the iliofibularis muscle of the energetic terrestrial monitor lizard (932 capillaries mm^{-2}) was $1\frac{1}{2}$ times greater than that of the less energetic desert iguana *Dipsosaurus dorsalis* (Mutungi 1990). The limitations of the reptilian pulmonary system is well evident during exercise where aerobic metabolism may account for less than half the energy expenditure (Gatten 1974; Ruben 1976). At maximal activity, anaerobic metabolism provided 86% of total energy consumption in the water snake, *Natrix rhombifera* (Gratz and Hutchison 1977). These limitations may reside in the fact that the physiological diffusing capacity (DLo_2) of the reptilian lung at a body temperature of 20–23 °C is an order of magnitude smaller than that of a mammal (Crawford et al. 1976; Glass and Johansen 1981). The mass-specific DLo_2 of the tissue barrier in the tegu was 0.88 ml O_2 min^{-1} kg^{-1} mm Hg^{-1}, while it was 0.75 in the monitor (Perry 1981). The reptiles, being the first fully adapted terrestrials, are important in discerning the final stratagems adopted for effective colonization of the dry land. The evolution of the appendages capable of effective body support and locomotion and also that of an impermeable skin to prevent excessive water loss are but some of the important developmental changes that made the reptiles once dominate the dry land. They are still an unspent force.

13.3 Mammals

The beginning of the Tertiary period saw the mammals take over and eclipse the reptiles as the dominant terrestrial vertebrate group (Romer 1967). A drop in the relatively higher temperatures that had prevailed in the last part of the Cretaceous caught the tenuously insulated reptiles unprepared. The mammals which had been slowly evolving during the dinosaur epoch (Mesozoic era) and were better prepared in having evolved endothermy and acquired a hair cover took advantage of the reptilian demise. The appearance of the placental mammals formed the pinnacle of the mammalian evolution. Endothermy, homoiothermy and agility on land are but some of the factors that have influenced pulmonary design in mammals. Commensurate with their generally high metabolic demands, the mammalian lungs are structurally well adapted for gas exchange in the following principal respects: (1) they have a relatively thin blood-gas barrier (BGB), (2) have a profuse pulmonary blood capillary network and (3) the branching of the airway system has resulted in an extensive respiratory surface area. Correspondingly, the lungs are well ventilated with air and well perfused with blood. The development, structure and function of the mammalian respiratory system is better understood than in other animals (see Burri 1984; Weibel 1984a,b for recent reviews). Some of these features will be mentioned in so far as they will serve to highlight how the mammalian respiratory system resembles or differs from the other systems presented in this account.

The mammalian lung arises as a diverticulum of the foregut and gradually grows in size to definitively occupy the thoracic cavity. In the process, the lung is interfaced with the heart via the pulmonary arteries and veins (the arc which will form the functional circuit) and to the systemic arterial vascular system which forms the nutritive pathway. Morphologically, the mammalian lungs are similar except for fine differences like bronchiolar bifurcation, lobulation and topographic relationships between bronchi and blood vessels. Remarkable uncharacteristic paedomorphic features were, however, recently observed in the lung of the fossorial rodent, the naked mole rat *(Heterocephalus glaber)* by Maina et al. (1992). Dichotomous branching of the bronchial system presents an inverted tree design. For every branching, the airway diameter is reduced by a factor of the cube root of 2 (Weibel 1963), an engineering design of branching tubes characterized by minimal fluid or air-flow resistance. Recently, Weibel (1991) attempted to interpret bronchial bifurcation in terms of the mathematical concept of fractal geometry (Mandelbrot 1983), a feature which may explain how the large respiratory surface areas are ventilated and perfused at very low energetic cost. Every day, the human lung is ventilated with about 12 000 l of air and perfused with half as much blood (Randall et al. 1981; Burri 1985), there being 200 ml of blood in the lung at any one time. The dimensions of the airways determine the ventilatory flow and hence the energetic cost of breathing. Resistance to flow through a pipe is directly proportional to length but inversely proportional to the radius to the fourth power in laminar flow and to the power 4.75 in turbulent flow. As the cross-sectional area increases along the pulmonary airways (Schreider and Raabe 1981), flow velocity and resistance decrease with distance. Cilia and subepithelial lymphatics ascertain bronchiolar clearance and patency. Airway branching (Fig. 15a) intensifies the subdivision of the parenchyma enhancing the respiratory surface area affordable within the constraints of the volume of the lung and, more importantly, that of the thoracic cavity. The respiratory surface area (RSA) in man (Gehr et al. 1978) of 143 m^2, which is packed in only about 4 l of lung volume, is equivalent to the size of a tennis court, while in the horse (with a value of 2457 m^2) (Gehr and Erni 1980) the RSA is 17 times the area of a tennis court. The alveoli (Fig. 15a) constitute the terminal components of the respiratory tree and their numbers and dimensions are species- and size-dependent (Tenney and Remmers 1963). The human lung (Weibel 1963) has about 300 million alveoli. About 2×10^3 alveoli are attached to the air-ducts of each acinus (Weibel 1984a). The shrew and the bat have the smallest alveoli (about 30 μm in diameter) while in the placid sirenians (dugongs and manatees) the diameters are about 1 mm (Tenney and Remmers 1963). Empirically, alveolar diameter is a measure of the intensity of subdivision of the parenchyma and hence maximization of the respiratory area. It scales inversely with VO_2 (Tenney and Remmers 1963; Geelhaar and Weibel 1971), the smaller animals having a greater surface area per unit volume of exchange tissue (Table 4 and Fig. 19). The alveoli are rather polymorphic air-cells (Assimacopoulos et al. 1976) which are defined by an interalveolar septum, which is variably perforated by interalveolar pores (pores of Kohn) (Fig. 15b) which may facilitate collateral ventilation. The alveolar surface is basically lined by type I and type II cells (Fig. 15b, d) and the rather rare type III brush cells (Weibel

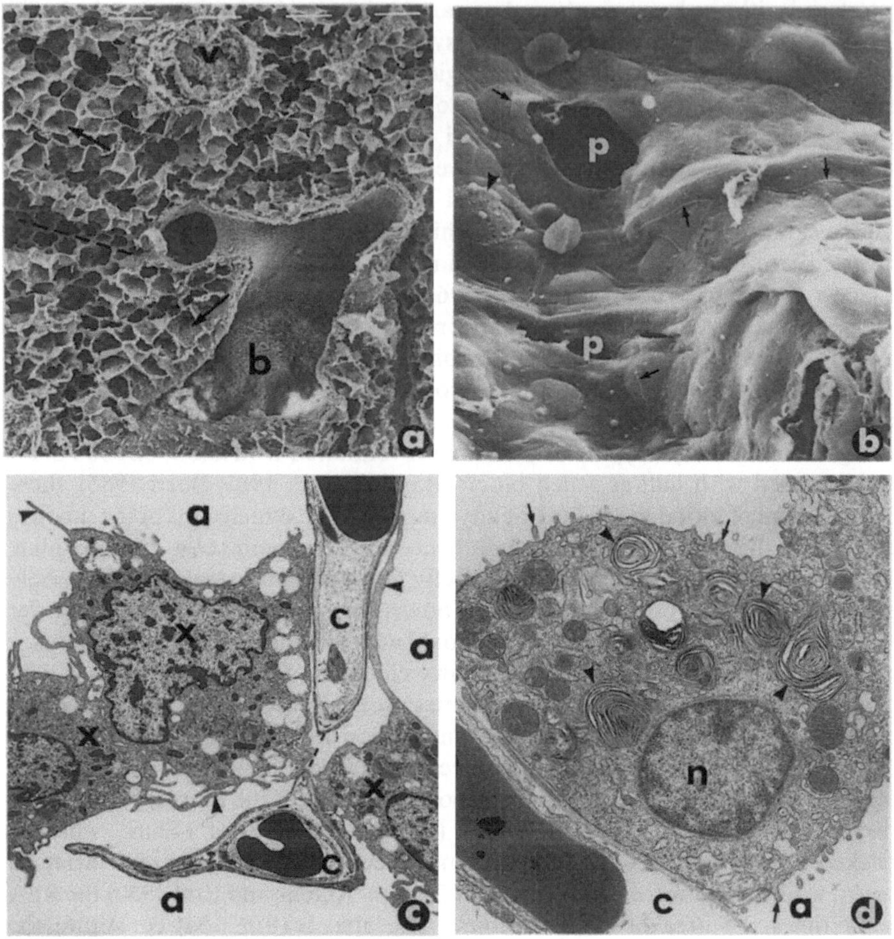

Fig. 15a–d. The mammalian lung. **a** Bronchi (*b*) conduct air to the terminal gas-exchange components, the alveoli (*arrows*), the *dashes* showing a respiratory bronchiole and (*v*) a large blood vessel, × 220; **b** surface of an alveolus showing interalveolar pores (*p*), a granular pneumocyte (*arrowhead*) and intercellular junctions of the smooth (type I) pneumocytes (*arrows*), × 1,100; **c** sectional view of adjacent alveoli (*a*), blood capillaries (**c**), macrophages (*x*) with pseudopodia (*arrowheads*); *dashed line* an interalveolar pore, × 8,630; **d** granular (type II) pneumocyte containing numerous osmiophilic lamellated bodies (*arrowheads*) lying on a blood capillary (*c*); *n* an eccentrically located nucleus; *arrows* microvilli, × 5,415. **a** and **d** from Maina 1985b, courtesy Oxford University Press, **b** Maina 1987b, courtesy of Cambridge University Press, **c** Maina 1988b, courtesy of Wiley/Alan R. Liss Inc.)

1973, 1984a, b; Dormans 1985). The total number of pulmonary pneumocytes scales linearly with body size (Gehr and Crapo 1988). The parenchyma of the human lung is made up of 230×10^6 cells (Gehr and Crapo 1988), the type I cells constituting 8% of them. The type I cells have thin and extensive cytoplasmic flanges (Fig. 15b) with an area of 5098 μm^2 and a volume of 1764 μm^3 (Crapo et al. 1982): type I cells cover about 90% of the alveolar surface. Type II cells cover an area of only 889 μm^2 (Crapo et al. 1982; Gehr and Crapo 1988) and overall only 0.1% of the alveolar surface. The type II cells (Fig. 15d) secrete the surfactant, the surface-active factor. Secretion of the surfactant appears to take place during the resting stage of an animal, while formation and accumulation occurs when it is active (Ishii et al. 1989). The suggested chemosensory, absorptive and contractile roles of the type III cells (Dormans 1985) are only speculative. Alveolar macrophanges (derived from the monocyte cells of the bone marrow and blood) are transient defensive cells of the alveolar surface (Fig. 15c). Their numerical density is extremely variable ranging from 2.3–4.3% of all cells of the lung parenchyma in rats, baboons, dogs and human non-smokers to as high as 19%in smokers (Crapo et al. 1982). The interstitium which constitutes 41% of the thickness of the BGB (Meban 1980) contains supportive elements such as collagen and elastic tissue which serve as the fibrous skeleton of the lung and provide mechanical support and elasticity to the lung (Weibel 1986). The epithelium and the endothelium respectively constitute 31 and 28% of the volume of the BGB (Meban 1980).

Mammals span an extensive weight range from a 2.5-g shrew to the about 150-tonne whale, a remarkable factorial difference of about 60 million. The mammalian lungs exhibit adaptive and allometric (size-dependent) features. Further, the lungs are inherently malleable to respond to altered functional demands. There is an unequivocal correlation between factors like body size, physical activity and ambient PO_2 with the morphometry of the mammalian lung (Weibel 1972; Maina et al. 1991). The alveolar surface area of the lung of the Etruscan shrew (*Suncus etruscus*), the smallest extant and most highly metabolically active mammal (Fons and Sicart 1976) is 64 $cm^2 g^{-1}$ (Gehr et al. 1980), while the corresponding value in an indolent animal like the rabbit (*Oryctolagus cuniculus*) is only 15 $cm^2 g^{-1}$ (Gehr et al. 1981). The anatomical pulmonary diffusing capacity (DLo_2), the most comprehensive estimator of the structural capacity of a respiratory organ in gas exchange, of the Japanese waltzing mouse (*Mus wagneri*), a pathologically hyperactive animal which consumes 80% more O_2 than a normal white mouse (Geelhaar and Weibel 1971), has a DLo_2 55% greater than a normal one. Burri and Weibel (1971) observed that rats raised in a hypoxic environment had a higher DLo_2 (0.72 ml O_2 min^{-1} mm Hg^{-1} 100 g^{-1}) than the controls (0.60) while those raised in a hyperoxic environment interestingly had a decreased DLo_2 (0.51 ml O_2 min^{-1} mm Hg^{-1} 100 g^{-1}). Lungs of genetically adapted high altitude rats had smaller alveoli and a higher respiratory surface area than the sea-level dwelling ones (Pearson and Pearson 1976). Experimental lobectomy results in compensatory overgrowth (Rannels and Rannels 1988), the remaining lobes in young rats attaining the same DLo_2 as the control animals (Burri and Sehovic 1979). Studying groups of animals of equivalent body sizes but different aerobic capacities, Weibel et al. (1987) observed that DLo_2 in the athletic species (dogs and ponies) which had a VO_2max 2.5 times higher than the lesser energetic ones (goats and calves) was 1.5 times higher in the athletic species. Prolonged exposure to cold causes increased VO_2 which results in accelerated lung development (Lechner and Banchero 1980). The mass-specific DLo_2 (DLo_2W^{-1}) of the dasyurid marsupial, the kowari (*Dasyuroides byrnei*) (4.5 ml O_2 min^{-1} mm Hg^{-1} kg^{-1}), an animal from a group with a standard metabolic rate generally lower than that of the placental mammals (McNab 1986a), is remarkably low (Hallam et al. 1989). The DLo_2 W^{-1} of 6.48 ml O_2 min^{-1} mm Hg^{-1} kg^{-1} in the horse is two times higher than the steer's 3.18 (Constantinopol et al. 1989). Correspondingly, the structural parameters of the lung are about two times larger in the horse than in the steer, features largely attributable to the larger lung volume in the horse. Bats, the only flying mammals and a taxon with potential for a high aerobic capacity (Thomas 1987), have lungs which are morphometrically superior to those of the non-flying mammals (Maina and Nicholson 1982; Maina et al. 1982a, 1991; Maina and King 1984; Maina 1986). While retaining the basic mammalian pulmonary design (Maina 1985b, 1986), the chiropteran lung has undergone distinct modifications which include a large lung volume, extensive respiratory surface area, thin blood gas barrier (BGB) and large pulmonary capillary blood volume. These adaptational features are essentially similar to those found in the shrew (Gehr et al. 1980) but fall far short of the bat ones. In *Epomophorus wahlbergi* (Maina et al. 1982a), the DLo_2 W^{-1} of 20 ml O_2 min^{-1} mm Hg^{-1} kg^{-1} is five times greater than the average value of 3.73 in the shrews (Gehr et al. 1980). Allometric comparison of pulmonary morphometric variables in terrestrial mammals [$Y = aW^b$], birds and bats (Mains and Settle 1982; Maina and King 1984; Maina 1989e; Maina et al. 1991) and terrestrial mammals and birds (Maina et al. 1989a) indicate that birds have superior lungs to those of non-flying mammals, while bats have lungs which are better specialized than those of birds and non-flying mammals (Figs. 16–21). While in both birds and mammals the principal pulmonary structural parameters, i.e. lung volume (Fig. 16), volume of the pulmonary capillary blood (Fig. 17), surface area of the BGB (Fig. 18) and total morphometric diffusing capacity, DLo_2 (Fig. 21) scale linearly with body weight (Gehr et al. 1981; Maina et al. 1989a), the surface density (S_V) of the BGB in the parenchyma (p), i.e. S_VBGB, p (Fig. 19) and harmonic mean thickness (τht) of the BGB (Fig. 18) scale weakly with weight in both birds and mammals. The thickness of the BGB in the bowhead whale, *Balaena mysticetus* (Henk and Haldiman 1990) is only about 0.35 μm, a value similar to that of other much smaller mammals (Table 4). An inverse scaling between τht and DLo_2 ($b = -0.01$) was observed in reptiles by Duncker and Guntert (1985a). In the 33 mammalian species reported by Gehr et al. (1981), in which the body mass ranged from 2.6 g in a shrew to 700 kg in a Swiss cow, a mass factorial difference of about 270000, τht ranged from 0.26 μm in a shrew to 0.62 μm in human being, a difference of a factor of only about 2. S_VBGB, p is a relative indicator of the degree of subdivision of the gas exchange tissue into terminal air-spaces and changes relatively less with body mass (Table 4). Such sizes have an important effect on the design and mode of ventilation of the lung. This suggests that the sizes of the terminal gas exchange components and the thickness of the BGB in the air-breathing vertebrates may be somewhat optimized to meet individual O_2

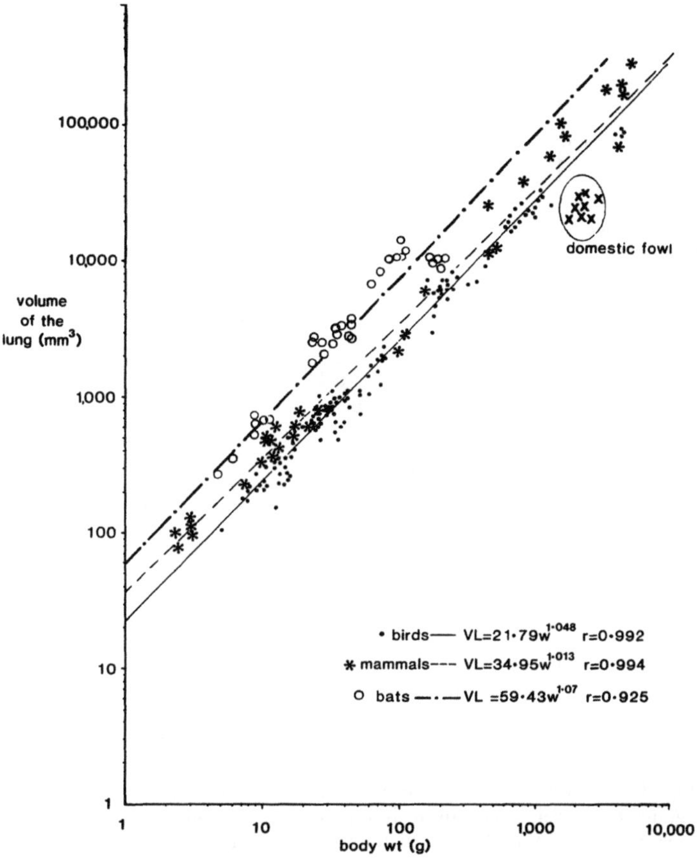

Fig. 16. A double logarithmic plot of the volume of the lung (*VL*) in birds, terrestrial (non-flying) mammals and bats against body weight (*W*). *VL* and *W* scale about proportionately in the three groups of animals. The correlation coefficients (*r*) in the three groups of animals (between *VL* and *W*) are very strong. Bats have significantly larger lungs than non-flying mammals which in turn have larger lungs than birds. The domestic species of birds like the domestic fowl have smaller weight-specific lungs (VLW^{-1}) than the wild (flying) ones. (Maina and King 1984, courtesy of Company of Biologists, UK)

needs. Discordance of scaling of DLo_2 was observed by Weibel (1984a) whereby No_2 scaled linearly with body mass (b = 0.95) and to VO_2 (b = 0.77). This suggests that large animals have greater DLo_2 in excess of their demands, a feature which runs counter to the expected tenets of symmorphosis (Sect. 3). While other factors may be involved, it appears possible (Weibel 1983, 1989, 1990) that larger animals have a lower alveolar partial pressure of O_2 due to slower diffusion of O_2 in their longer pulmonary acini. Lindstedt (1984) has further pointed out that as pulmonary erythrocyte transit time must scale disproportionately to body mass, DLo_2 must scale linearly to body mass to

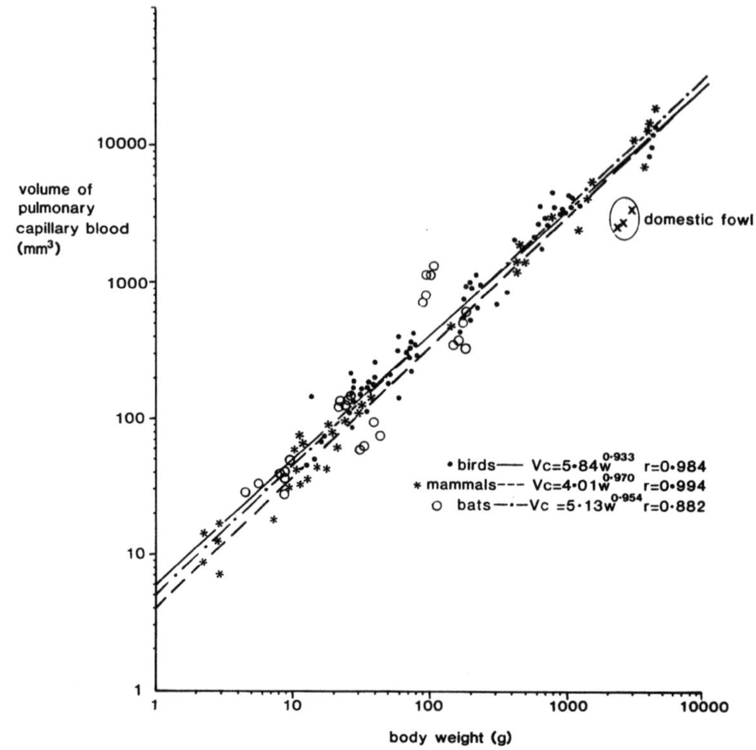

Fig. 17. A double logarithmic plot of the volume of the pulmonary capillary blood (*Vc*) against body weight (*W*) in birds, terrestrial (non-volant) mammals and bats. Birds in general have a larger *Vc* than terrestrial mammals and bats. *Vc* scales very strongly with *W* in the three groups of animals and the correlation coefficients (*r*) between *Vc* and *W* are notably strong. (Maina and King 1984, courtesy of Company of Biologists, UK)

ascertain adequate O_2 delivery in all mammals. Due to large cross-sectional areas, the transport of O_2 by mass flow in an acinus may be slower than by diffusion (Gomez 1965; Paiva 1985), a feature which would create a greater acinus stratification of PO_2 in the larger lungs. As $VO_2 = DLo_2PO_2$ ($PO_2 = PAO_2 - PCO_2$, where PAO_2 is the alveolar and PCO_2 the mean pulmonary capillary O_2 tension and PO_2 the partial pressure gradient or the driving force for O_2), DLo_2 can be varied by adjusting either VO_2 or PO_2. The PO_2 in the energetic animals (dogs and ponies) of 19 mm Hg was observed to be about 1.6 times greater than that in goats and calves (12 mm Hg) (Weibel et al. 1987). The high VO_2max in foxes (3.6 ml O_2 s^{-1} kg^{-1}) is largely attributed to the high PO_2 of about 40 mm Hg (Longworth et al. 1989). When compared with the physiological diffusing capacity (DLo_2p), the morphometric (DLo_2m) value has been observed to be consistently greater by a factor of 2 (Weibel et al. 1983; Weibel 1984a). The current explanation for this difference is that DLo_2 presents

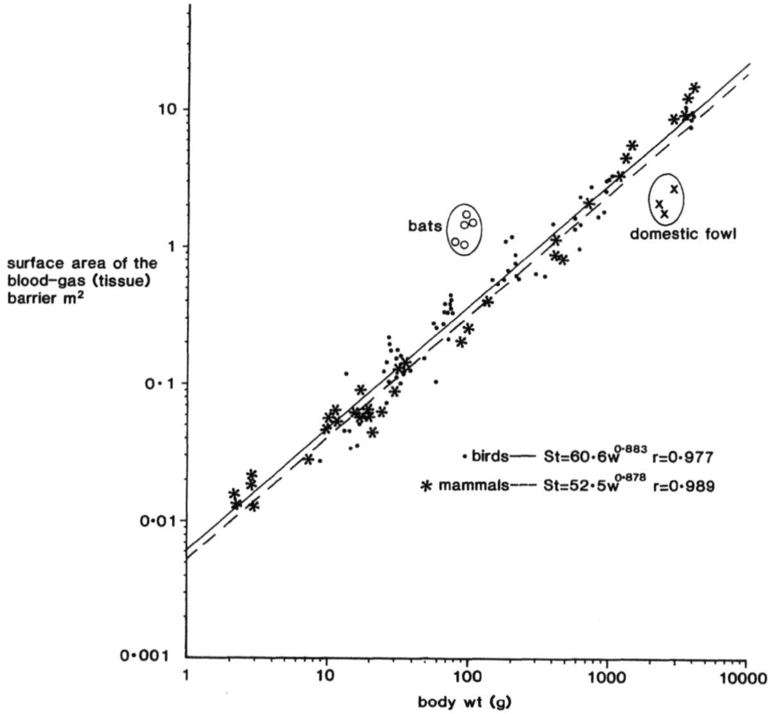

Fig. 18. Double logarithmic plot of the pulmonary respiratory surface area (*Sa*) against body weight (*W*) in birds and non-flying mammals. Birds have a higher surface area than non-flying mammals. $Sa\ W^{-1}$ in non-flying birds like the domestic fowl is notably lower than in non-flying mammals and volant birds, the bats having remarkablly high values. (Maina et al. 1989a, courtesy of the Royal Society of London)

the maximal O_2 transfer capacity of the lung when the ventillation–perfusion ratio (V/Q) is unity and a positive PO_2 (alveolar capillary driving force) exists across the entire BGB. This state rarely if ever prevails in .life but may be approached at VO_2max. The functional reserve of the lung may be viewed as a safety margin which facilitates gas transfer in all possible circumstances the animal may be exposed to. In *Gallus gallus*, a differential factor of about 2 between the morphometric difusing capacity (DLo_2m) (Abdalla et al. 1982) and the physiological diffusing capacity (DLo_2p) (Scheid and Piiper 1970; Meyeret al. 1977) was observed while in the lungless salamander, *Desmoganathus fuscus*, DLo_2m and DLo_2p (of the skin) are virtually identical (Gatz et al. 1975; Piiper and Scheid 1975; Piiper et al. 1976). Biological safety designs of the order of 1.5–2 are well known in other body systems (Alexander 1981).

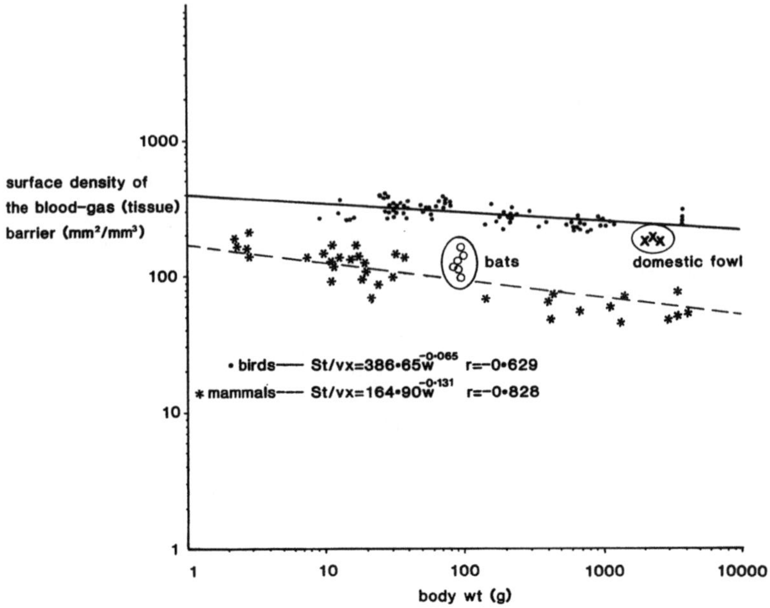

Fig. 19. A double logarithmic plot of the surface density of the blood-gas barrier in the parenchyma (S_VBGB, p) against body weight (W) in birds and non-flying mammals. Birds have a higher (S_VBGB, p) than non-flying mammals and the parameter scales negatively and very weakly with W in both groups of animals. (S_VBGB, p) is a relative indicator of the subdivision of the parenchyma and hence the size of the terminal respiratory air-spaces. (Maina et al. 1989a, courtesy of Royal Society of London)

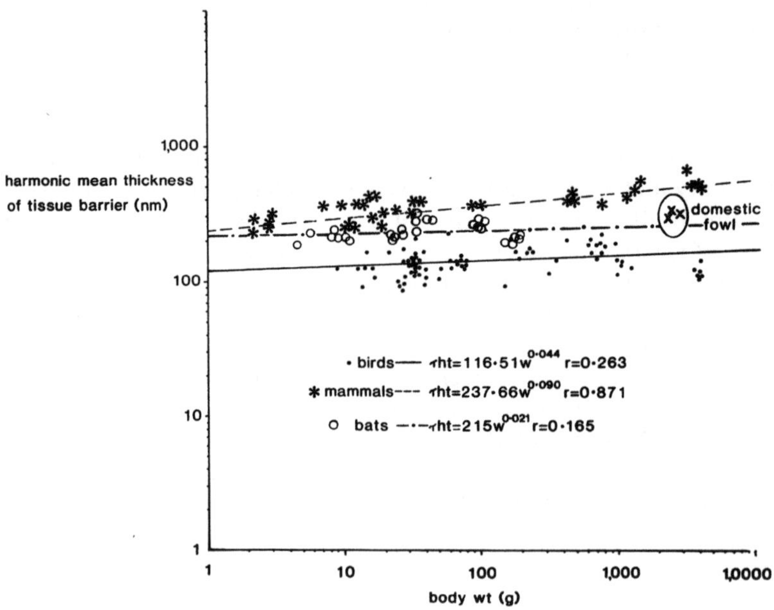

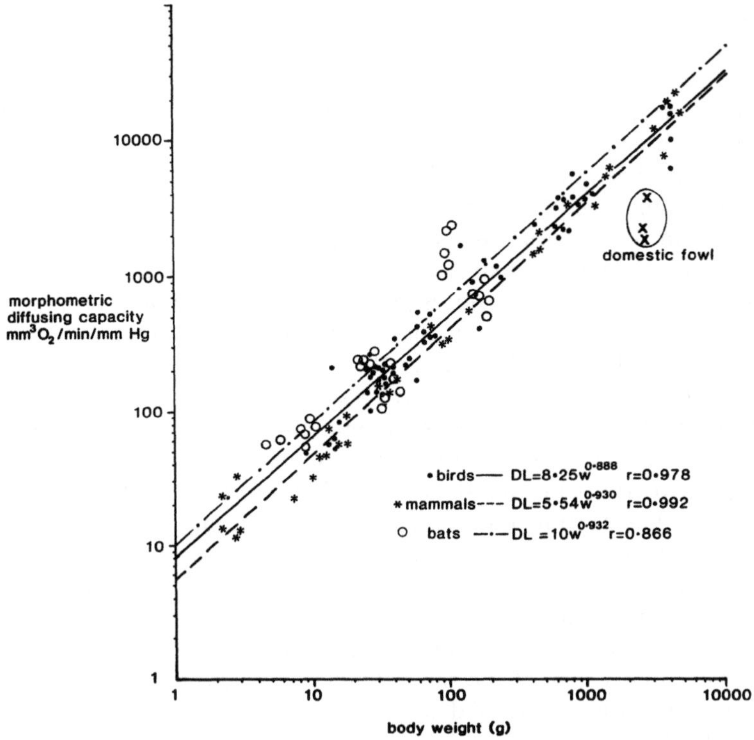

Fig. 21. A double logarithmic plot of the total morphometric pulmonary diffusing capacity (DLo_2) against body weight (W) in birds, non-flying mammals and bats. The parameter scales very strongly with W and the correlation coefficients (r) are strong between the two parameters. DLo_2 is the most comprehensive estimator of the structural capacity of the lung for gas exchange. Bats have higher DLo_2 than birds and non-flying mammals. (Maina and King 1984, courtesy of Company of Biologists, UK)

13.4 Birds

The oldest convincingly bird-like creature is the late Jurassic *Archaeopteryx lithographica* which has been aged at approximately 150 million years (de Beer 1954). The ancestry of Archaeopteryx and whether it was the immediate progenitor of modern birds is still debatable. *Archaeopteryx* was a close relative of the dinosaurs and probably arose either from the thecodonts or from the

Fig. 20. A double logarithmic plot of the harmonic mean thickness (τht) of the blood-gas barrier (BGB) against body weight (W) in birds, non-flying mammals and bats. Non-flying mammals have a thicker BGB than birds. τht scales very weakly with W in the three groups of animals and the correlation coefficients (r) are similarly weak except in the non-flying mammals. The weak scaling of τht with W is an indication of the optimization of this parameter. (Maina and King 1984, courtesy of Company of Biologists, UK) coelurosaurs, *Compsognathus* or *Ornitholestes*, which were bipedal fast-running reptiles (Ostrom 1974; Raath 1985). The close phylogenetic affinity between reptiles and birds accounts for their still evident anatomical similarities, including those of the respiratory system (Perry 1988). Arguably, *Archaeopteryx* is said to have been a heterothermic endotherm (Regal 1985), an adaptive feature which may have been essential for flight. The evolution of flight, a highly energetically demanding type of locomotion (Bartholomew and Lighton 1986; Butler 1991) but with distinct selective advantages, more than anything else should have greatly influenced the structure and function of the avian respiratory system. The system clearly was not a prerequisite to attainment of flight, as bats with a typical but refined mammalian respiratory system (Maina et al. 1982a, 1991; Maina and King 1984; Lechner 1985; Maina 1985b, 1986) and the insects with a tracheal system (Sect. 12.3) fly reasonably well, if not better, than birds. The VO_2 in the hovering Allen's hummingbird *(Selasphorus sasin)* of 70–80 ml $g^{-1} h^{-1}$ is the highest value reported in a vertebrate (Lasiewski 1963). So restrictive are the demands of flight that among the 9000 or so extant species there is less morphological variation than among the much fewer number of species in single taxa like Amphibia, Reptilia and Mammalia.

The lung air-sac system of birds is divided into two distinct regions namely the lung, a compact gas exchanger and the voluminous expansile air-sacs. The respiratory system constitutes 10–34% of the total body volume, the lungs constituting 1–3% of the same (Duncker 1972; Duncker and Guntert 1985a). The lungs are rather solid organs located at the dorsal part of the thoracic cavity, the cavum pulmonale, where they are firmly attached to the thoracic wall and ventrally subtended by the underlying membranous horizontal septum (King 1966; Duncker 1971, 1972, 1978a, b) and constitute 11–22% of the volume of the respiratory system (Dubach 1981; Duncker and Guntert 1985a, b). A stratified maze of bronchial system (Fig. 22a) pervades the lung and is continued to air-sacs (King 1966; Duncker 1971). The air-sacs are capacious transparent membranous air-bags which are contained in body spaces in some cases extending to pneumatize bones (King 1966). The parabronchi constitute 55–90% of the lung volume, and the air and blood capillaries (Fig. 22) 50–75% of the exchange tissue (Maina et al. 1982b; Powell and Mazzone 1983; Duncker and Guntert 1985a:). There are 300–500 parabronchi in both lungs of the domestic fowl (King and Cowie 1969). The parabronchial mantle ranges in diameter from 200–500 μm (Duncker 1972). Air capillaries have a width of 3–14 μm (Duncker 1972; Maina 1982a; Maina et al. 1982c; Duncker and Guntert 1985a), a size which is only possible in a non-compliant lung due to the enormous forces of surface tension which must exist on the gas-liquid interface of such small air spaces. Though during respiration the lung was estimated to change in volume on average by a mere 1.4% (Jones et al. 1985), on frozen goose lungs, Powell and Mazzone (1983) and Maina (1987c) obseved pleats and folds of the air and blood capillaries which, if not artifactual, would suggest that capillary volumes and surface areas may change under physiologic conditions and imply that the lungs may be finitely distensible. Macklem et al. (1979) observed that the parabronchi

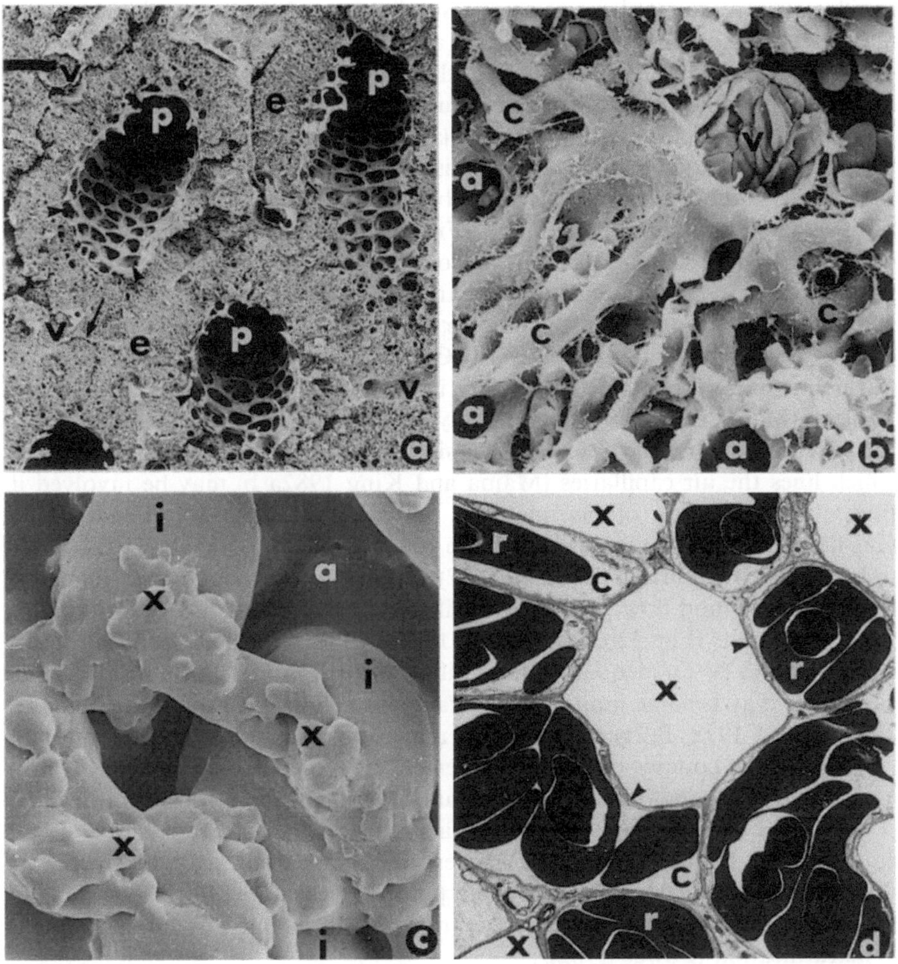

Fig. 22a–d. The avian lung. **a** and **b** are preparations from the lung of the emu (*Dromaius novaehollandiae*). **a** Parabronchial lumen (*p*), exchange tissue (*e*), blood vessels (*v*), interparabronchial septa (*arrows*), and atria (*arrowheads*), × 68; **b** large blood vessel (*v*) in the exchange tissue giving rise to blood capillaries (*c*) which intimately interdigitate with air capillaries (*a*), × 4010; **c** latex cast preparation of the lung of the domestic fowl *Gallus domesticus* showing atria (*a*), infundibula (*i*) and anastomosing air capillaries (*x*), × 230, **d** sectional view of the lung of *G. domesticus* showing air-capillaries (*x*), blood capillaries (*c*) containing red blood cells (*r*) and blood-gas barrier (*arrowheads*), × 3412. **c** from Maina 1988a, courtesy of Wiley/Alan R. Liss Inc. **d** Maina et al. 1989a, courtesy of the Royal Society of London)

and secondary bronchi of the duck lungs have a limited compliance which would potentially result in possible changes in lung volume during respiration but are not powerful enough to lead to a significant collapse of air capillaries. Experimental compression of the duck lung failed to produce any substantial collapse of the air capillaries (Macklem et al. 1979). The minuscule air and blood capillaries have resulted from intense subdivision of the gas-exchange tissue, maximizing the respiratory surface area and pulmonary capillary blood volume in the lung. The total blood volume in the avian lung makes up 15–36% of the lung volume, 58–80% of which is found in the capillaries (Duncker 1973; Dubach 1981; Abdalla et al. 1982; Duncker and Guntert 1985a, b). In mammals, the pulmonary blood capillary volume is 2.5–3 times less than that of a bird of similar size and only 20% of it is in the blood capillaries (Weibel 1963). Though birds have a smaller lung volume (Fig. 16) and a smaller volume of the parenchyma than mammals due to the lower volume density of the parenchyma (Table 4), birds have significantly greater ($P < 0.05$) respiratory surface area compared with mammals (Maina 1989e; Maina et al. 1989a, 1991). The surface density (S_V) of the blood-gas barrier (BGB) in the parenchyma (p), i.e. S_VBGB, p which is on average ten times higher in birds than in mammals (Fig. 19 and Table 4) (Maina et al. 1989a) empirically expresses the degree of subdivision of the parenchyma and the much smaller size of the air-capillaries. The surfactant which lines the air capillaries (Maina and King 1982a, b) may be involved in preventing exudation of plasma onto the respiratory surface (Fedde 1980). Gas-flow in the air capillaries, which are 200–500 μm long (Powell 1982), is said to take place by diffusion (Zeuthen 1942), but due to the unique arrangement of the capillary blood-flow to the air capillaries – two intimately interdigitating respiratory units (Fig. 22b, c, d) (Maina 1982a, 1988a) – during exercise ventilation-perfusion (V/Q) inequalities and diffusive air-flow resistance inside the air capillaries may become limiting in overall gas exchange (Crank and Gallanger 1978; Scheid 1978; Burger 1980). This may, however, be compensated for by factors like the counter-current arrangement of the gas exchange units (Piiper and Scheid 1973) and the slower pulmonary capillary blood transit time in birds (Powell and Mazzone 1983), which facilitates more time for blood – gas equilibration. Diffusive limitation in exercise may also be avoided if the blood-flow in the pulmonary capillaries were pulsatile, a feature which would create a convective process and enhance gas transport in the air capillaries, reducing O_2 stratification (Powell 1982).

Though the mechanical ventilation of the respiratory system in birds, except for the absence of a diaphragm, is essentially similar to that of mammals (Scheid 1979), the pattern of air-flow and the mode of gas exchange in the avian lung is remarkably different and more complicated than the mammalian one. Some important functional aspects, like the mechanism of unidirectional rectification of air-flow in the lung, gas-flow in the air capillaries and enervation and control of respiration are still enigmatic (Fedde 1980; Powell and Scheid 1989; Scheid and Piiper 1989). Recent studies suggest that aerodynamic valving, which is attributable to the geometry of the bronchial system, synchronized ventilation by the air-sacs, gas velocity and density and gas convective inertial forces, may explain the nature of intrapulmonary air-flow in the lung (Brackenbury 1972; Banzett et al. 1987; Butler et al. 1988; Kuethe 1988; Wang et al. 1988). Anatomical valves are lacking in the avian lung (King and Payne 1960). Most recently, however, Banzett et al. (1991) and Wang et al. (1992) described a constriction at the caudal end of the primary bronchus, which they have termed "segmentum accelerans" and associated it with aerodynamic valving by accelerating the incoming air-stream past the ventrobronchial openings. The periparabronchial (gas exchange) tissue is ventilated in a continuous and unidirectional manner (by synchronized bellows action of the air-sacs) (Brackenbury 1971; Bretz and Schmidt-Nielsen 1972) in contrast to the oscillatory pattern of the mammalian lung. This may maintain a high O_2 gradient and eliminate CO_2 buildup in the exchange tissue (Piiper and Scheid 1973; Scheid 1979). Intermittent ventilatory flow would have a greater effect on gas-exchange efficiency in the avian lung than in the mammalian one with its greater alveolar space (Piiper and Scheid 1973). In birds with a well-developed neopulmo, however, unidirectionality of air-flow in the parabronchi does not appear essential in achieving efficient gas exchange (Piiper 1985). The blood-flow in the exchange tissue is centrifugal (i.e. from the periphery towards the parabronchial lumen), while the air diffuses in opposite direction (i.e. centripetally) (Abdalla and King 1975; West et al. 1977; Maina 1988a, 1993). The spatial relationship between the flow of the deoxygenated blood and that of the bulk parabronchial air is, however, essentially perpendicular (Fig. 6. and 22a): the first arrangement constitutes a counter-current system and the second a cross-current one. The cross-current system was conclusively experimentally demonstrated by Scheid and Piiper (1972) on experimental reversal of the direction of air-flow which did not affect the partial pressures of gases in the inhaled air and blood, which should have occurred had the system been counter-current and hence converted to a co-current one. The counter-current arrangement has not been considered to be a true one in nature but an auxiliary one because the flow is only convective in the blood capillaries (Scheid 1979). The actual contribution of this system to the overall efficiency of the avian lung has not been established (Piiper and Scheid 1973) but it may be significant in gas exchange at high altitude flight and in hypoxic environments. The cross-current arrangement increases the interaction between air and blood along the parabronchial length enhancing gas-exchange efficiency. Birds remove more O_2 from inhaled air than mammals of equivalent body weight (Lasiewski and Calder 1971). From the additive effect of the serial multicapillary system of the cross-current design, in exercise or in a hypoxic environment, the PO_2 in the mixed end arterial blood (PAO_2) exceeds the PO_2 in the end expired air (PEO_2), while $PACO_2$ in turn exceeds $PECO_2$ (Scheid and Piiper 1970; Piiper and Scheid 1973; Powell et al. 1978; Bech and Johansen 1980) (Sect. 9). In resting normoxic birds, $PACO_2$ is less than $PECO_2$ but PAO_2 is usually not greater than PEO_2 due to V/Q mismatching and diffusive resistances or a combination of both factors (Powell 1982). As a mark of this efficiency, the $PACO_2$ in the resting chicken was estimated at 33 mm Hg (Kawashiro and Scheid 1975) and in exercising emu at 30 mm Hg (Jones et al. 1983), values much lower than the average ones of 40 mm Hg in the mammalian lung. In the mute swan (*Cygnus olor*), a pronounced positive PCO_2 difference between end-tidal gas (42.3 mm Hg) and mixed venous blood (32.5 mm Hg) averaging 9.8 mm Hg was observed (Bech and Johansen 1980). PO_2 in the exhaled air may drop to as low as 15% in birds and only to 17% in mammals (Tucker 1972; Schmidt-Nielsen 1975). In the "uniform pool" design of mammals (Piiper and Scheid 1973), arterial PO_2 is generally lower than alveolar PO_2 and only under ideal conditions, where V/Q inequalities are abolished, would the two values potentially equilibrate (Fig. 2D). Overlapping of PO_2 is only compatible in the counter-current system of gills (Fig. 2C) and the cross-current one (Fig. 2E) of birds (Sect. 9). The lung air-sac system of birds is said to be the most efficient respiratory system to have evolved among the air-breathing vertebrates (Weibel 1973; Tenney 1979). Birds in general have a lower breathing frequency than mammals (Lasiewski and Calder 1971; Schmidt-Nielsen 1975). Since the VO_2std in birds (non-passerine species) is similar to that of mammals (Lasiewski and Dawson 1967), it follows (as has been experimentally proven) that the avian respiratory system must be more efficient in extracting more O_2 than a mammalian one. The unique spatial designs, the cross-current and counter-current systems and an efficient mode of pulmonary ventilation and perfusion, are but some of the features which contribute to the high avian pulmonary efficiency.

Morphometrically, the avian lung is well adapted for O_2 uptake (Dubach 1981; Duncker and Guntert 1985a, b; Maina 1989e; Maina et al. 1989a). Whereas when compared with terrestrial mammal birds have smaller lungs by a factor of 27% (Fig. 16), a greater pulmonary capillary blood volume by 22% (Fig. 17), a more extensive respiratory area by a factor of 15% (Fig. 18) and a thinner BGB by a factor of 57% (Fig. 20) (Maina 1989e; Maina et al. 1989a). These specializations give the avian lung a DLo_2 22% higher than in mammals (Fig. 21). Adaptively, bats have superior lungs to those of birds (Maina et al. 1982a; Maina 1986, 1989e; Maina and King 1984; Maina et al. 1991). As expected, pulmonary variables in reptiles are remarkably inferior to those of birds and mammals (Duncker and Guntert 1985a). In birds, the magnitude of the pulmonary morphometric parameters correspond to their individual aerobic capacities (Maina 1982b): the small and energetic passeriform birds have better specialized lungs than the generally larger and less active non-passeriforms (Maina 1984). The emu *Dromaius novaehollandiae* (Maina and King 1989), a flightless bird which leads a relatively inactive life, has poorly adapted lungs, and diving birds like the penguins (Welsch and Aschauer 1986; Maina and King 1987) and the razorbills and guillemots (Maina 1987b) have specializations like a large pulmonary capillary blood volume. The thinnest BGBs in the air-breathing vertebrates have been reported in the rock martin (*Hirundo fuligula*) (0.090 μm), house sparrow (*Passer domesticus*) (0.096 μm) (Maina 1984) and the violet-eared hummingbird (*Colibri coruscans*) (0.099 μm) (Dubach 1981), while the thickest were in domestic species like the chicken (*Gallus gallus* var. *domesticus*) (0.34 μm) and the guinea fowl (*Numida meleagris*) (0.320 μm) (Abdalla and Maina 1981; Abdalla et al. 1982). The morphometric parameters in the lung of the domestic fowl are lower than those of the red jungle fowl, its wild-form progenitor (Vidyadaran et al. 1987, 1990); comparatively, when normalized with body weight the lung of the domestic fowl is 20% smaller, the BGB 28% thicker, and DLo_2 25% lower. These differences may have come about from the long-standing intense breeding of the domestic fowl for early maturity with concomitant deterioration of supporting pulmonary parameters. The smallest and more energetic species of birds have a more extensive respiratory surface area (Maina 1982b) with *Hirundo* having 87 cm^2 (Maina 1984) and *Colibri* 90 $cm^2\,g^{-1}$ (Dubach 1981): the value in *Gallus* is only 11 cm^2g^{-1} (Abdalla et al. 1982). Morphometric adaptations and spatial arrangement of the structural components of the avian lung are but some of its specializations for gas exchange. Other important features which contribute to its efficiency are a high cardiac output and a large tidal volume (Hart and Roy 1966; Lasiewski and Calder 1971). Interplay of these factors cumulatively led to an exceptionally highly efficient lung with remarkable functional reserves. It may explain the observations made by Bech et al. (1985) and Koteja (1986) that cold exposure did not result in any perceivable respiratory distress in birds. Brackenbury and Amaku (1990) noted that experimental isolation of the thoracic and abdominal air-sacs, a procedure which effectively removes 70% of the total air capacity, did not have a detrimental effect on the efficiency of the respiratory system and the birds could ran at three times the pre-exercise metabolic rate. The O_2 extraction coefficient, FIo_2 (the ratio of the amount of O_2 used to that available in air) was the same in exercising Pekin ducks in both normoxic and hypoxic conditions (Kiley et al. 1985). During flight, at ambient temperatures between 12–22 °C, the FIo_2 remains unchanged in the crows (Bernstein 1976) and starlings (Torre-Bueno 1978). The exceptionally long and deep dives (Kooyman et al. 1982) and high altitude flights made by some birds (Laybourne 1974; Black et al. 1978) are attributable to the great functional reserves of the respiratory system. The mute swan (Bech and Johansen 1980) has a high FIo_2 of 33%, an unusually large tidal volume, long breath holds, and the V/Q ratio which is large in most birds (Hastings and Powell 1986) is near unity (i.e. 1.06). In the bar-headed goose (*Anser indicus*), a champion high altitude flyer, neither ventilation nor pulmonary gas transfer were limiting in a bird exercising in a hypoxic environment (7% O_2) (Fedde et al. 1989). Shams and Scheid (1989) have, however, suggested that higher tolerance to hypoxia in birds (compared with mammals) is not due to a higher efficiency of the parabronchial compared with alveolar gas exchange, but may lie in the ability of the birds to tolerate lower arterial PCO_2 levels than mammals.

14 Conclusions

Although this account is by no means exhaustive in either depth or extent of such a vast subject, some salient conclusions can be drawn:

1. The terms advanced, successful or primitive when used in the wide field of natural history, more so in respiratory morphology and physiology without specifically defining the particular adaptive aspect considered, can often be misleading. The fancied phylogenetic tree which places mammals at the apex of

the animal kingdom is essentially contributive to such inferences. Highly perfected animals like birds and mammals are adapted to occupy only one environment (largely aerial). While they reap maximum benefit from such mastery, they are adaptively inflexible and hence vulnerable to changes. The more highly evolved an animal supposedly is, the narrower are its limits of tolerance to ambient changes. With time, it will have acquired a very firm hold on its habitat. For example, while the aquatic turtles and freshwater molluscs have undergone successful reversed evolution and adapted very well to aquatic life, it is difficult to envisage birds and mammals reverting to such a life. Bimodal breathers can derive benefits from two totally different habitats and hence occupy habitats entirely alien to the so called advanced animals. Such organisms are well prepared for survival in case of unpropitious environmental changes in any of their alternative ecological niches. Teleologically, those animals that have retained bimodal breathing may have purposely done so “to play it safe”.

2. The marked pulmonary morphological adverseness, and in some instances anachronism, exhibited by the extant organisms indicates the divergent responses to different external forces in animals at different phylogenetic levels of development.

3. The malleability of the respiratory system to ambient changes and the morphological and functional diversity of respiratory organs can be attributed to lack of ubiquitous tissue structure(s), a feature which otherwise may not be tenable in the phylogenetically wide cross-section of extant animals. Exposure of blood or haemolymph to air over a thin barrier (in any tissue) is essentially all that is required for gas exchange to occur.

4. The evolution of the respiratory systems and processes has been implicitly parsimonius in nature. It is clearly most expedient to modify an already existing organ than attempt the drastic and frequently risky process of evolving an entirely new one particularly when this entails (usually due to lack of space) total extirpation of the primary one. The common stratagem which appears to have been adopted is that of an organism retaining the primary respiratory organ(s) as the secondary one(s) evolved at a different site or a gradual phasing out and concomitant replacement of the original one. During this time, the skin and buccal cavity may have served as “bridging” respiratory organs. In the course of this transition, reversed evolution could easily have occurred. The change from a unimodal water-breathing fish to a bimodal stage and subsequently to a unimodal air-breather appears to be the pathway taken in the evolution of aerial-breathing and terrestriality. It shows a clear and gradual de-emphasis of the gills as respiratory organs: the coexistence stage is presented at the bimodal respiratory stage.

5. Since gas exchange occurs through wet surfaces (as O_2 and CO_2 have to dissolve in water to diffuse across a respiratory barrier), strictly the so-called air-breathers have not totally delinked the “aquatic umbilical cord”. The need for moisture on the respiratory surface is crucial and in fully terrestrial animals has called for distinct adaptations of the gas conducting passages to humidify and warm the inhaled air and on exhalation minimize water loss from the

respiratory microcosm. The amphibians and other animals which use the skin to a significant extent for gas exchange cannot survive for an extended period in a highly xeric environment.

6. Most of the pulmonary parameters generally scale proportionately with body mass (BM). However, the thickness of the blood – gas barrier in particular allometrically varies very little with increasing body mass. This indicates the significance of barrier thickness to gas exchange and hence the need for its optimization. Such refinement of a parameter in a wide cross-section of animals is a reflection of their adaptation to the same potential functional constraints during their evolutionary development.

7. Despite the appreciable differences between the lungs of the contemporary vertebrates, on closer scrutiny, there appears to be logical and fundamental principles on which the designs of the lungs are based. Conversion of one lung type to another could occur at any level of the developmental hierarchy, such flexibility being less likely the more refined the organ is. The words of L.J. Handerson befit the final words in this account, and I quote:

> .... after a half a century, there is no longer room for doubt that the fitness of organic beings for life in the world has been won in whole or in part by an almost infinite series of adaptations of life to its environment, whereby, through a corresponding series of transformations, present complexity has grown out of former simplicity.

Acknowledgments. I am grateful to Prof. (Emeritus) A.S. King for guiding me in the formative stages of my research career and exciting me to "look beyond birds ": My wife and our three children Ndegwa, Wanjiru and Kireru have been an unfailing inspiration to me through their support and understanding.

References

Abdalla MA, King AS (1975) The functional anatomy of the pulmonary circulation of the domestic fowl. Respir Physiol 23: 267–290

Abdalla MA, Maina JN (1981) Quantitative analysis of the exchange tissue of the avian lung (Galliformes). J Anat 133: 677–680

Abdalla MA, Maina JN, King AS, King DZ, Henry J (1982) Morphometrics of the avian lung. I. The domestic fowl *Gallus gallus* var. *domesticus.* Respir Physiol 47: 267–278

Abdel-Magid AM (1971) The ability of *Clarias lazera* (Pisces) to survive without air-breathing. J Zool (Lond) 163: 63–72

Agassiz LD (1857) Contributions to the natural history of the United States, vol I. Little Brown, Boston

Alexander McN (1981) Factors of safety in the structure of animals. Sci Prog (Oxf) 67: 109–130

Alexander McN (1982) Optima for animals. Arnold, London

Al-Wassa AH, Innes AJ, Whiteley NM, Taylor EW (1989) Aerial and aquatic respiration in the ghost crab *Ocypode saratan.* I. Fine structure of respiratory surfaces, their ventilation and perfusion, oxygen consumption and carbon dioxide production. Comp Biochem Physiol 94A: 755–764

Andersen HT (1961) Physiological adjustments to prolonged diving in American alligator, *Alligator mississipiensis.* Acta Physiol Scand 53: 23–45

Anderson DT (1973) Embryology and phylogeny in annelids and arthropods. Pergamon Press, Oxford
Anderson JF (1970) Metabolic rates of spiders. Comp Biochem Physiol 33: 51–72
Anderson JF, Prestwich KN (1980) Scaling subunit structure in book lungs of spiders (Araneae). J Morphol 165: 167–174
Anderson JF, Prestwich KN (1982) Respiratory gas exchange in spiders. Physiol Zool 55: 72–90
Aschoff J, Pohl H (1970) Rhythmic variations in energy metabolism. Fed Proc 29: 1541–1552
Assimacopoulus A, Guggenheim R, Kapanci Y (1976) Changes in alveolar capillary configuration at different levels of lung inflation in the rat. Lab Invest 34: 10–21
Awati PR, Tembe VB (1956) *Buthus tamulus* (Fabr), the Indian scorpion: morphology, anatomy and bionomics. Zool Monogr Univ (Bombay) 2: 1–62
Bailey L (1954) The respiratory currents in the tracheal system of the adult bee. J Exp Biol 31: 589–595
Bakker RT (1975) Dinosaur renaissance. Sci Am 232: 58–78
Banzett RB, Butler PJ, Nations CS, Barnas GM, Lehr JL, Jones HJ (1987) Inspiratory aerodynamic valving in goose lungs depends on gas density and velocity. Respir Physiol 70: 287–300
Banzett RB, Nations CS, Wang N, Fredberg JJ, Butler JP (1991) Pressure profiles show features essential to aerodynamic valving in geese. Respir Physiol 84: 295–309
Barnhart MC (1986) Respiratory gas tensions and gas exchange in active and dormant land snails, *Otala lactea*. Physiol Zool 59: 733–745
Barrel J (1916) Influence of Silurian-Devonian climates on the rise of air-breathing vertebrates. Bull Geol Soc Am 27: 387–436
Bartels H, Welsch U (1984) Freeze fracture of the turtle lung 2. Rod shaped particles in the plasma membrane of a mitochondria-rich pneumocyte in *Pseudemys* (*Chrysemys*) *scripta*. Cell Tissue Res 236: 453–457
Bartholomew GA, Barnhart CM (1984) Tracheal gases, respiratory gas exchange, body temperature and flight in some tropical cicadas. J Exp Biol 111: 131–144
Bartholomew GA, Lighton JRB (1986) Oxygen consumption during hover feeding in free ranging Anna hummingbird. J Exp Biol 123: 191–199
Bartlett D, Mortola JP, Doll EJ (1986) Respiratory mechanics and control of the ventilatory cycle in the garter snake. Respir Physiol 64: 13–27
Baskin DG, Detmers PA (1976) Electron microscopic study of the gill bars of Amphioxus (*Brachiostoma californiense*) with special reference to neurociliary control. Cell Tissue Res 166: 167–178
Baudrimont A (1955) Organisation générale du poumon et structure des alvéoles pulmonaires des vertèbres (amphibiens, reptiliens mammifères) considerés dans leurs rapports avec la mécanique respiratoire, la circulation pulmonaire fonctionelle et l'activité métabolique de ces animaux. Arch Anat Histol Embryol 38: 99–136
Baumgarten-Schumann D, Piiper J (1968) Gas exchange in the gills of resting unaesethetized dogfish (*Scyliorhinus stellaris*). Respir Physiol 5 317–325
Bayne BL, Bayne CJ, Carefoot TC, Thompson RJ (1976) The physiological ecology of *Mytilus californicus* Conrad. 1. Metabolism and energy balance. Oecologia 22: 211–228
Beadle LC (1957) Respiration in the African swampworm *Alma emini* (Mich). J Exp Biol 34: 1–10
Bech C, Johansen K (1980) Ventilation and gas exchange in the mute swan, *Cygnus olor*. Respir Physiol 39: 285–295
Bech C, Rautenberg W, May B (1985) Ventilatory oxygen extraction during cold exposure in the pigeon (*Columba livia*). J Exp Biol 116: 499–502
Belkin DA (1963) Anoxia tolerance in reptiles. Science 139: 492–493
Belkin DA (1964) Variations in heart rate during voluntary diving in the turtle *Pseudemys concinna*. Copeia 2: 321–330
Belkin DA (1968) Aquatic respiration and under water survival of two fresh water turtle species. Respir Physiol 4: 1–14

Bell PB, Stark-Vancs VI (1983) Scanning electron microscopic study of the microarchitecture of the lung of the giant salamander *Amphiuma tridactylum*. Scan Electr Microsc 1983: 449–456
Benerjee V (1966) Blood of some Indian vertebrates. Naturwissenschaften 53: 1370
Bennett AF (1972) A comparison of activities of metabolic enzymes in lizards and rat. Comp Biochem Physiol B 42: 637–647
Bennett AF (1988a) Structural and functional determinates of metabolic rate. Am Zool 28: 699–708
Bennett MB (1988b) Morphometric analysis of the gills of the European eel *Anguilla anguilla*. J Zool (Lond) 215: 549–560
Bennett AF, Dawson WR (1976) Metabolism. In: Gans C, Dawson WR (eds) Biology of the reptilia, vol 5. Academic Press, New York, pp 127–223
Bennett AF, Naggy KA (1977) Energy expenditure in free ranging lizards. Ecology 58: 697–700
Bennett AF, Ruben JA (1979) Endothermy and activity in vertebrates. Science 206: 649–654
Bennett FM, Tenney SM (1982) Comparative mechanics of mammalian respiratory system. Respir Physiol 49: 131–140
Berg T, Steen JB (1965) Physiological mechanisms for aerial respiration in the eel. Comp Biochem Physiol 15: 469–484
Berg T, Steen JB (1966) Regulation of ventilation in eels exposed to air. Comp Biochem Physiol 18: 511–516
Bernstein MH (1976) Ventilation and respiratory evaporation in the flying crow, *Corvus ossifragus*. Respir Physiol 26: 371–382
Bertin L (1958) Organes de la respiration aerienne. In: Grasse PD (ed) Traité de Zoologie, vol 13. Massonet Cie, Paris, pp 1363–1398
Bishop IR, Foxon GEH (1968) The mechanism of breathing in the South American lungfish, *Lepidosiren paradoxa*; a radiological study. J Zool (Lond) 154: 262–271
Black CP, Tenney SM, Kroonenburg M (1978) Oxygen transport during progressive hypoxia in bar-headed geese (*Anser indicus*) acclimated to sea level and 5600 m. In: Piiper J (ed) Respiratory function in birds, adult and embryonic. Springer, Berlin Heidelberg New York, pp 78–83
Black EC, Fry FEJ, Black VS (1954) The influence of carbon dioxide on the utilization of oxygen by some fresh water fish. Can J Zool 32: 408–420
Bliss DE (1968) Transition from water to land in decapod crustacea. Am Zool 8: 355–392
Boresi AP (1955) A refinement of the theory of the buckling of rings under uniform pressure. J Appl Mech 22: 95–102
Boutilier RG (1990) Control and co-ordination of gas exchange in bimodal breathers. In: Boutilier RG (ed) Advances in comparative and environmental physiology, vol 6. Vertebrate gas exchange from environment to cell. Springer, Berlin Heidelberg New York, pp 279–345
Bouverot P, Douguet D, Sebert P (1979) Role of the arterial chemoreceptors in the ventilatory adaptations to hypoxia in awake Pekin ducks. J Comp Physiol 133B: 177–186
Brackenbury JH (1971) Air-flow dynamics in the avian lung as determined by direct and indirect methods. Respir Physiol 13: 319–329
Brackenbury JH (1972) Lung air-sac anatomy and respiratory pressures in the bird. J Exp Biol 57: 543–550
Brackenbury JH, Amaku J (1990) Effects of combined abdominal and thoracic air sac occlusion on respiration in domestic fowl. J Exp Biol 152: 93–100
Brett JR (1964) The respiratory metabolism and swimming performance of the young sockeye salmon. J Fish Res 21: 118–126
Bretz WL, Schmidt-Nielsen K (1972) The movement of gas in the respiratory system of the duck. J Exp Biol 56: 57–65
Bridges CR, Kester P, Scheid P (1980) Tracheal volume in the pupa of the saturniid moth *Hyalophora cecropia* determined with inert gases. Respir Physiol 40: 281–291
Brooks RE (1970) Ultrastructure of the physostomatous swimbladder of rainbow trout *Salmo gairdneri*. Z Zellforsch Mikrosk Anat 106: 473–483

Browman MW, Kramer DL (1985) *Pangasius sutchi* (Pangasiidae), an air breathing catfish uses the swimbladder as an accessory organ. Copeia 1985: 994–998
Buck JB (1948) The anatomy and physiology of the light organs in fire-flies. Ann NY Acad Sci 49: 397–482
Buck JB (1962) Physical aspects of insect respiration. Annu Rev Entomol 7: 27–56
Buck JB, Keister M (1958) Cyclic CO_2 release in diapausing pupae. II. Tracheal anatomy, volume and PCO_2, blood volume, interburst CO_2 release rate. J Insect Physiol 1: 327–340
Burger R (1980) Respiratory gas exchange and control in the chickens. Poult Sci 59: 2654–2665
Burggren WW, Johansen K (1986) Circulation and respiration in lungfishes (Dipnoi). J Morphol (Suppl) 1: 217–236
Burggren WW, McMahon BR (1983) An analysis of scaphognathite pumping performance in the crayfish *Orconectes virilis*: compensatory changes to acute and chronic exposure. Physiol Zool 56: 309–318
Burggren WW, McMahon BR (1988) Circulation. In: Burggren BB, McMahon BR (eds) Biology of the crabs. Cambridge University Press, Cambridge, pp 298–332
Burggren WW, Moalli R (1984) "Active" regulation of cutaneous gas exchange by capillary recruitment in amphibians: experimental evidence and a revised model for skin respiration. Respir Physiol 55: 379–392
Burggren WW, McMahon BR, Costerton JW (1974) Branchial water and blood flow patterns and the structure of the gill of the crayfish *Procambarus clarkii*. Can J Zool 52: 1511–1518
Burggren WW, Glass ML, Johansen K (1978) Intrapulmonary variation of gas partial pressure and ventilation inequalities in chelonian reptiles. J Comp Physiol 126: 203–209
Burggren WW, Dupre RK, Wood SC (1987) Allometry of red cell oxygen binding and hematology in larvae of the salamander *Ambryostoma tigrinum*. Respir Physiol 70: 73–84
Burnett LE, McMahon BR (1987) Gas exchange, haemolymph acid base status and the role of branchial water stores during air exposure in three littoral crab species. Physiol Zool 60: 27–36
Burri PH (1984) Lung development and histogenesis. In: Fishman AP, Fischer AB (eds) Handbook of physiology; respiration, vol 4. American Physiological Society Press, New York, pp 1–46
Burri PH (1985) Morphology and respiratory function of the alveolar unit. Int Arch Allergy Appl Immunol 76: 1–12
Burri PH, Sehovic S (1979) The adaptive responses of the rat lung after bilobectomy. Am Rev Respir Dis 119: 769–777
Burri PH, Weibel ER (1971) Morphometric estimation of pulmonary diffusion capacity. II. Effect of PO_2 on the growing lung. Adaptation of the growing rat lung to hypoxia and hyperoxia. Respir Physiol 11: 247–264
Bursell E (1970) An introduction to insect physiology. Academic Press, London
Butler PJ (1991) Exercise in birds. J Exp Biol 160: 233–262
Butler PJ, Metcalfe JD (1983) Control of respiration and circulation. In: Rankin JC, Pitcher TJ, Duggan R (eds) Control processes in fish physiology. Croom Helm, Beckenham, pp 41–65
Butler PJ, Banzett RB, Fredberg JJ (1988) Inspiratory valving in avian bronchi: aerodynamic considerations. Respir Physiol 72: 241–256
Calder WA (1984) Size, function and life history. Harvard University Press, Cambridge, Mass
Callahan PS (1972) The evolution of insects. Holliday House New York
Cameron JN, Mecklenburg TA (1973) Aerial gas exchange in the coconut crab *Birgus latro* with some notes on *Gecaroidea lalandii*. Respir Physiol 19: 245–261
Carey FG (1982) Warm fish. In: Taylor CR, Johansen K, Bolis L (eds) Animal physiology. Cambridge University Press, Cambridge, pp 216–233
Carey FG, Teal JM (1969) Regulation of body temperature by the blue-fin tuna. Comp Biochem Physiol 28: 205–213
Carey FG, Teal JM, Kanwisher JW, Lawson KD (1971) Warm-blooded fish. Am Zool 11: 135–145
Carter GS (1931) Aquatic and aerial respiration in animals. Biol Rev 6: 1–35

Carter GS (1935) The fresh waters of the rain forest areas of British Guiana. J Linn Soc (Lond) 39: 147–193
Carter GS (1955) The papyrus swamps of Uganda. Heffer, Cambridge
Carter GS (1957) Air-breathing. In: Brown ME (ed) The physiology of fishes. Academic Press, London, pp 65–79
Carter GS (1967) Structure and habitat in vertebrate evolution. Washington University Press, Seattle
Carter GS, Beadle LC (1931) The fauna of swamps of Paraguayan Chaco in relation to its environment. II. Respiratory adaptations in the fishes. J Linn Soc (Lond) 37: 327–368
Casey TM, May ML, Morgan KR (1985) Flight energetics of euglossine bees in relation to morphology and wing stroke frequency. J Exp Biol 116: 271–289
Cheatum EP (1934) Limnological investigations on respiration, annual migratory cycle and other related phenomena in freshwater pulmonate snails. Trans Am Microsc Soc 53: 348–407
Cloud P (1974) Evolution of ecosystems. Am Sci 62: 54–66
Coates CW (1937) Slowly the lungfish gives up its secrets. Bull NY Zool Soc 40: 25–34
Collip JP (1921) A further study of the respiratory processes in *Mya arenaria* and other marine Mollusca. J Biol Chem 49: 297–310
Constantinopol M, Jones JH, Weibel ER, Taylor CR, Lindholm A, Karas RH (1989) Oxygen transport during exercise in large mammals. II. Oxygen uptake by the pulmonary gas exchanger. J Appl Physiol 67: 871–878
Cope ED (1894) On the lungs of Ophidia. Am Philos Soc 33: 217–224
Copeland DE (1968) Fine structure of salt and water uptake in the land crab *Gecarcinus lateralis*. Am Zool 8: 417–432
Coulson RA, Hernandez T (1980) Oxygen debt in reptiles: relationship between the time required for repayment and metabolic rate. Comp Biochem Physiol 65A: 453–457
Courtice GP (1981) The effect of temperature on bimodal gas exchange and the respiratory ratio in the water dragon, *Physognathus lesuerii*. Comp Biochem Physiol 68A: 437–441
Cowles RG (1947) Comments by readers. Science 105: 282
Cragg P (1975) Respiration and body weight in the reptilian genus Lacerta: a physiological, anatomical and morphometric study. PhD Thesis, University of Bristol
Crank WD, Gallanger RR (1978) Theory of gas exchange in the avian parabronchus. Respir Physiol 35: 9–25
Crapo JD, Barry BE, Gehr P, Bachofen M, Weibel ER (1982) Cell number and cell characteristics of the normal human lung. Am Rev Respir Dis 125: 332–337
Crawford EC, Gatz RN, Magnussen H, Perry SF (1976) Lung volumes, pulmonary blood flow and carbon monoxide diffusing capacity of turtles. J Comp Physiol 107: 169–178
Crompton AW, Taylor CR, Jagger AJ (1978) Evolution of homoiotherm in mammals. Nature 272: 333–336
Czopek J (1955) Vascularization of the respiratory surfaces in *Leiopema hochstetteri* (Fitzinger) and *Xenopus laevis* (Daudin). Acta Anat 25: 346–360
Czopek J (1959) Vascularization of respiratory surfaces in *Salamandra salamandra* L. in ontogeny. Bull Acad Pol Sci B 7: 473–477
Czopek J (1961) Vascularization of the respiratory surfaces of some Plethodontidae. Zool Pol 11: 131–148
Czopek J (1962a) Vascularization of respiratory surfaces in some caudata. Copeia 1962: 576–587
Czopek J (1962b) Smooth muscle in the lungs of some urodeles. Nature 193: 798
Czopek J (1965) Quantitative studies on the morphology of respiratory surfaces in amphibians. Acta Anat 62: 296–323
Dahr E (1927) Studien über die Respiration der Landpulmonaten. Lunds Univ Aarsskr 23: 1–20
Das BK (1927) The bionomics of certain air-breathing fishes of India together with an account of the development of the air-breathing organs. Philos Trans R Soc Lond B 216: 183–219
Das BK (1940) Nature and causes of evolution and adaptation of the air-breathing fishes. Proc 27th Indian Sci Congr 2: 215–260

Davenport J (1985) Environmental stress and behavioural adaptation. Croom Helm, London
Dawson TJ, Hubert AJ (1970) Standard metabolism, body temperature and surface areas of Australian marsupials. Am J Physiol 218: 1233–1238
de Beer G (1954) *Archeopteryx lithographica*. Br Mus Nat Hist, London
de Fur PL, McMahon BR (1978) Respiratory responses of *Cancer productus* to air exposure. Am Zool 18: 605A
Dehadrai PV (1962) Respiratory function of the swimbladder of *Notopterus* (Lacepede). Nature (Lond) 185: 929
Dehadrai PV, Tripathi SD (1976) Environment and ecology of freshwater air-breathing teleosts. In: Hughes GM (ed) Respiration of amphibious vertebrates. Academic Press, London, pp 139–172
Dejours P (1976) Water versus air as the respiratory media. In: Hughes GM (ed) Respiration of amphibious vertebrates. Academic Press, London, pp 1–15
Dejours P (1979) Oxygen demand and gas exchange. In: Wood SC, Lenfant C (eds) Evolution of respiratory processes: a comparative approach. Dekker, New York, pp 1–49
Dejours P (1988) Respiration in water and air. Elsevier, Amsterdam
Dejours P, Garey WF, Rahn H (1970) Comparison of ventilatory and circulatory flow rates between animals in various physiological conditions. Respir Physiol 9: 108–117
Demoll R (1927) Untersuchungen über die Atmung der Insekten. Z Biol 87: 8–22
Denison RH (1941) The soft anatomy of *Bothriolepis*. J Palaeontol 15: 553–561
de Villiers CJ, Hodgson C (1987) The structure of the secondary gills of *Siphonaria capensis* (Gastropoda; Pulmonata). J Molluscan Stud 53: 129–138
Diaz H, Rodriguez G (1977) The branchial chamber in terrestrial crabs: a comparative study. Biol Bull Mar Biol Lab Woods Hole 153: 485–504
Dormans JAMA (1985) The alveolar type III cell. Lung 163: 327–335
Dubach M (1981) Quantitative analysis of the respiratory system of the house sparrow, budgerigar, and hummingbird. Respir Physiol 46: 43–60
Dubale MS (1951) A comparative study of the extent of gill surface in some representative Indian fishes. J Univ Bombay 19: 90–101
Dubale MS (1959) A comparative study of the oxygen carrying capacity of the blood in water and air-breathing teleosts. J Anim Morphol Physiol 6: 48–54
Duncker HR (1971) The lung air sac system of birds. A contribution to the functional anatomy of the respiratory system. Ergeb Anat Entwicklungsgesch 45: 1–171
Duncker HR (1972) Structure of avian lungs. Respir Physiol 14: 44–63
Duncker HR (1973) Der quantitative Aufbau des Lungen-Luftsacksystems der Vögel. Verh Anat Ges 67: 197–204
Duncker HR (1978a) Development of the avian respiratory systems. In: Piiper J (ed) Respiratory function in birds, adult and embryonic. Springer, Berlin Heidelberg New York, pp 260–273
Duncker HR (1978b) General morphological principles of amniotic lungs. In: Piiper J (ed) Respiratory function in birds, adult and embryonic. Springer, Berlin Heidelberg New York, pp 1–15
Duncker HR, Guntert MD (1985a) The quantitative design of the avian respiratory system-from hummingbird to the mute swan. In: Nachtigall W (ed) BIONA report 3, Publ Acad Wiss Lit Mainz. Fischer, Stuttgart, pp 361–378
Duncker HR, Guntert MD (1985b) Morphometric analysis of the avian respiratory system. In: Duncker HR, Gleischer G (eds) Vertebrate morphology, Forsch Zool 3. Fischer, Stuttgart, pp 383–387
Dunham DW, Gilchrist SL (1988) Behaviour. In: Burggren WW, McMahon BR (eds) Biology of the land crabs. Cambridge University Press, Cambridge, pp 97–138
Dunn JF, Davison W, Maloiy GMO, Hochachka PW, Guppy M (1981) An ultrastructural and histochemical study of the axial musculature in the African lungfish. Cell Tissue Res 220: 599–609
Dunson WA (1986) Aquatic respiration in *Trionyx spiniferus asper*. Herpetologia 16: 277–283
Edney EB (1960) Terrestrial adaptations. In: Waterman TH (ed) The physiology of Crustacea, vol 1. Academic Press, London, pp 367–393

Edwards GA, Ruska H, de Harven E (1958) The fine structure of insect tracheoblasts, tracheae and tracheoles. Arch Biol 69: 351–369
Else PL, Hubert AJ (1981) Comparison of the "mammal machine" and the "reptile machine": energy production. Am J Physiol 240: R3–R9
Else PL, Hubert AJ (1985a) An allometric comparison of the mitochondria of mammalian and reptilian tissues: the implications for the evolution of endothermy. J Comp Physiol B 156: 3–11
Else PL, Hubert AJ (1985b) Mammals: an allometric study of metabolism at tissue and mitochondrial level. Am J Physiol 248: R415–421
Emery SH, Szczepanski A (1986) Gill dimensions in pelagic elasmobranch fishes. Biol Bull 171: 441–449
Emilio MG, Shelton G (1974) Gas exchange and its effect on blood gas concentrations in the amphibian, *Xenopus laevis*. J Exp Biol 60: 567–579
Eshky AA, Atkinson RJ, Taylor AC (1988) Effects of temperature on oxygen consumption and heart rate in the semiterrestrial crab *Ocypode saratan* (Forskal). Mar Behav Biol 13: 341–358
Fange R (1953) The mechanism of gas transport in the euphysoclist swimbladder. Acta Physiol Scand (Suppl 30) 110: 1–33
Fange R (1966) Physiology of the swimbladder. Physiol Rev 46: 299–322
Fange R (1976) Gas exchange in the swimbladder. In: Hughes GM (ed) Respiration of amphibious vertebrates. Academic Press, London, pp 189–211
Fange R (1983) Gas exchange in fish swimbladder. Physiol Biochem Pharmacol 97: 111–158
Farber J, Rahn H (1970) Gas exchange between air and water and the ventilation pattern in the electric eel. Respir Physiol 9: 151–161
Farber J, Tenney SM (1971) The pouch gas of the Virginia opossum (*Didelphis virginiana*). Respir Physiol 11: 335–345
Farhi L, Rahn H (1960) Dynamic changes in carbon dioxide stores. Anaesthesiology 21: 604–614
Farley RD (1990) Regulation of air and blood flow through the booklungs of the desert scorpion, *Paruroctonus mesaensis*. Tissue Cell 22: 547–569
Farrelly CA, Greenaway P (1987) The morphology and vasculature of the lungs and gills of the soldier crab *Mictyris longicarpus*. J Morphol 193: 285–304
Farrelly CA, Greenaway P (1992) Morphology and ultrastructure of the gills of the terrestrial crabs (Crustacea, Gecarcinidae and Grapsidae): Adaptations for air-breathing. Zoomorphology 112: 39–49
Fedde MR (1980) Structure and gas flow pattern in the avian respiratory system. Poult Sci 59: 2642–2653
Fedde MR, Orr JA, Shams H, Scheid P (1989) Cardiopulmonary function in exercising bar-headed geese during normoxia and hypoxia. Respir Physiol 77: 239–262
Fincke T, Paul R (1989) Book lung function in arachnids. III. The function and control of the spiracles. J Comp Physiol 159B: 433–441
Fish GR (1956) Some aspects of the respiration of six species of fish from Uganda. J Exp Biol 33: 186–195
Fishman AP, Pack AI, Delany RG, Galante RJ (1986) Estivation in *Protopterus*. J Morphol (Suppl 1) 1: 237–248
Florkin M, Jeuniaux C (1974) Haemolymph composition. In: Rockestein M (ed) The physiology of insects, vol V, 2nd edn. Academic Press, New York, pp 256–307
Foelix RF (1979) Biologie der Spinnen. Thieme, Stuttgart
Foelix RF (1982) Biology of spiders. Harvard University Press, Cambridge, MA
Fons R, Sicart R (1976) Contribution à la connaissance du métabolisme énergétique chez deux Crocidurinae: *Suncus etruscus* (Savi, 1822) et *Crocidura russula* (Hermann, 1780), Insectivora, Soricidae. Mammalia 40: 229–311
Forster RE (1964) Diffusion of gases. In: Fenn WO, Rahn H (eds) Handbook of respiration, vol 1, sect 3. Respiration. Am Physiol Soc, Washington, pp 839–872
Fox HM (1921) Methods of studying the respiratory exchange in small aquatic organisms, with particular reference to the use of flagellates as an indicator for oxygen consumption. J Gen Physiol 3: 365–573

Fraenkel G, Hertford GVB (1938) The respiration of insects through the skin. J Exp Biol 15: 266–280
Fuhrmann O (1912) Le genre *Typhlonectes.* Mem Soc Neuchatel Sci Nat 5: 112–138
Full RJ (1987) Locomotion energetics of the ghost crab. I. Metabolic cost and endurance. J Exp Biol 130: 137–153
Fung YB, Sobin S (1977) Pulmonary alveolar blood flow. In: West JB (ed) Bioengineering aspects of the lung. Dekker, New York, pp 267–359
Gans C (1970) Strategy and sequence in the evolution of the external exchangers of ectothermal vertebrates. Forma Functio 3: 61–104
Gans C (1971) Respiration in early tetrapods – the frog is a red herring. Evolution (Lancaster-Pa) 24: 723–734
Gans C (1976) Ventilatory mechanisms and problems in some amphibious aspiration-breathers (Chelydra, Caiman-Reptilia). In: Hughes GM (ed) Respiration of amphibious vertebrates. Academic Press, London, pp 357–374
Gans C (1978) Ventilation mechanisms: problems in evaluating the transition to birds. In: Piiper J (ed) Respiratory function in birds, adult and embryonic. Springer, Berlin Heidelberg New York, pp 16–22
Gans C (1979) Momentarily excessive construction as the basis for protoadaptation. Evolution 33: 227–233
Gans C (ed) (1986) The biology of the lungfishes. J Morphol (Suppl 1) 1: 1–383
Gans C, Northcutt RG (1983) Neural crest and the origin of vertebrates: a new head. Science 220: 268–274
Garland T, Huey RB (1987) Testing symmorphosis: Does structure match functional requirements? Evolution 41: 1404–1409
Gatten RE (1974) Effects of temperature and activity on aerobic and anaerobic metabolism and heart rate in turtles *Pseudemys scripta* and *Terrapene ornata.* Comp Biochem Physiol 48A: 619–648
Gatten RE (1987) Activity metabolism of anuran amphibians: tolerance to dehydration. Physiol Zool 60: 576–585
Gatz RN, Crawford EC, Piiper J (1974) Respiratory properties of the blood of a lungless and gill-less salamander *Desmognathus fuscus.* Respir Physiol 20: 33–41
Gatz RN, Crawford EC, Piiper J (1975) Kinetics of inert gas equilibration in an exclusively skin-breathing salamander, *Desmognathus fuscus.* Respir Physiol 24: 15–29
Gee JH (1976) Buoyancy and aerial respiration: factors influencing the evolution of reduced swimbladder volume of some Central American catfishes (Trichomycteridae, Callichthyidae, Loricariidae, Astroblepidae). Can J Zool 54: 1030–1037
Geelhaar A, Weibel ER (1971) Morphometric estimation of pulmonary diffusing capacity. III. The effect of increased oxygen consumption in Japanese waltzing mice. Respir Physiol 11: 354–366
Gehlbach FR, Gordon R, Jordon JB (1973) Aestivation of salamander, *Siren intermedia.* Am Midl Nat 89: 455–463
Gehr P, Crapo JD (1988) Morphometric analysis of the gas exchange region of the lung. In: Gardner DE, Crap JD, Massaro EJ (eds) Toxicology of the lung. Raven Press, New York, pp 1–42
Gehr P, Erni H (1980) Morphometric estimation of pulmonary diffusion capacity in two horse lungs. Respir Physiol 41: 199–210
Gehr P, Bachofen, M, Weibel ER (1978) The normal human lung: ultrastructure and morphometric estimation of diffusion capacity. Respir Physiol 32: 121–140
Gehr P, Sehovic S, Burri PH, Claasen H, Weibel ER (1980) The lung of shrews: morphometric estimation of diffusion capacity. Respir Physiol 40: 33–47
Gehr P, Mwangi DK, Amman A, Maloiy GMO, Taylor CR, Weibel ER (1981) Design of the mammalian respiratory system. V. Scaling morphometric pulmonary diffusing capacity to body mass: wild and domestic animals. Respir Physiol 44: 61–86
George JC, Shah RV (1956) Comparative morphology of the lung in snakes with remarks on the evolution of the lung in reptiles. J Anim Morphol Physiol 3: 1–7
George JC, Shah RV (1965) Evolution of air sacs in Sauropsida. J Anim Morphol Physiol (Bombay) 12: 255–263

Ghiretti F (1966) Respiration. In: Wilbur KM, Yonge CM (eds) Physiology of Mollusca, vol 2. Academic Press, New York, pp 175–298
Ghiretti F, Ghiretti M (1975) Respiration. In: Fretter VV, Peake J (eds) Pulmonates, vol V. Academic Press, London, pp 33–52
Girgis S (1961) Aquatic respiration in the common Nile turtle, *Trionyx triunguis* (Forskal). Comp Biochem Physiol 3: 206–217
Glass ML, Johansen K (1981) Pulmonary diffusing capacity in reptiles (relations to temperature and oxygen uptake). J Comp Physiol 107: 169–178
Glass ML, Wood SC (1983) Gas exchange control of breathing in reptiles. Physiol Rev 63: 232–260
Gleeson TT (1991) Patterns of metabolic recovery from exercise in amphibians and reptiles. J Exp Biol 160: 187–207
Gomez DM (1965) A physical mathematical study of lung function in normal subject and in patients with obstructive pulmonary disease. Med Thorac 22: 275–294
Goncalves AA, Sawaya P (1978) Oxygen uptake by *Typhlonectes compressicaudatus* related to body weight. Comp Biochem Physiol 61A: 141–143
Goniakowska-Witalinska L (1973) Metabolism, resistance to hypotonic solutions and ultrastructure of erythrocytes of five amphibian species. Acta Biol (Cracov) 16: 113–134
Goniakowska-Witalinska L (1978) Ultrastructure and morphometric study of the lung of the European salamander, *Salamandra salamandra* L. Cell Tissue Res 191: 343–356
Goniakowska-Witalinska L (1980a) Ultrastructural and morphometric changes in the lung of newt *Triturus cristatus carnifex* Laur during ontogeny. J Anat 130: 571–585
Goniakowska-Witalinska L (1980b) Scanning and transmission electron microscopic study of the lung of the newt, *Triturus alpestris* Laur. Cell Tissue Res 205: 133–145
Goniakowska-Witalinska L (1982) Development of the larval lung of *Salamandra salamandra* L. Anat Embryol 164: 113–137
Goniakowska-Witalinska L (1985) The structure of the lung of the tree frog *Hyla arborea* L. In: Duncker HR, Fleischer G (eds) Vertebrate morphology. Fischer, Giessen, pp 375–377
Goniakowska-Witalinska L (1986) Lung of the tree frog, *Hyla arborea* L. A scanning and transmission electron microscopic study. Anat Embryol 174: 379–389
Gordon MS, Boetius I, Evans DH, McCarthey R, Oglesby LC (1969) Aspects of the physiology of terrestrial life in amphibious fishes. I. The mudskipper *Periophthalmus sobrinus*. J Exp Biol 50: 141–149
Gordon MS, Gabaldon DJ, Yip AYW (1985) Exploratory observations on the microhabitat selection within the intertidal zone by the Chinese mudskipper *Periophthalmus cantonensis*. Mar Biol 85: 209–215
Gould SJ, Lewontin RC (1979) The spandrels of San Macro and the Panglossian paradigm: a critique of adaptationist programme. Proc R Soc Lond B 205: 581–598
Graham JB (1973) Terrestrial life of the amphibious fish *Mnierpes macrocephalus*. Mar Biol 37: 83–91
Graham JB (1974) Aquatic respiration in the sea-snake *Pelamis platurus*. Respir Physiol 21: 1–7
Graham JB (1976) Respiratory adaptations of marine air breathing fishes. In: Hughes GM (ed) Respiration of amphibious vertebrates. Academic Press, London, pp 165–187
Graham JB (1983) The transition of air-breathing in fishes. II. Effects of hypoxia acclimation on the bimodal gas exchange of *Ancistrus chagresi* (Locariidae). J Exp Biol 102: 157–173
Graham JB, Baird TA (1982) The transition to air-breathing in fishes. I. Environmental effects on the facultative air-breathing of *Ancistrus chagresi* and *Hypostomus plecoatomus* (Loricariidae). J Exp Biol 96: 53–67
Graham JB, Baird TA (1984) The transition to air-breathing in fishes. III. Effects of body size and aquatic hypoxia on the aerial gas exchange in the swamp eel *Synbranchus marmoratus*. J Exp Biol 108: 357–375
Graham JB, Kramer DL, Pineda E (1977) Respiration of the air-breathing fish *Piabucina festae*. J Comp Physiol 122B: 295–310
Graham JB, Baird TA, Stockmann W (1987) The transition to air-breathing in fishes. IV. Impact of branchial specializations for air-breathing on the aquatic respiratory mechanisms and ventilatory costs of the swamp eel *Synbranchus marmoratus*. J Exp Biol 129: 83–106

Gratz RK, Hutchison HV (1977) Energetics for activity in the diamond-back watersnake, *Natrix rhombifera*. Physiol Zool 50: 99–114
Gratz RN, Crawford EC, Piiper J (1974) Metabolic and heart rate response of the plethodontid salamander *Desmognathus fuscus* to hypoxia. Respir Physiol 20: 43–49
Gray IE (1954) Comparative study of the gill area of marine fishes. Biol Bull Mar Biol Lab (Woods Hole) 107: 219–225
Gray IE (1957) A comparative study of the gill area of crabs. Biol Bull Mar Biol Lab (Woods Hole) 112: 34–42
Green SL (1984) Ultrastructure and enervation of the swimbladder of *Tetractenos glaber* (Tetraodontidae). Cell Tissue Res 237: 277–284
Greenaway P (1984) The relative importance of the gills and lungs in the gas exchange of amphibious crabs of the genus *Holthuisana*. Aust J Zool 8: 355–392
Greenaway P (1988) Ionic and water balance. In: Burggren WW, McMahon BR (eds) Biology of the land crabs. Cambridge University Press, Cambridge, pp 211–248
Greenaway P, Farrelly C (1990) Vasculature of the gas exchange organs in air-breathing brachyurans. Physiol Zool 63: 117–139
Greenaway P, Morris S, McMahon BR (1988) Adaptations to a terrestrial existence by the robber crab *Birgus latro*. II. In vivo respiratory gas exchange and transport. J Exp Biol 149: 493–509
Greenwood PH (1986) The natural history of the African lungfishes. J Morphol (Suppl) 1: 163–179
Gregor DS, Prior DJ (1986) Modification of cardiac activity in response to dehydration in the terrestrial slugs, *Limax maximus*. J Exp Biol 237: 185–190
Grigg GC (1965a) Studies on the Queensland lungfish, *Neoceratodus forsteri* (Krefft). I. Anatomy, histology, and functioning of the lung. Aust J Zool 13: 243–253
Grigg GC (1965b) Studies on the Queensland lungfish, *Neoceratodus forsteri* (Krefft). III. Aerial respiration in relation to habits. Aust J Zool 13: 413–421
Guimond RV, Hutchison HV (1968) The effect of temperature and photoperiod on gas exchange in the leopard frog, *Rana pipiens*. Comp Biochem Physiol 27: 177–195
Guimond RW, Hutchison HV (1972) Pulmonary, brachial and cutaneous gas exchange in the mud puppy, *Necturus maculosus maculosus* (Rafinesque). Comp Biochem Physiol 42A: 367–392
Guimond RW, Hutchison HV (1973) Trimodal gas exchange in the large aquatic salamander *Siren lacertina* L. Comp Biochem Physiol 46A: 249–268
Guimond RW, Hutchison HV (1976) Gas exchange of giant salamanders. In: Hughes GM (ed) Respiration in amphibious vertebrates. Academic Press, London, pp 313–338
Gutman WF, Bonik K (1981) Kritische Evolutionstheorie. Gerstenberg, Hildesheim
Hakim A, Munshi JSD, Hughes GM (1978) Morphometrics of the respiratory organs of the Indian green snake headed fish *Channa punctata*. J Zool (Lond) 184: 519–543
Hallam JF, Dawson TJ, Holland RAB (1989) Gas exchange in the lung of a dasyurid marsupial: Morphometric estimation of diffusion capacity and blood oxygen uptake kinetics. Respir Physiol 77: 309–322
Hardley NF, Hill RD (1969) Oxygen consumption of the scorpion *Centruroides sculpturatus*. Comp Biochem Physiol 33: 51–72
Harms JW (1932) Die Realisation von Genen und die konsekutive Adaptation. II. *Birgus latro* L. als Landkrebs und seine Beziehungen zu den Coenabiten. Z Wiss Zool 140: 167–290
Hart JS, Roy OZ (1966) Respiratory and cardiac responses to flight in pigeons. Physiol Zool 39: 291–306
Hartnool RG (1988) Evolution, systematics and geographic distribution. In: Burggren WW, McMahon BR (eds) Biology of land crabs. Cambridge University Press, New York, pp 6–54
Harvey EN (1928) The oxygen consumption of luminous bacteria. J Gen Physiol 11: 469–475
Harvey HW (1960) The chemistry and fertility of sea waters. Cambridge University Press, London
Hastings RH, Powell FL (1986) Physiological dead space and effective parabronchial ventilation in ducks. J Appl Physiol 60: 85–91

Hazelhoff EH (1939) Über die Ausnutzung des Sauerstoffs bei verschiedenen Wassertieren. Z Vgl Physiol 26: 306: 327
Heath JE, Heath M (1982) Energetics of locomotion in endothermic insects. Annu Rev Physiol 44: 133–143
Heinrich B (1977) Why have some animals evolved to regulate a high body temperature. Am Nat 111: 623–640
Heinrich B (1981) Insect thermoregulation. Wiley, New York
Heinrich B (1982) Thermoregulation in endothermic insects. Science 185: 747–756
Heinrich B, Bartholomew GA (1972) Temperature control in flying moths. Sci Am 226: 71–77
Heisler N (1982) Intracellular and extracellular acid-base regulation in the tropical freshwater teleost fish *Synbranchus marmoratus* in response to transition from water-breathing to air-breathing. J Exp Biol 99: 9–28
Henk WG, Haldiman JT (1990) Microanatomy of the lung of the bowhead whale *Balaena mysticetus*. Anat Rec 226: 187–197
Herreid CF (1980) Hypoxia in invertebrates. Comp Biochem Physiol 67A: 311–320
Herreid CF, Full RJ (1988) Energetics and locomotion. In: Burggren WW, McMahon BR (eds) Biology of the land crabs. Cambridge University Press, Cambridge, pp 333–377
Herreid CF, Lee LW, Spampata R (1981) How do spiders breathe? Am Zool 21: 917
Herrnkind WF (1983) Movement patterns and orientation. In: Vernberg FJ, Vernberg WB (eds) Biology of Crustacea, vol 7. Academic Press, London, pp 41–105
Heusner AA (1982) Energy metabolism and body size. I. Is the 0.75 mass exponent of Kleiber's equation a statistical artifact? Respir Physiol 48: 1–12
Heusner AA (1983) Body size, energy metabolism and the lung. J Appl Physiol 54: 867–873
Hexter SH (1982) Lung book microstructure in *Tegenaria* sp. Bull Br Arachnol Soc 5: 323–326
Hill DE (1977) Some observations on the physiology of living *Lyssomanes viridis* which should apply to the Araneae in general. Peckhamia 1: 41–44
Hillman SS (1987a) Dehydrational effects of cardiovascular and metabolic capacity of two amphibians. Physiol Zool 60: 608–613
Hillman SS (1987b) The roles of oxygen delivery and electrolyte levels in the dehydrational death of *Xenopus laevis*. J Comp Physiol 128: 169–178
Hillman SS, Withers PC, Hedrick MS, Kimmel PB (1985) The effects of erythrocythemia on blood viscosity, maximal systemic oxygen transport capacity and maximal rates of oxygen consumption in an amphibian. J Comp Physiol 155: 577–581
Hills BA, Hughes GM, Koyama T (1982) Oxygenation and deoxygenation kinetics of red cells in isolated lamellae of fish. J Exp Biol 98: 269–275
Hinton HE (1966) Plastron respiration in marine insects. Nature 209: 200–221
Hlastala MP, Standaert TA, Pierson DJ, Luchtel DL (1985) The matching of ventilation and perfusion in the lung of the tegu lizard *Tupinambis nigropunctus*. Respir Physiol 60: 277–294
Hochachka PW (1979) Cell metabolism, air-breathing, and the origins of endothermy. In: Wood SC, Lenfant C (eds) Evolution of respiratory processes: a comparative approach. Dekker, New York, pp 253–288
Hochachka PW, Guppy M (1987) Metabolic arrest and the control of biological time. Harvard University Press, Cambridge (Mass)
Hochachka PW, Fields J, Mustafa T (1973) Animal life without oxygen: basic biochemical mechanisms. Am Zool 13: 543–555
Hodge AJ (1955) Studies on the structure of muscle. III. Phase contrast and electron microscopy of Dipteran flight muscle. J Biophys Biochem Cytol 1: 361–380
Holeton GF (1970) Oxygen uptake and circulation by a haemoglobinless Antarctic fish (*Chaenocephalus aceratus* Lonnberg) compared with 3 red-blooded Antarctic fish. Comp Biochem Physiol 34: 457–471
Hora SL (1922) Structural modifications in the fish. Q Rev Biol 21: 46–58
Hora SL (1933) Respiration in fishes. J Bombay Nat Hist Soc 36: 538–560
Hora SL (1935) Physiology, bionomics and evolution of the air-breathing fishes of India. Trans Natl Inst Sci India 1: 1–16
Howell BJ (1970) Acid-base balance in transition from water to air-breathing. Fed Proc 29: 1130–1134

Hubert AJ (1978) The evolution of energy metabolism in mammals. In: Schmidt-Nielsen K, Bolis L, Taylor CR (eds) Comparative physiology: primitive mammals. Cambridge University Press, Cambridge, pp 13–46
Hughes GM (1966a) Dimensions of fish gills in relation to their function. J Exp Biol 45: 177–195
Hughes GM (1966b) Evolution between air and water. In: de Reuck AVS, Porter R (eds) CIBA Foundation Symposium on development of the lung. Churchill, London, pp 64–80
Hughes GM (1967) Experiments on the respiration of the trigger fish (*Balistes capriscus*). Experientia 23: 1077
Hughes GM (1970) Ultrastructure of the air breathing organs of some lower vertebrates. Proc 17th Int Congr Electr Microsc (Grenoble) 3: 599
Hughes GM (1972a) Distribution of oxygen tensions in the blood and water along the secondary lamella of icefish. J Exp Biol 56: 481–492
Hughes GM (1972b) Morphometrics of fish gills. Respir Physiol 14: 1–26
Hughes GM (1973) Ultrastructure of the lung of *Neoceratodus* and *Lepidosiren* in relation to the lung of the vertebrates. Folia Morphol (Prague) 21: 155–161
Hughes GM (1978) Some features of gas transfer in fish. Bull Inst Math Appl 14: 39–43
Hughes GM (1980) Ultrastructure and morphometry of the gills of *Latimeria chalumnae* and a comparison with the gills of associated fishes. Proc R Soc Lond B 208: 309–328
Hughes GM (1981) Fish gills – past, present and future. Biol Bull India 3: 69–87
Hughes GM (1984) General anatomy of the gills. In: Randall DJ (ed) Fish physiology, vol XA. Academic Press, London, pp 1–72
Hughes GM, Hills BA (1971) Oxygen tension distribution in water and blood at the secondary lamella of the dogfish gill. J Exp Biol 55: 399–408
Hughes GM, Iwai T (1978) A morphometric study of the gills in some Pacific deep sea fishes. J Zool (Lond) 184: 155–170
Hughes GM, Morgan M (1973) The structure of fish gills in relation to their respiratory function. Biol Rev 48: 419–475
Hughes GM, Munshi JSD (1968) Fine structure of the respiratory surfaces of an air-breathing fish, the climbing perch *A. testudineus* (*Bloch*). *Nature* 219: 1382–1384
Hughes GM, Munshi JSD (1973a) Fine structure of the respiratory organs of the climbing perch *Anabas testudineus* (Pisces: Anabantidae). J Zool (Lond) 170: 201–225
Hughes GM, Munshi JSD (1973b) Nature of the air-breathing organs of the Indian fishes *Channa, Amphipnous, Clarias*, and *Saccobranchus* as shown by electron microscopy. J Zool (Lond) 170: 245–270
Hughes GM, Singh BR (1970) Respiration in air-breathing fish the climbing perch *Anabas testudineus* (Bloch). I. Oxygen uptake and carbon dioxide release into air and water. J Exp Biol 53: 265–280
Hughes GM, Singh BN (1971) Gas-exchange with air and water in an air-breathing catfish *Saccobranchus fossilis*. J Exp Biol 55: 667–682
Hughes GM, Vergara GA (1978) Static pressure-volume curves for the lung of the frog (*Rana pipiens*). J Exp Biol 76: 149–165
Hughes GM, Weibel ER (1976) Morphometry of fish lungs. In: Hughes GM (ed) Respiration of amphibious vertebrates. Academic Press, London, pp 213–232
Hughes GM, Weibel ER (1978) Visualization of layers lining the lung of the South American lungfish (*Lepidosiren paradoxa*) and a comparison with the frog and rat. Tissue Cell 10: 343–353
Hughes GM, Wright DE (1970) A comparative study of the ultrastructure of the water-blood pathway in the secondary lamellae of teleost and elasmobranch fishes – benthic forms. Z Zellforsch Mikrosk Anat 104: 478–493
Hughes GM, Knights B, Scammell CA (1969) The distribution of oxygen partial pressure and hydrostatic pressure changes within the branchial chambers in relation to gill ventilation of the shore crab, *Carcinus maenas* L. J Exp Biol 51: 203–220
Hughes GM, Dube SC, Munshi JSD (1973) Surface area of the respiratory organs of the climbing perch, *Anabas testudineus* (Pisces: Anabantidae). J Zool (Lond) 170: 227–243

Hughes GM, Singh BR, Guha G, Dube SC, Munshi JSD (1974a) Respiratory surface area of an air-breathing siluroid fish *Saccobranchus* (= *Heteropneustes*) *fossilis* in relation to body size. J Zool (Lond) 172: 215–232
Hughes GM, Singh BR, Thakur RN, Munshi JSD (1974b) Areas of the air-breathing surfaces of *Amphipnous cuchia* (Ham). Proc Indian Natl Sci Acad 40: 379–392
Hunter WR (1953) The condition of the mantle cavity in two pulmonate snails in Loch Lomond. Proc R Soc Edinb 65B: 143–165
Hutchison VH (1968) Relation of body size and surface area to gas exchange in anurans. Physiol Zool 41: 65–85
Hutchison VH, Dowling HG, Vinegar A (1966) Thermoregulation in a brooding female Indian python, *Python morulus bivittatus*. Science 151: 694–696
Hutchison VH, Whitford WG, Kohl M (1968) Relation of body size and surface area to gas exchange in anurans. Physiol Zool 41: 65–85
Hyman LH (1967) The invertebrates, vol XI. Mollusca I. McGraw-Hill, New York
Inger RF (1957) Ecological aspects of the origin of the tetrapods. Evolution 11: 373–376
Innes AJ, Taylor EW (1986) A functional analysis of pulmonary, cutaneous and brachial gas exchange in the Trinidad mountain crab: a comparison with other land crabs. J Physiol 373: 46P
Innes AJ, Forster ME, Jones MB, Mardsen ID, Taylor HH (1986) Bimodal respiration, water balance and acid base regulation in a high shore crab *Cyclograpsus lavaux* H. J Exp Mar Biol Ecol 100: 127–145
Ishii Y, Hasegawas S, Uchiyama Y (1989) Twenty four hour variations in subcellular structures of the rat type II alveolar epithelial cell: A morphometric study at the electron microscopic level. Cell Tissue Res 256: 347–353
Ishimatsu A, Itazawa Y (1983) Difference in blood oxygen levels in the outflow vessels of the heart of an air-breathing fish *Channa argus*: do separate streams exist in a teleost heart? Comp Physiol 149: 435–440
Ishimatsu A, Itazawa Y, Takeda T (1979) On the circulatory systems of the snake head, *Channa maculata* and *C. argus* with reference to bimodal breathing. Jpn J Ichthyol 26: 167–180
Jakubowski M (1963) The structure and vascularization of the skin of the river bullhead (*Cottus gobio* L.) and black sea turbot (*Rhombus aeoticus*, Pall). Acta Biol Cracov Ser Zool 6: 159–175
Jaskinski A (1973) Air-blood barrier in the respiratory intestine of the pond loach, *Misgurnas fossalis*. Acta Anat 86: 376–393
Jesse MJ, Shub J, Fishman AP (1967) Lung and gill ventilation on the African lungfish. Respir Physiol 3: 267–287
Jeuken M (1957) A study of the respiration of *Misgurnus fossilis* (L.), the pond loach. Thesis, Univ Leiden
Johansen K (1966) Air-breathing in the teleost *Synbranchus marmoratus*. Comp Biochem Physiol 18: 383–395
Johansen K (1970) Air-breathing in fish. In: Hoar WS, Randall DJ (eds) Fish physiology, vol 4. Academic Press, New York, pp 361–411
Johansen K (1982) Respiratory gas exchange of vertebrate gills. In: Houlihan DF, Rankin JC, Shuttleworth TJ (eds) Gills. Cambridge University Press, Cambridge, pp 99–128
Johansen K, Hanson D (1968) Functional anatomy of the hearts of lungfishes and amphibians. Am Zool 8: 191–210
Johansen K, Lenfant K (1967) Respiratory function of in the South American lungfish, *Lepidosiren paradoxa* (Fitz). J Exp Biol 46: 205–218
Johansen K, Lenfant C (1968) Respiration in the African lungfish, *Protopterus aethiopicus*. II. Control of breathing. J Exp Biol 49: 453–468
Johansen K, Lenfant C (1972) A comparative approach to the adaptability of oxygen affinity. In: Roth M, Astrup P (eds) Oxygen affinity and red cell acid-base status. Proc A Benzon Symp IV. Munksgaard, Copenhagen, pp 750–780
Johansen K, Reite O (1967) Effects of acetylcholine and biogenic amines on pulmonary smooth muscle in African lungfish *P. aethiopicus*. Acta Physiol Scand 71: 248–252

Johansen K, Reite OB (1968) Influence of acetylcholine and biogenic amines on pulmonary smooth muscle in the African lungfish, *Protopterus aethiopicus*. Acta Physiol Scand 71: 248–252
Johansen K, Lenfant C, Griggs CG (1967) Respiratory control in the lungfish, *Neoceratodus forsteri* (Krefft). Comp Biochem Physiol 20: 835–854
Johansen K, Lenfant C, Hanson D (1968a) Cardiovascular dynamics in the lungfishes. Z Vgl Physiol 59: 157–186
Johansen K, Lenfant C, Schmidt-Nielsen K (1968b) Gas exchange and control of breathing in the electric eel *Electrophorus electricus*. Z Vgl Physiol 61: 137–163
Johansen K, Hanson D, Lenfant C (1970) Respiration in primitive air-breather *Amia calva*. Respir Physiol 9: 162–174
Johansen K, Maloiy GMO, Lykebboe G (1975) A fish in extreme alkalinity. Respir Physiol 24: 159–162
Johansen K, Lomholt JP, Maloiy GMO (1976a) Importance of air and water breathing in relation to size of the African lungfish *Protopterus amphibius* (Peters). J Exp Biol 65: 395–399
Johansen K, Lykkeboe G, Weber RE, Maloiy GMO (1976b) Respiratory properties of blood in awake and aestivating lungfish, *Protopterus amphibius*. Respir Physiol 27: 335–345
Johansen K, Lykkeboe G, Kornerup S, Maloiy GMO (1980) Temperature insensitive oxygen binding in blood of the tree frog *Chiromantis petersi*. J Comp Physiol 136: 71–76
Johnston IA, Bernard LM, Maloiy GMO (1983) Aquatic and aerial respiration rates, muscle capillary supply and mitochondrial volume density in the air-breathing catfish (*Clarias mossambicus*) acclimated either to air or hypoxic water. J Exp Biol 105: 317–338
Jones DR, Schwarzfeld T (1974) The oxygen cost to the metabolism and efficiency of breathing in trout *Salmo gairdneri*. Respir Physiol 21: 241–254
Jones HD (1983) The circulatory systems of gastropods and bivalves. In: Saleudin W, Wilbur K (eds) The Molluscs. Academic Press, New York, pp 189–238
Jones JH, Grubb B, Schmidt-Nielsen K (1983) Panting in the emu causes arterial hypoxemia. Respir Physiol 54: 189–195
Jones JH, Effman P, Scheid P (1985) Lung volume changes during respiration in ducks. Respir Physiol 59: 15–25
Kardong KV (1972) Morphology of the respiratory system and its musculature in different snake genera (part I), *Crotalus* and *Elaphe*. Gegenbaurs Morphol Jahrb 117: 285–302
Kawashiro T, Scheid P (1975) Arterial blood gases in undisturbed resting birds: measurements in chicken and duck. Respir Physiol 23: 337–342
Kemp A (1986) The biology of the Australian lungfish, *Neoceratodus forsteri* (Krefft 1870). J Morphol (Suppl 1) 1: 181–198
Kiley JP, Faraci FM, Fedde MR (1985) Gas exchange during exercise in hypoxic ducks. Respir Physiol 59: 105–115
King As (1966) Structural and functional aspects of the avian lung and its air-sacs. Int Rev Gen Exp Zool 2: 171–267
King AS, Cowie AF (1969) The functional anatomy of the bronchial muscle of bird. J Anat 105: 323–326
King AS, Payne DC (1960) Does air circulate in the avian lung? Anat Rec 136: 223
Kirsch R, Nonnotte G (1977) Cutaneous respiration in three freshwater teleosts. Respir Physiol 29: 339–354
Kjellesvig-Waering EN (1986) A restudy of the fossil scorpionida of the world. Palaeontogr Am 55: 1–287
Klauber LM (1956) Rattlesnakes. University of California, Berkeley
Kleiber M (1961) The fire of life. Wiley, New York
Klemm RD, Gatz RM, Westfall TA, Fedde MR (1979) Microanatomy of the lung parenchyma of a tegu lizard *Tupinambis nigropunctus*. J Morphol 161: 257–280
Klika E, Lelek A (1967) A contribution to the study of the lungs of *Protopterus annectens* and *Polypterus senegalensis*. Folia Morphol 15: 168–175
Knight JH (1984) Studies on the biology and biochemistry of *Pholas dactylus*. PhD Thesis, University of London

Knight JH, Knight R (1986) The blood vascular system of the gills of *Pholas dactylus* L. (Mollusca, Bivalvia, Eulamellibrachia). Philos Trans R Soc Lond B 313: 509–523
Kooyman GL, Davis RW, Croxall JP, Costa DP (1982) Diving depths and energy requirements of king penguins. Science 217: 726–727
Koteja P (1986) Maximum cold induced oxygen consumption in the house-sparrow *Passer domesticus* L. Physiol Zool 59: 43–48
Krogh A (1904) On the cutaneous and pulmonary respiration of the frog. Skand Arch Physiol 15: 328–419
Krogh A (1919) The rate of diffusion of gases through animal tissues with some remarks on the coefficient of invasion. J Physiol (Lond) 52: 391–408
Krogh A (1922a) Studien Über Tracheen Respiration, II. Über Gasdiffusion in den Tracheen. Pfluegers Archiv Gesamte Physiol Menschen Tiere 179: 97–112
Krogh A (1922b) Studien Über der Tracheen Respiration, III. Die Kombination von mechanischer Ventilation mit Gasdiffusion nach Versuchen an Dytiscuslarven. Pfluegers Archiv Gesamte Physiol Menschen Tiere 179: 113–120
Krogh A (1941) The comparative physiology of respiratory mechanics. University Pennsylvania Press, Philadelphia
Krogh A (1959) The anatomy and physiology of the capillaries. Hafner, New York
Kuethe DO (1988) Fluid mechanical valving of air-flow in bird lungs. J Exp Biol 136: 1–12
Lahiri S, Szidon JP, Fishman AP (1970) Potential respiratory and circulatory adjustments to hypoxia in the African lungfish. Fed Proc 29: 1141–1148
Lasiewski RC (1963) The energetic cost of small size in hummingbirds. Proc XIII Int Ornithol Congr 10: 1095–1103
Lasiewski RC, Calder WR (1967) A re-examination of the relation between standard metabolic rate and body weight in birds. Condor 69: 13–23
Lasiewski RC, Calder WA (1971) A preliminary allometric analysis of respiratory variables in resting birds. Respir Physiol 11: 152–166
Lasiewski RC, Dawson WR (1967) A re-examination of the relationship between standard metabolic rate and body weight in birds. Condor 69: 12–23
Lauder G, Liem K (1983) The evolution and interrelationships of the actinopterygian fishes. Bull Mus Comp Zool 150: 95–197
Laurent P (1982) Structure of vertebrate gills. In: Houlihan DF, Rankin JC, Shuttleworth TJ (eds) Gills. Cambridge University Press, Cambridge, pp 25–43
Laurent P, Perry SF (1991) Environmental effects on fish gill morphology. Physiol Zool 64: 4–25
Laybourne RC (1974) Collision between a vulture and an aircraft at an altitude of 37 000 ft. Wilson Bull 86: 461–462
Lechner AJ (1985) Pulmonary design in a microchiroptera bat *Pipistrellus pipistrellus* during hibernation. Respir Physiol 59: 301–312
Lechner AJ, Banchero N (1980) Lung morphometry in guinea pigs acclimated to cold during growth. J Appl Physiol Respir Environ Excercise Physiol 48: 886–891
Lenfant C, Johansen K (1966) Respiratory function in the elasmobranch *Squalus suckleyi* G. Respir Physiol 1: 13–29
Lenfant C, Johansen K (1967) Respiratory adaptations in selected amphibians. Respir Physiol 2: 247–260
Lenfant C, Johansen K (1968) Respiration in an African lungfish, *Protopterus aethiopicus*: respiratory properties of blood and normal patterns of breathing and gas exchange. J Exp Biol 49: 437–457
Lenfant C, Johansen K, Grigg GC (1966) Respiratory properties of blood and pattern of gas exchange in the lungfish *Neoceratodous forsteri* (Krefft). Respir Physiol 2: 1–21
Lenfant C, Johansen K, Hanson D (1970) Bimodal gas exchange and ventilation-perfusion relationship in lower vertebrates. Fed Proc 29: 1124–1129
Levi HW (1967) Adaptations of respiratory systems of spiders. Evolution 21: 571–583
Lewis SV, Potter IC (1982) A light and electron microscope study of the gills of the lampreys (*Geotria australis*) with particular reference to the water-blood pathway. J Zool (Lond) 198: 157–176

Lewontin RG (1979) Fitness, survival and optimality. In: Horn DH, Mitchell R, Stairs GR (eds) Analysis of ecological systems. Ohio University Press, Columbus, pp 3–21
Liem KF (1967) Functional morphology of the integumentary, respiratory and digestive systems of synbranchoid fish *Monopterus albus*. Copeia 2: 375–380
Liem KF (1980) Air ventilation in advanced teleosts: Biomechanical and evolutionary aspects. In: Ali MA (ed) Environmental physiology of fishes. Plenum Press, London, pp 57–91
Liem KF (1981) Larvae of air-breathing fishes as counter-current flow devices in hypoxic environments. Science 211: 1177–1179
Liem KF (1984) The muscular basis of aquatic and aerial ventilation in the air-breathing teleost fish *Channa*. J Exp Biol 113: 1–18
Lillywhite HB (1987) Temperature, energetics and physiological ecology. In: Seigel RA, Novak SS, Collins JT (eds) Snakes: Ecology and evolutionary biology. Macmillan, New York, pp 422–477
Lindstedt SL (1984) Pulmonary transit time and diffusing capacity in mammals. Am J Physiol 246: 384–388
Lindstedt SL, Calder WA (1981) Body size, physiological time and longevity of homeothermic animals. Q Rev Biol 56: 1–16
Linzen B, Gallowitz P (1975) Enzyme activity patterns in muscles of the lycosid spider *Cupiennius salaei*. J Comp Physiol 143: 331–338
Locke MJ (1958a) The structure of insect trachea. Q J Microsc Sci 98: 487–492
Locke MJ (1958b) Coordination of growth in the tracheal system of insects. Q J Microsc Sci 99: 373–391
Locke MJ (1958c) The formation of tracheae and tracheoles in *Rhodnius prolixus*. Q J Microsc Sci 99: 29–46
Lomholt JP, Johansen K (1979) Hypoxia acclimation in carp – How it affects O_2 uptake, ventilation, and O_2 extraction from water. Physiol Zool 52: 38–49
Lomholt JP, Johansen K, Maloiy GMO (1975) Is aestivating lungfish the first vertebrate with suctional-breathing. Nature (Lond) 257: 787–788
Longworth KE, Jones JH, Bicudo JEPW, Taylor CR, Weibel ER (1989) High rate of oxygen consumption in exercising foxes: large PO_2 difference drives diffusion across the lung. Respir Physiol 77: 263–276
Loudon C (1989) Tracheal hypertrophy in mealworms: design and plasticity in oxygen supply systems. J Exp Biol 147: 217–235
Loveridge JP (1970) Observations on nitrogenous excretion and water relations of *Chiromantis xerapherina* (Amphibia, Anura). Arnoldia 5: 1–6
Loveridge JP (1976) Strategies of water conservation in the Southern African frogs. Zool Afr 11: 319–333
Lovtrup S (1977) The phylogeny of vertebrata. Wiley, London
Low WP, Ip YK, Lane DJ (1990) A comparative study of the gill morphology of the mudskippers – *Periophthalmus chrysospilos, Boleophthalmus boddaerti* and *Periophthalmodon schlosseri*. Zool Sci 7: 29–38
Luchtel DL, Kardong KV (1981) Ultrastructure of the lung of the rattlesnake, *Crotalus viridis oreganus*. J Morphol 169: 29–47
Macchin J (1974) Water relationships. In: Fretter VV, Peake J (eds) Pulmonates, vol V. Academic Press, London, pp 105–163
Mackay AR, Gelperin A (1972) Pharmacology and reflex responsiveness of the heart in the giant garden slug *Limax maximus*. Comp Biochem Physiol 43A: 877–896
Macklem PT, Bouvent P, Scheid P (1979) Measurements of the distensibility of the parabronchi in duck lungs. Respir Physiol 38: 23–35
MacMillen RE (1983) Adaptive physiology of heteromyid rodents. Great Basin Nat Mem 7: 65–76
Maina JN (1982a) A scanning electron microscopic study of the air and blood capillaries of the lung of the domestic fowl *Gallus domesticus*. Experientia 35: 614–616
Maina JN (1982b) A comparison of the lungs of two species of birds of different exercise capacities. J Anat 134: 604–605

Maina JN (1984) Morphometrics of the avian lung. 3. The structural design of the passerine lung. Respir Physiol 55: 291–309
Maina JN (1985a) The comparative morphometry of the respiratory organs of the East African air-breathing fish: the catfish *Clarias mossambicus* and the lungfish *Protopterus aethiopicus*. In: Karson B, Warren R, Wasen J (eds) Proc IV Europ Sympo Stereol. Chalmers Univ Press, Goteborg, pp 101–104
Maina JN (1985b) A scanning and transmission electron microscopic study of the bat lung. J Zool (Lond) 205: 19–27
Maina JN (1986) The structural design of the bat lung. Myotis 23: 71–77
Maina JN (1987a) The morphology of the lung of the African lungfish, *Protopterus aethiopicus*: a scanning electron microscopic study. Cell Tissue Res 250: 191–196
Maina JN (1987b) The morphology and morphometry of the adult normal baboon lung (*Papio anubis*). J Anat 150: 229–245
Maina JN (1987c) Morphometrics of the avian lung. 4. The structural design of the charadrii-form lung. Respir Physiol 68: 99–119
Maina JN (1988a) Scanning electron microscopic study of the spatial organization of the air and blood conducting components of the avian lung *Gallus gallus domesticus*. Anat Rec 222: 145–153
Maina JN (1988b) The morphology and morphometry of the normal lung of the adult vervet monkey *Cercopithecus aethiops*. Am J Anat 183: 258–267
Maina JN (1989a) The morphology of a tropical terrestrial slug *Trichotoxon copleyi* – Ver-dicourt (Mollusca: Gastropoda: Pulmonata). A scanning and transmission electron micro-scopic study. J Zool (Lond) 217: 355–366
Maina JN (1989b) A scanning and transmission electron microscopic study of the tracheal-air sac system in a grasshopper (*Chrotogonus senegalensis*) Kraus – (Orthoptera: Acri-didae: Pygomorphinae). Anat Rec 223: 393–405
Maina JN (1989c) The morphology of the lung of the East African tree frog *Chiromantis petersi* with observations on the skin and the buccal cavity as secondary gas exchange organs: a TEM and SEM study. J Anat 165: 29–43
Maina JN (1989d) The morphology of the lung of the black mamba *Dendroaspis polylepis* (Reptilia: Ophidia: Elapidae): a scanning and transmission electron microscopic study. J Anat 167: 31–46
Maina JN (1989e) The morphometry of the avian lung. In: King AS, McLelland J (eds) Form and function in birds, vol 4. Academic Press, London, pp 307–368
Maina JN (1990a) A study of the morphology of the gills of an extreme alkalinity and hyperosmotic adapted teleost *Oreochromis alcalicus grahami* (Boulenger) with particular emphasis on the ultrastructure of the chloride cells and their modifications with water dilution: a SEM and TEM study. Anat Embryol 181: 83–98
Maina JN (1990b) The morphology of the gills of the freshwater crab *Potamon niloticus* – Ortman (Crustacea: Brachyura: Potamonidae): scanning and transmission electron microscopic study. J Zool (Lond) 221: 499–515
Maina JN (1990c) The morphology and morphometry of the lung of the lesser bush-baby *Galago senegalensis*. J Anat 172: 129–144
Maina JN (1991) A morphometric analysis of chloride cells in the gills of the teleosts *Oreochromis alcalicus* and *Oreochromis niloticus* and a description of presumptive area excreting cells in *Oreochromis alcalicus*. J Anat 175: 131–145
Maina JN (1993) Perspectives on the structure and function in birds. In: Rosskopf W (ed) Petrak's diseases of cage and aviary birds. Lea-Ferbiger, New York (in press)
Maina JN, King AS (1982a) The thickness of the avian blood-gas barrier. Qualitative and quantitative observations. J Anat 134: 553–562
Maina JN, King AS (1982b) Morphometrics of the avian lung. 2. The wild mallard (*Anas platyrhynchos*) and greylag goose (*Anser anser*). Respir Physiol 50: 299–313
Maina JN, King AS (1984) The Structural functional correlation in the design of the bat lung: A morphometric study. J Exp Biol 111: 43–63
Maina JN, King AS (1987) A morphometric study of the lung of a humboldt penguin (*Spheniscus humboldti*). Zentralblatt Veterinaermed C Anat Histol Embryol 16: 293–297

Maina JN, King AS (1989) The lung of the emu, *Dromaius novaehollandiae*: a microscopic and morphometric study. J Anat 163: 67–74
Maina JN, Maloiy GMO (1985) The morphometry of the lung of the lungfish (*Protopterus aethiopicus*): its structural-functional correlations. Proc R Soc (Lond) B 244: 399–420
Maina JN, Maloiy GMO (1986) The morphology of the respiratory organs of the African air breathing catfish (*Clarias mossambicus*) (Peters): a light, and electron microscopic study, with morphometric observations. J Zool (Lond) 209: 421–445
Maina JN, Maloiy GMO (1988) A scanning and transmission electron microscopic study of the lung of a caecilian *Boulengerula taitanus*. J Zool (Lond) 215: 739–751
Maina JN, Nicholson T (1982) The morphometric diffusing capacity of a bat *Epomophorus wahlbergi*. J Physiol 325: 36–37
Maina JN, Settle G (1982) Allometric comparison of some morphometric parameters of avian and mammalian lungs. J Physiol 330: 28
Maina JN, King AS, King DZ (1982a) A morphometric analysis of the lungs of a species of bat. Respir Physiol 50: 1–11
Maina JN, Abdalla MA, King AS (1982b) Light microscopic morphometry of the lungs of 19 avian species. Acta Anat 112: 264–270
Maina JN, Howard CV, Scales L (1982c) The length densities and size distributions of the air and blood capillaries of the paleopulmo and neopulmo regions of the avian lung. Acta Stereol (Yugoslavica) 2: 101–107
Maina JN, King AS, Settle G (1989a) An allometric study of the pulmonary morphometric parameters in birds, with mammalian comparison. Philos Trans R Soc (Lond) 326: 1–57
Maina JN, Maloiy GMO, Warui CN, Njogu EK, Kokwaro ED (1989b) A scanning electron microscopic study of the morphology of the reptilian lung: the savanna monitor lizard (*Varanus exanthematicus*) and the pancake tortoise (*Malacochersus tornieri*). Anat Rec 224: 514–522
Maina JN, Thomas SP, Hyde DM (1991) A morphometric study of bats of different size: correlations between structure and function of the chiropteran lung. Philos Trans R Soc Lond B 333: 31–50
Maina JN, Maloiy GMO, Makanya AN (1992) The morphology and morphometry of two East African mole rats *Tachyoryctes splendens* and *Heterocephalus glaber*. Zoomorphology 112: 167–179
Maitland DP (1986) Crabs that breath air with their legs – *Scopimera* and *Dotilla*. Nature 319: 493–495
Maitland DP (1987) A highly complex invertebrate lung: the gill chambers of the soldier crab *Mictyris longicarpus*. *Naturwissenschaften* 74: 293–295
Maitland DP (1990a) Carapace and branchial water circulation and water related behaviours in the semiphore crab *Heloecius cordiformis* (Decapoda: Brachura: Ocypode). Mar Biol 105: 275–286
Maitland DP (1990b) Aerial respiration in the semiphore crab, *Holoecius cordiformis*, with or without branchial water. Comp Biochem Physiol 95A: 267–274
Mandelbrot BB (1983) The fractal geometry of nature. Freeman, New York
Mangum CP (1976) Primitive respiratory adaptations. In: Newell RC (ed) Adaptations to the environment: essays on the physiology of marine animals. Butterworth, London, pp 191–278
Mangum CP (1982) The functions of gills in several groups of invertebrate animals. In: Houlihan DF, Rankin JC, Shuttleworth TJ (eds) Gills. Cambridge University Press, Cambridge, pp 77–97
Marcus H (1937) Lungen. In: Bolk L, Goppert E, Kallius E, Lubosch W (eds) Handbuch der vergleichenden Anatomie der Wirbeltiere III. Urban Schwarzenberg, Berlin, pp 127–224
Martin KM, Hutchison VH (1979) Ventilatory activity in *Amphiuma tridactylum* and *Siren lacertina*. J Herpetol 13: 427–434
Mass JA (1939) Über die Atmung von *Helix pomatia*. Z Vgl Physiol 26: 605–610
Maynard-Smith J (1978) Optimization theory in evolution. Annu Rev Ecol Syst 9: 31–56
McCutcheon FH, Hall FG (1937) Haemoglobin in Amphibia. J Cell Comp Physiol 9: 191–197
McDonald DG, Caudek V, Ellis R (1991) Gill design in fresh water fishes: interrelationships among gas exchange, ion regulation, and acid base regulation. Physiol Zool 64: 103–123

McMahon BR (1981) Oxygen uptake and acid base balance during activity in decapod crustaceans. In: Harreid CF, Fortner CR (eds) Locomotion and energetics of arthropods. Plenum Press, New York, pp 299–335
McMahon BR, Burggren WW (1979) Respiration and adaptation to the terrestrial habitation the land hermit crab *Coenobita clypeatus*. J Exp Biol 79: 265–281
McMahon BR, Burggren WW (1981) Acid-base balance following temperature acclimation in land crabs. J Exp Biol 218: 45–52
McMahon BR, Burggren WW (1988) Respiration. In: Burggren WW, McMahon BR (eds) Biolgy of the crabs. Cambridge University Press, Cambridge, pp 249–297
McMahon BR, Wilkens JL (1984) Ventilation, perfusion and oxygen uptake. In: Mantel LH (ed) The biology of Crustacea, vol 5. Academic Press, New York, pp 289–372
McMahon BR, McDonald DG, Wood CM (1979) Ventilation, oxygen uptake, and haemolymph oxygen transport following enforced exhaustive activity in the Dungenese crab *Cancer magister*. J Exp Biol 80: 271–285
McNab B (1966) The metabolism of fossorial rodents: a study of convergence. Ecology 47: 712–733
McNab B (1984) Physiological convergence amongst ant-eating and termite-eating mammals. J zool (Lond) 203: 485–510
McNab B (1986a) Food habits, energetics, and the reproduction of marsupials. J Zool (Lond) 208: 595–614
McNab B (1986b) The influence of food habits on the energetics of the eutherian mammals. Ecol Monogr 56: 1–19
Meban C (1973) The pneumocytes in the lung of *Xenopus laevis*. J Anat 114: 235–244
Meban C (1980) Thicknesses of the air-blood barriers in vertebrate lungs. J Anat 131: 299–307
Meyer M, Burger RE, Scheid P, Piiper J (1977) Analyse des échanges gaseux dans les poumons des oiseaux. J Physiol (Paris) 73: 9A
Milani A (1894) Beiträge zur Kenntnis der Reptilienlunge. II. Zool Jahrb Abt Anat Ontog Tiere 7: 545–592
Mill PJ (1972) Respiration in the invertebrates. MacMillan St Martins Press, London
Miller PL (1966) The supply of oxygen to the active flight muscles of some large beatles. J Exp Biol 45: 285–304
Miller PL (1974) Respiration-aerial gas transport. In: Rockstein M (ed) The physiology of insects, 2nd edn. Academic Press, New York, pp 346–402
Milner AR (1988) The relationships and origins of living amphibians. In: Benton MJ (ed) The phylogeny and classification of tetrapods, vol 1 Amphibians, reptiles and birds. Clarendon Press, Oxford, pp 59–102
Milsom WK (1984) The interrelationship between pulmonary mechanics and spontaneous breathing pattern in the tokay lizard *Gekko gecko*. J Exp Biol 113: 203–214
Milsom WK, Johansen K (1975) The effect of buoyancy induced lung volume changes on respiratory frequency in a chelonian (*Caretta caretta*). J Comp Physiol 98: 157–160
Mitchell HA (1964) Investigation of the cave atmosphere of a Mexican bat colony. J Mammal 45: 568–577
Moore SJ (1976) Some spider organs as seen by the scanning electron microscope, with special reference to the book lung. Bull Br Arachnol Soc 3: 177–187
Morberly WR (1968) The metabolic responses of the common iguana, *Iguana iguana*, to activity under restraint. Comp Biochem Physiol 27: 21–37
Morris S, Taylor HH, Greenaway P (1991) Adaptations to a terrestrial existence by the robber crab *Birgus latro*. VII. The branchial chamber and its role in urine reprocessing. J Exp Biol 161: 315–331
Morton JE (1958) Molluscs. Hutchinson University Library, London
Moussa TA (1956) Physiology of the accessory respiratory organs of the teleost, *Clarias lazera* (C & V). J Exp Zool 136: 419–454
Moussa TA (1957) Morphology of the accessory air-breathing organs of the teleost, *Clarias lazera* (C & V). J Morphol 98: 125–160
Moy-Thomas JA (1971) Palaeozoic fishes, 2nd edn. Saunders, Philadelphia
Muir BS, Hughes GM (1969) Gill dimensions for three species of tunny. J Exp Biol 51: 271–285

Muir BS, Kendall JI (1968) Structural modifications in the gills of tunas and some oceanic fishes. Copeia 1968: 388–398
Munshi JSD (1960) The structure of the gills of certain fresh water teleosts. Indian J Zootomy 1: 135–174
Munshi JSD (1961a) On the accessory respiratory organs of *Heteropneustes fossilis* (Bloch). Proc R Soc Edinb B 68: 128–146
Munshi JSD (1961b) The accessory respiratory organs of *Clarias batrachus* (Linn). J Morphol 109: 115–140
Munshi JSD (1968) The accessory respiratory organs of *Anabas testudineus* (Bloch) (Anabantidae, Pisces). Proc Linn Soc Lond (Zool) 179: 107–126
Munshi JSD (1976) Gross and fine structure of the respiratory organs of air-breathing fishes. In: Hughes GM (ed) Respiration of amphibious vertebrates. Academic Press, London, pp 73–104
Munshi JSD, Hughes GM (1986) Scanning electron microscopic study of the respiratory organs of juvenile and adult climbing perch *Anabas testudineus*. Jpn J Ichthyol 33: 39–45
Munshi JSD, Singh BN (1968) On the respiratory organs of *Amphipnous cuchia* (Ham. Buch). J Morphol 124: 423–444
Munshi JSD, Weibel ER, Gehr P, Hughes GM (1986) Structure of the respiratory air-sac of *Heteropneustes fossilis* (Bloch) (Heteropneustidae, Pisces): an electron microscope study. Proc Indian Natl Sci Acad 52: 703–713
Munshi JSD, Olson KR, Ghosh TK, Ojha J (1990) Vasculature of the head and respiratory organs in an obligate air-breathing fish, the swamp eel *Monopterus* (= *Amphipnous*) *cuchia*. J Morphol 203: 181–201
Mutungi G (1990) Histochemistry, innervation, capillary density, and mitochondrial volume of the red and white muscle fibers isolated from a lizard, *Varanus exanthematicus*. Can J Zool 68: 476–481
Naggy KA, Odell DK, Seymour RS (1972) Temperature regulation by the inflorescence of *Philodendron*. Science 178: 1195–1197
Neckvasil NP, Olson KR (1986) Extraction and metabolism of circulating catecholamines by trout gills. Am J Physiol 19: R526–531
Newell RC (1976) Adaptations in intertidal life. In: Newell RC (ed) Adaptations to environment: essays on the physiology of marine animals. Butterworths, London, pp 1–82
Newell RC, Courtney WAM (1965) Respiratory movements in *Holothuria forskali* Delle Chiaje. J Exp Biol 42: 45–57
Nicoll PA (1954) The anatomy and behaviour of the vascular systems in *Nereis virens* and *Nereis limbata*. Biol Bull Mar Biol Lab (Woods Hole) 106: 69–82
Nilsson S (1983) Autonomic nerve function in the vertebrates. Springer, Berlin Heidelberg New York
Noble GK (1925) The integumentally, pulmonary and cardiac modifications correlated with increased cutaneous respiration in the Amphibia: solution to the "hairly frog" problem. J Morphol 40: 341–416
Noble GK (1929) The adaptive modifications of the arboreal tadpoles of *Hoplophryne* and torrent tadpoles of *Staurois*. Bull Am Mus Nat Hist 58: 291–334
Noble GK (1931) The biology of Amphibia. McGraw-Hill, New York
Ojha J, Munshi JSD (1974) Morphometric studies on the gill and skin dimensions in relation to body weight in freshwater mud-eel *Macrognathus aculeatum* (Bloch). Zool Anz (Jena) 193: 364–381
Ojha J, Singh SK (1986) Scanning electron microscopy of the gills of a hill stream fish, *Danio dangila* (Ham). Arch Biol (Brux) 97: 455–467
Okada Y, Ishiko S, Daido S, Kim J, Ikeda S (1962) Comparative morphology of the lung with special reference to the alveolar epithelial cells. I. lung of Amphibia. Acta Tuberc Jpn 11: 63–72
Olson KR (1981) Morphology and vascular anatomy of the gills of a primitive air-breathing fish, the bow-fin (*Amia calva*). Cell Tissue Res 218: 499–517
Olson KR, Munshi JSD, Ghosh TK, Ojha J (1986) Gill microcirculation of the air-breathing climbing perch, *Anabas testudineus* (Bloch): relationships with the accessory respiratory organs and systemic circulation. Am J Anat 176: 305–320

Olson KR, Munshi JSD, Ghosh TK, Ojha J (1990) Vascular organization of the head and respiratory organs of the air breathing catfish, *Heteropneustes fossilis*. J. Morphol 203: 165–179
O'Mahoney P (1984) Heart rate of the land crab *Gecarcinus lateralis* during aquatic and aerial respiration. Comp Biochem Physiol 78A: 621–624
Ostrom JH (1974) *Archaeopteryx* and the origin of flight. Q Rev Biol 49: 27–47
Packard GC (1974) The evolution of air-breathing in Paleozoic gnathostome fishes. Evolution 28: 320–325
Paiva M (1985) Theoretical studies of gas mixing in the lung. In: Engel LA, Paiva M (eds) Gas mixing and distribution in the lung. Dekker, New York, pp 221–285
Parker HW (1940) The Percy Sladen Trust Expedition to Lake Titicaca, Amphibia Ser 3. Trans Linn Soc (Lond) 12: 203–216
Pattle RE (1976) Lung surfactant in the evolutionary tree. In: Hughes GM (ed) Respiration of amphibious vertebrates. Academic Press, London, pp 233-255
Pattle RE, Schock C, Creasey JM, Highes GM (1977) Surpellic films, lung surfactant, and their cellular origin in newt, caecilian and frog. J Zool (Lond) 182: 125–136
Paul R, Fincke T (1989) Book lung function is arachnids. II. Carbon dioxide release and its relations to respiratory surface, water loss and heart frequency. J Comp Physiol 159B: 419–432
Paul R, Fincke T, Linzen B (1987a) Respiration in tarantula *Eurypelma californicum*: the effects of hydraulic force generation. J Comp Physiol 158B: 673–687
Paul R, Fincke T, Linzen B (1987b) Respiration in tarantula: evidence for diffusion lung. J Comp Physiol 157B: 209–217
Paul R, Fincke T, Linzen B (1989) Book lung function in arachnids. I. Oxygen uptake and respiratory quotient during rest, activity and recovery-relations to gas transport in the haemolymph. J Comp Physiol 159B: 409–418
Pearson OP, Pearson AK (1976) A stereological analysis of the ultrastructure of the lungs of wild mice living at low and high altitude. J Morphol 150: 359–368
Pelseneer P (1935) Essai d éthologie zoologique. Acad R Belg Cl Sci Publ Fond, Agathon Potter, pp 1–662
Perry SF (1978) Quantitative anatomy of the lungs of the red-eared turtle, *Pseudemys scripta elegans*. Respir Physiol 35: 245–262
Perry SF (1981) Morphometric analysis of pulmonary structure: methods for evaluation of unicameral lungs. Microscopie 38: 278–293
Perry SF (1983) Reptilian lungs: functional anatomy and evolution. Adv Anat Embryol Cell Biol, vol 79, Springer, Berlin Heidelberg New York, pp 1–81
Perry SF (1988) Functional morphology of the lungs of the Nile crocodile *Crocodylus niloticus*: non-respiratory parameters. J Exp Biol 143: 99–117
Perry SF (1989) Mainstreams in the evolution of vertebrate respiratory structures. In: King AS, McLelland J (eds) Form and function in birds, vol 4. Academic Press, London, pp 1–67
Perry SF (1990) Recent advances and trends in the comparative morphometry of vertebrate gas exchange organs. In: Boutilier RG (ed) Advances in comparative and environmental physiology, vol 6. Springer, Berlin Heidelberg New York, pp 45–71
Perry SF, Duncker HR (1978) Lung architecture, volume and static mechanics in five species of lizards. Respir physiol 34: 61–81
Perry SF, Duncker HR (1980) Interrelationship of static mechanical factors and anatomical structure in lung ventilation. J Comp Physiol 138: 321–334
Perry SF, Bauer AM, Russel AP, Alston JT, Maloney JE (1989a) Lungs of the gecko *Rhacodactylus leachianus* (Reptilia: Gekkonidae): a correlative gross anatomical and light and electron microscopic study. J Morphol 199: 23–40
Perry SF, Darian-Smith C, Alston J, Limpas CJ, Maloney JE (1989b) Histological structure of the lungs of the loggerhead turtle, *Caretta caretta*, before and after hatching. Copeia 1989 (4): 1000–1010
Peters HM (1978) On the mechanism of air-ventilation in anabantoids (Pisces: Toleostei). Zoomorphology 89: 93–124
Peters RH (1983) The ecological implications of body size. Cambridge University Press, Cambridge (UK)

Phleger CF, Saunders BS (1978) Swimbladder surfactants of Amazon air-breathing fishes. Can J Zool 56: 946–952
Pickard WF (1974) Transitional regime diffusion and the structure of the insect tracheolar system. J Insect Physiol 20: 947–956
Piiper J (1985) Air-flow pathway in the avian respiratory tract. In: Nachtigall W (ed) Bird flight BIONA report 3. Fischer, Stuttgart, pp 235–349
Piiper J, Scheid P (1972) Maximum gas transfer efficacy of models for fish gills, avian lungs and mammalian lungs. Respir Physiol 14: 115–124
Piiper J, Scheid P (1973) Gas exchange in avian lungs: models and experimental evidence. In: Bolis L, Schmidt-Nielsen K, Maddrell SHP (eds) Comparative physiology. Elsevier North-Holland, Amsterdam, pp 161–185
Piiper J, Scheid P (1975) Gas transport efficiency of gills, lungs and skin: theory and experimental data. Respir Physiol 23: 209–221
Piiper J, Scheid P (1980) Blood-gas equilibration in lungs. In: West JB (ed) Pulmonary gas exchange, vol 3. Academic Press, New York, pp 131–171
Piiper J, Schumann D (1967) Efficiency of O_2 exchange in the gills of the dogfish, *Scyliorhinus stellaris*. Respir Physiol 2: 135–148
Piiper J, Gatz RN, Crawford EC (1976) Gas transport in *Desmognathus fuscus*. In: Hughes GM (ed) Respiration in amphibious vertebrates. Academic Press, London, pp 339–356
Pisam M, Rambourg A (1991) Mitochondria rich cells in the gill epithelium of teleost fishes: an ultrastructural approach. Int Rev Cytol 130: 191–232
Poczopko P (1959) changes in blood circulation in *Rana escutenta* L. while diving. Zool Pol 10: 29–43
Pohunkova H (1967) The ultrastructure of the lung of the snail *Helix pomatia*. Folia Morphol 15: 250–257
Pohunkova H (1969) Lung ultrastructure of the Arachnida-Arachnoidea. Folia Morphol (Prague) 17: 309–361
Pohunkova H, Hughes GM (1985a) Structure of the lung of clawed toad (*Xenopus laevis* Daudin). Folia Morphol 23: 385–390
Pohunkova H, Hughes GM (1985b) Ultrastructure of the lungs of the garter-snake. Folia Morphol 23: 254–258
Portner HO, Heisler N, Grieshaber MK (1985) Oxygen consumption and mode of energy production in the intertidal worm *Sipunculus nudus* L.: definition and characterization of the critical PO_2 for an oxyconformer. Respir Physiol 59: 361–377
Potter GE (1927) Respiratory function of swimbladder in *Lepidosteus*. J Exp Zool 49: 45–52
Pough FH (1980) Blood oxygen transport and delivery in reptiles. Am Zool 31: 455–456
Pough FH, Taigen TL, Stawart MM, Brussard PF (1983) Behavioural modification of evaporative water loss by a Puerto Rican frog. Ecology 60: 608–613
Powell FL (1982) Diffusion in avain lungs. Fed Proc 41: 53–55
Powell FL, Mazzone RW (1983) Morphometrics of rapidly frozen goose lungs. Respir Physiol 51: 319–332
Powell FL, Scheid P (1989) Physiology of gas exchange in the avian respiratory system. In: King AS, McLelland J (eds) Form and function in birds, vol 4. Academic Press, London, pp 393–437
Powell FL, Fedde MR, Gratz RK, Scheid P (1978) Ventilatory responses to CO_2 in birds. I. measurements in anaethetized duck. Respir Physiol 35: 349–359
Powers LW, Bliss DE (1983) Terrestrial adaptations. In: Vernberg FJ, Vernberg VB (eds) The biology of Crustacea, vol 8. Academic Press, New York, pp 271–331
Prasad MS, Mishra AP, Singh BR (1982) Development of the air-breathing organs in *Colisa fasciatus*. Arch Ital Anat Embriol 87: 43–53
Precht H (1939) Die Lungenatmung der Süsswasserpulmonaten. Z Vgl Physiol 26: 696–739
Pringle JWS (1983) Insect flight. Carolina Biological Readers, Burlington, NC
Prior DJ, Hume M, Varga D, Hess SD (1983) Physiological and behavioural aspects of water balance and respiratory function in terrestrial slug, *Limax maximus*. J Exp Biol 104: 111–127

Qasim SZ, Qayyum A, Garg KK (1960) The measurement of carbon dioxide produced by bair-breathing fishes and evidence of respiratory function of accessory respiratory organs. Proc Indian Acad Sci 52: 19–26
Raath MA (1985) The theropod *Syntarsus* and its bearing on the origin of birds. In: Hecht MK et al. (eds) The beginning of birds. Freunde des Jura Museums Eichstatt, Willibaldsburg, pp 219–227
Rahn H, Howell BJ (1976) Bimodal gas exchange. In: Hughes GM (ed) Respiration of amphibious vertebrates. Academic Press, London, pp 271–285
Rahn H, Paganelli CV (1968) Gas exchange in gas gills of diving insects. Respir Physiol 5: 145–164
Rahn H, Rahn KB, Howell BJ, Gans C, Tenney SM (1971) Air-breathing of the garfish (*Lepisosteus osseus*). Respir Physiol 11: 285–307
Randall DJ (1982) The control of respiration and circulation in fish during exercise and hypoxia. J Exp Biol 100: 275–288
Randall DJ, Burggren WW, Farrell AP, Haswell MS (1981) The evolution of air-breathing vertebrates. Cambridge University Press, Cambridge
Rannels DE, Rannels SR (1988) Compensatory growth of the lung following partial pneumonectomy. Exp Lung Res 14: 157–182
Raubal FR (1987) Gill surface area and its components in the yellow-fin bream, *Acanthopagrus australis* (Gunther) (Pisces: Sparidae). Aust J Zool 35: 25–34
Ready NE (1983) Wing development in hemimetabolous insects. PhD Thesis, Univ California, Irvine
Redmond JR (1955) The respiratory function of haemocyanin in Crustacea. J Cell Comp Physiol 46: 209–247
Regal PJ (1985) Common sense reconstruction of the biology of fossils. In: Hecht MK et al. (eds) The beginnings of birds. Freunde des Jura Museums Eichstatt, Willibaldsburg, pp 67–74
Reite OB, Maloiy GMO, Aasehaug B (1974) pH, salinity and temperature tolerance of Lake Magadi Tilapia. Nature 247: 315
Richards AG, Anderson TF (1942) Further electron microscope studies on arthropod trachea. J NY Entomol Soc 50: 245–247
Richards AG, Korda FH (1950) Studies on arthropod cuticle. IV. an electron microscope survey of the intima of arthropod tracheae. Ann Entomol Soc Am 43: 49–71
Robin ED, Vester JW, Murdaugh HV, Millen JE (1964) Prolonged anaerobiosis in a vertebrate: anaerobic metabolism in the freshwater turtle. J Cell Comp Physiol 63: 287–297
Rodl S (1987) Raster- und transmissionselektronenmikroskopische Untersuchungen an der Buchlunge des Kaiserskorpions *Pandinus imperator*. Diplomathesis, Univ Munich, Munich
Romer AS (1946) The early evolution of fishes. Q Rev Biol 21: 33–69
Romer AS (1967) Major steps in vertebrate evolution. Science 158: 1629–1637
Romer AS (1970) The vertebrate body, 4th edn. Saunders, Philadelphia
Romer AS (1972) Skin breathing – primary or secondary? Respir Physiol 14: 183–192
Rose FL, Zambernard P (1966) Cardiac glycogen depletion in *Amphiuma means* during induced anoxia. J Morphol 120: 391–396
Rosen R (1968) Optimality principles in biology. Butterworths, London
Rosen DE, Forey PL, Gardiner BG, Petterson C (1981) Lungfishes, tetrapods paleontology and plesiomorphology. Bull Am Mus Nat Hist 167: 159–276
Ross LG (1978) The innervation of the resorptive structures in the swimbladder of a physoclist fish, *Polachius virens* (L). Comp Biochem Physiol 61C: 385–388
Ruben J (1976) Aerobic and anaerobic metabolism during activity in snakes. J Comp Physiol 109: 147–157
Rudd JT (1954) Vertebrates without erythrocytes and blood pigment. Nature (Lond) 173: 848–850
Rudd JT (1965) The ice-fish. Sci Am 213: 108–114
Runham NW, Hunter PJ (1970) Terrestrial slugs. Hutchinson University Library, London
Rzoska J (1974) The Upper Nile swamps, a tropical wet land study. Freshwater Biol 4: 1–30

Sacca R, Burggren WW (1982) Oxygen uptake in air and water in the air-breathing reed-fish *Calamoichthyes calabaricus*: role of skin, gills and lungs. J Exp Biol 97: 179–186
Satchell GH (1971) Circulation in fishes. In: Brain PW, Hughes GM, Salt G, Willmer EN (eds) Cambridge monographs in experimental biology 18. Cambridge University Press, Cambridge, pp 1–131
Satchell GH (1976) The circulatory system of air-breathing fish. In: Hughes GM (ed) Respiration of amphibious vertebrates. Academic Press, London, pp 105–123
Saunders RL (1962) The irrigation of the gills in fishes. 2. efficiency of oxygen uptake in relation to respiratory flow activity and concentrations of oxygen and carbon dioxide. Can J Zool 40: 817–862
Saupe M (1940) Anatomie und Histologie der Schwimmblase des Feussbarsches (*Perca fluviatilis*) mit besonderer Berücksichtigung des Ovals. Z Zellforsch 30: 1–34
Sawaya P (1946) Sobre abiologia de alguns peixes de respiracao aerea. Univ Sao Paulo Fac Filosof Cien Letras Zool 11: 225–286
Saxena DB (1959) Extent of gill surface in teleost *Rita rita*. Proc 1st All India Congr Zool 105: 108–168
Saxena DB (1960) On the asphyxiation and influence of CO_2 on respiration of air-breathing fish, *Heteropneustes fossilis* and *Clarias batrachus*. J Zool (Lond) 12: 114–124
Saxena DB (1962) A comparative study of the gill surface of the fresh water fishes, *Lobeo rohita*, *Ophicephalus striatus* and *Anabas testudineus*. Ichthyologica 1: 59–68
Sexena DB (1963) A review of ecological studies and their importance in the physiology of air-breathing fishes. Ichthyologica 2: 116–128
Scheid P (1978) Analysis of gas exchange between air capillaries and blood capillaries in avian lung. Respir Physiol 32: 27–49
Scheid P (1979) Mechanisms of gas exchange in birds. Rev Physiol Biochem Pharmacol 86: 137–186
Scheid P (1985) Gas exchange in the avian parabronchial lung. In: Nachtigall W (ed) Bird flight, BIONA report 3. Fischer, Stuttgart, pp 131–171
Scheid P, Piiper J (1970) Analysis of gas exchange in the avian lung: theory and experiments in the domestic fowl. Respir Physiol 9: 246–262
Scheid P, Piiper J (1972) Cross current gas exchange in avian lungs: effects of reversed parabronchial air flow in ducks. Respir Physiol 16: 304–312
Scheid P, Piiper J (1989) Respiratory mechanics and air flow in birds. In: King AS, McLelland J (eds) Form and function in birds, vol 4. Academic Press, London, pp 369–391
Scheid P, Hook C, Bridges CR (1981) Diffusion in gas exchange of insects. Fed Proc 41: 2143–2145
Schidlowski M (1975) Archean atmosphere and evolution of the terrestrial O_2 budget. In: Windley BF (ed) The early history of the earth. Wiley, London, pp 67–103
Schlotte E (1932) Morphologie und Physiologie der Atmung bei wasser, schlammm- und landlebenden Gobiiformes. Z Wiss Zool 140: 1–113
Schmalhausen II (1968) The origin of terrestrial vertebrates. Academic Press, London
Schmidt PT (1942) A fish with feather-like gills. Copeia 1942: 98–100
Schmidt-Nielsen K (1975) Recent advances in avian respiration. In: Peaker M (ed) Avian Physiology. Academic Press, London, pp 33–47
Schmidt-Niesen K (1984) Scaling: Why is animal size so important. Cambridge University Press, Cambridge
Scholander PF (1954) Secretion of gases against high pressures in the swimbladder of deep sea fishes. II. the rete mirabile. Biol Bull Mar Biol Lab (Woods Hole) 107: 247–259
Schreider JP, Raabe OG (1981) Structure of the human respiratory acinus. Am J Anat 162: 221–232
Schuurmans-Stekhoven JH (1920) Über die Atmung der Schnecken *Limax agrestis* und *Helix pomatia*. Tijdscrned Dierk Vereen 18: 1–43
Seed R (1983) Structural organization, adaptive radiation and classification of molluscs. In: Hochachka PW (ed) The Mollusca, vol 1. Academic Press, New York, pp 1–53
Shafiq SA (1963) Electron microscopy of the development of tracheoles in *Drosophila melanogaster*. Q J Microsc Sci 104: 135–140

Shams H, Scheid P (1989) Efficiency of parabronchial gas exchange in deep hypoxia: measurements in the resting duck. Respir Physiol 77: 135–146
Shelton G, Croghan PC (1988) Gas exchange and its control in steady state systems: the consequences of evolution from water to air-breathing in the vertebrates. Can J Zool 66: 109–123
Shoemaker VH, Balding D, Raubal R, McClanahan LL (1972) Uricotelism and low evaporative water loss in a South American frog. Science 175: 1018–1020
Singh BN (1976) Balance between aquatic and aerial respiration. In: Hughes GM (ed) Respiration of amphibious vertebrates. Academic Press, London, pp 125–164
Singh BN, Hughes GM (1971) Respiration of air-breathing catfish *Clarias batrachus* (Linn). J Exp Biol 55: 421–434
Singh BR, Thakur RN (1975) A histochemical study on the respiratory epithelium of an eel-fish, *Amphipnous cuchia* (Ham). Acta Histochem 54: 161–167
Singh BR, Sheel M, Mishra AP (1981) Development of the air-breathing organ in the catfish *Anabas testudineus* and the catfish *Heteropneustes fossilis*. Zool Anz 206: 215–226
Singh BR, Mishra AP, Sheel M, Singh I (1982a) Development of the air-breathing organ in the catfish *C. batrachus* (Linn). Zool Anz 208: 100–111
Singh BR, Singh RP, Mishra AP (1982b) Development of the air-breathing organ in the snake headed fish, *Channa punctatus* (Bloch). Zool Anz 208: 428–439
Singh BR, Mishra AP, Singh I (1984) Development of air breathing organ in the mud-eel *Amphipnous cuchia* (Ham). Zool Anz (Jena) 213: 395–407
Singh BR, Prasad MS, Yadav AN (1991) The morphology of the respiratory epithelium of the air-breathing organ of the swamp-eel, *Monopterus albus* (Zuiew): a SEM study. Zool Anz 226: 27–32
Smatresk NJ (1988) Control of the respiratory mode of air-breathing fishes. Can J Zool 148: 88–152
Smith DG, Gannon BJ (1978) Selective control of brachial arch perfusion in an air-breathing Amazon fish *Hoplerythrinus unitaeniatus*. Can J Zool 56: 959–964
Smith DG, Rapson NL (1977) Differences in pulmonary microvascular anatomy between *Bufo marinus* and *Xenopus laevis*. Cell Tissue Res 178: 1–15
Smith DS (1961) The structure of insect fibrillar flight muscle: a study made with special reference to membrane systems of the fiber. J Biophys Biochem Cytol 10: 123–158
Smith HW (1930) Metabolism of the lungfish, *Protopterus aethiopicus*. J Biol Chem 88: 97–130
Smith RE (1956) Quantitative relations between liver mitochondria metabolism and total body weight in mammals. Ann NY Acad Sci 62: 403–422
Smith SL (1982) Introduction to fish physiology. TFH, Neptune, NJ
Solomon SE, Purton M (1984) The respiratory epithelium of the lung in the green turtle (*Chelonia mydas*). J Anat 139: 353–370
Spengel JW (1904) Ueber Schwimmblasen, Lungen und Kiementaschen der Wirbeltiere. Zool Jahrb (Suppl) 7: 727–749
Staendaert T, Johansen K (1974) Cutaneous gas exchange in snakes. J Comp Physiol 89: 313–320
Stahl WR (1967) Scaling of respiratory variables in mammals. J Appl Physiol 22: 453–460
Stark-Vancs V, Bell PB, Hutchison VH (1984) Morphological and physiological basis for pulmonary ventilation in *Amphiuma tridactylum*: an ultrastructural study. Cell Tissue Res 238: 1–12
Steen JB (1963) The physiology of the swimbladder in the eel *Anguilla vulgaris*, III. the mechanisms of gas secretion. Acta Physiol Scand 59:221–241
Steen JB (1971a) Comparative physiology of respiratory mechanics. Academic Press, New York
Steen JB (1971b) The swimbladder as a hydrostatic organ. In: Hoar WS, Randall DJ (eds) Fish physiology. Academic Press, London, pp 414–440
Steffensen JF, Lomholt JP, Johansen K (1981) The relative importance of skin oxygen uptake in the naturally buried plaice, *Pleuronectes platessa*, exposed to graded hypoxia. Respir Physiol 44: 269–276
Stevens ED (1972) Some aspects of gas exchange in tuna. J Exp Biol 56: 809–823

Stevens ED, Holeton GF (1978) The partitioning of oxygen uptake from air and from water by the large obligate air-breathing teleost, piracucu (*Arapaima gigas*). Can J Zool 56: 974–976
Stevens ED, Randall DJ (1967) Changes in gas compositions in blood and water during moderate swimming activity in rainbow trout. J Exp Biol 46: 329–337
Stinner JN (1982) Functional anatomy of the lung of the snake *Pituophis melanoleucus*. Am J Physiol 243: R251–257
Stinner JN, Shoemaker VH (1987) Cutaneous gas exchange and low evaporative water loss in the frogs *Phyllomedusa sauvagei* and *Chiramantis xeraphelina*. J Comp Physiol 157B: 423–427
Stone PA, Dobie JL, Henry RP (1992a) Cutaneous surface area and bimodal respiration in soft-shelled turtles (*Trionyx spiniferus*), stinkpot (*Sternotherus odoratus*) and mud-turtles (*Kinosternon subrubrum*). Physiol Zool 65: 311–330
Stone PA, Dobie JL, Henry RP (1992b) The effect of aquatic O_2 levels on diving and ventilatory behaviour in soft-shelled (*Trionyx spiniferus*), stinkpot (*Sternotherus odoratus*) and mud-turtles (*Kinosternon subrubrum*). Physiol Zool 65: 331–345
Storch V, Welsch U (1975) Über Bau und Funktion der Kiemen und Lungen von *Ocypode ceratophthalma* (Decapoda: Crustacea). Mar Biol 29: 363–371
Strazny F (1982) Morphometrisch bestimmte Sauerstoff-Diffusionskapazität und funktionelle Anatomie der Spinnenlungen in der Gattung *Tegenaria*. J Biophys Biochem Cytol 4: 475–478
Strazny F, Perry SF (1984) Morphometric diffusion capacity and functional anatomy of the booklungs in the spider *Tegenaria* sp (Agelenidae). J Morphol 182: 339–354
Strazny F, Perry SF (1987) Respiratory system: structure and function. In: Nentwig W (ed) Ecophysiology of spiders. Springer, Berlin Heidelberg New York, pp78–94
Swan H (1969) Metabolic torpor in *Protopterus aethiopicus*: An anti-metabolic agent from the brain. Am Nat 103: 247–258
Szarski H (1962) The origin of the Amphibia. Q Rev Biol 37: 189–241
Szarski H (1964) The structure of respiratory organs in relation to body size in Amphibia. Evolution 18: 118–126
Szarslo H (1977) Sarcopterygii and the origin of the tetrapods. In: Hecht MK, Goody PC, Hecht BM (eds) Major patterns of vertebrate evolution. Plenum Press, New York, pp 517–544
Tamura O, Moriyama T (1976) On the morphological feature of the gill of amphibious and air breathing fishes. Bull Fac Fish Nagasaki Univ 41: 1–8
Tamura O, Morii H, Yuzuriha M (1976) Respiration of the amphibious fishes *Periophthalmus cantonensis* and *Boleophthalmus chinensis* in water and land. J Exp Biol 65: 97–107
Tappan H (1974) Molecular O_2 evolution. In: Hayaishi O (ed) Molecular oxygen in biology. Elsevier, Amsterdam, pp 81–135
Taylor CR, Weibel ER (1981) Design of the mammalian respiratory system. I. problem and strategy. Respir Physiol 44: 1–10
Taylor CR, Weibel ER, Karas RH, Hoppelar H (1989) Matching structures and functions in the respiratory system: allometric and adaptive variations in energy demand. In: Wood SC (ed) Comparative pulmonary physiology: current concepts. Dekker, New York, pp 27–65
Taylor EW, Butler PJ (1978) Aquatic and aerial respiration in the shore crabs *Carcinus maenas* (L.), acclimated to 15 °C. J Comp Physiol 127: 315–323
Taylor HH, Greenaway P (1979) The structure of the gills and lungs of the arid zone crab *Holthuisana* (= *Austrothelphusa*) *transversa* (Mertens), (Sundathelphusidae: Brachyura) including observations on arterial vessels within gills. J Zool (Lond) 189: 359–384
Taylor HH, Greenaway P (1984) The role of gills and branchiostegites in gas exchange in bimodal-breathing crab *Holthuisana transversa*: evidence of a facultative change in the distribution of the respiratory circulation. J Exp Biol 11: 103–121
Teal JM, Carey FG (1967) Skin respiration and oxygen debt in the mudskipper, *Periophthalmus sobrinus*. Copeia 1967: 677–679
Tenney SM (1979) A synopsis of breathing mechanisms. In: Wood SC, Lenfant C (eds) Evolution of respiratory processes: a comparative approach. Dekker, New York, pp 51–106

Tenney SM (1985) Oxygen supply and limiting oxygen pressures in an insect larva. Respir Physiol 60: 121–134
Tenney SM, Boggs DF (1985) Comparative mammalian respiratory control. In: Fishman AP (ed) Handbook of physiology: the respiratory system II, sect 3. American Physiological Society, Bethesda, Maryland, pp 833–855
Tenney SM, Remmers JE (1963) Comparative morphology of the lung: diffusing area. Nature 197: 54–56
Tenney SM, Tenney JB (1970) Quantitative morphology of cold blooded lungs: Amphibia and Reptilia. Respir Physiol 9: 197–215
Tenney SM, Bartlett D, Farber JP, Remmers JE (1984) Mechanics of the respiratory cycle in the green turtle (*Chelonia mydas*). Respir Physiol 22: 361–368
Thomas SP (1987) The physiology of bat flight. In: Fenton MB, Racey P, Rayner MV (eds) Recent advances in the study of bats. Cambridge University Press, Cambridge, pp 75–99
Thorpe WH, Crisp DJ (1947) Studies on plastron respiration. II. the respiratory efficiency of the plastron in the *Aphelocheirus*. J Exp Biol 24: 270–303
Toews DP, Shelton G, Randall DJ (1971) Gas tension in the lungs and major blood vessels of the urodele amphibian, *Amphiuma tridactylum*. J Exp Biol 55: 47–61
Torre-Bueno J (1978) Respiration during flight in birds. In: Piiper J (ed) Respiratory function in birds, adult and embryonic. Springer Berlin Heidelberg New York, pp 89–94
Tucker VA (1972) Respiration during flight in birds. Respir Physiol 14: 75–82
Ultsch GR (1974) Gas exchange and metabolism in the Sirenidae (Amphibia: Caudata). I. Oxygen consumption of submerged sirenids as a function of body size and respiratory surface area. Comp Biochem Physiol 47A: 485–498
Ultsch GR (1976) Ecophysiological studies of some metabolic and respiratory adaptations of sirenid salamanders. In: Hughes GM (ed) Respiration in amphibious vertebrates. Academic Press, London, pp 287–312
Ultsch GR, Jackson DC (1982) Long term submergence at 3 °C of the turtle *Chrysemys picta bellii*, in normoxic and severely hypoxic water. I. Survival, gas exchange and acid base status. J Exp Biol 96: 11–28
Ultsch GR, Herbert CV, Jackson DC (1984) The comparative physiology of diving in North American freshwater turtles. I. submergence tolerance, gas exchange, and acid-base balance. Physiol Zool 57: 620–631
Valente D (1948) Mechanismo da respiracao de *Trichodactylus petropolitanus* (Goeldi). Bolm Fac Filos Cien Let (Zool) 13: 259–324
van Dam L (1935) On the utilization of oxygen by *Mya arenaria*. J Exp Biol 12: 86–94
van Dam L (1954) On the respiration in scallops (Lamellibrachia). Bio Bull Mar Biol Lab (Woods Hole) 107: 192–202
Vellard J (1951) Adaptation des Batraciens a la vie a grande hauteur. Trav Inst Fr Etudes And 3: 88–114
Verde MR (1951) The morphology and histology of the lung in snakes. J Univ Bombay 19: 79–89
Vidyadaran MK, King AS, Kassim H (1987) Deficient anatomical capacity of oxygen uptake of developing lung of the female domestic fowl when compared with the red jungle fowl. Schweiz Arch Tierheilkol 129: 225–237
Vidyadaran MK, King AS, Kassim H (1990) Quantitative comparisons of lung structure of adult domestic fowl and red-jungle fowl, with reference to broiler ascites. Avian Pathol 19: 51–58
Vinegar A, Hutchison VH (1965) Pulmonary and cutaneous gas exchange in the green frog. Zoologica 50: 47–53
Vogel GM (1980) Oxygen uptake and transport in the sabellid polychaete *Eudistylia vancouveri* (Kinburg). MA Thesis, College of William and Mary, Williamsburg, Virginia
von Brand T, Nolan MO, Mann ER (1948) Observations on the respiration of *Australorbis glabratus* and some other aquatic snails. Biol Bull 95: 199–223
von Raben K (1934) Veränderungen im Kiemendeckel und in den Kiemen einiger Brachyuren (Decapoden) im Verlauf der Anpassung an die Feuchtluftatmung. Z Wiss Zool 145: 425–461

Vosswinkel VR (1982) Das Blutgefässystem von *Helix pomatia* L. (Gastropoda, Pulmonata), 2. Licht-und elektronenmikroskopische Untersuchungen der verschiedenen Anteile des Systems. Zool Jahrb Anat 108: 341–374
Vyas AB, Laliwala SM (1976) Anatomical studies on the book lungs of the scorpion *Buthus tamulus* with a note on the respiratory mechanism. J Anim Morphol Physiol 23: 3–7
Wang N, Banzett RB, Butler JP, Fredberg JJ (1988) Bird lung models show that convective inertia effects inspiratory aerodynamic valving. Respir Physiol 73: 109–124
Wang N, Banzett RB, Nations CS, Jenkins F (1992) An aerodynamic valve in the avian primary bronchus. J Exp Zool 262: 441–445
Warner GF (1977) The biology of crabs. Elek Science, London
Wassnetzov W (1932) Über die Morphologie der Schwimmblase. Zool Jahrb Abt Anat Ont Tiere 56: 1–36
Weber RE (1978) Respiration. In: Mill PJ (ed) Physiology of annelids. Academic Press, London pp 369–392
Weibel ER (1963) Morphometry of the human lung. Springer, Berlin Heidelberg New York
Weibel ER (1970/71) Morphometric estimation of pulmonary diffusing capacity. I. Model and method. Respir Physiol 11: 54–75
Weibel ER (1972) Morphometric estimation of pulmonary diffusing capacity, V. Comparative analysis of mammalian lungs. Respir Physiol 14: 26–43
Weibel ER (1973) Morphological basis of alveolar-capillary gas exchange. Physiol Rev 53: 419–495
Weibel ER (1983) Is the lung built reasonably? The 1983 J Burns Amberson Lecture. Am Rev Respir Dis 128: 752–760
Weibel ER (1984a) The pathway for oxygen: structure and function of the mammalian respiratory system. Harvard University Press, Cambridge, Mass
Weibel ER (1984b) Lung cell biology. In: Fishman AP, Fisher AB (eds) Handbook of physiology: respiration, vol 4. American Physiological Society, Washington, pp 47–91
Weibel ER (1986) Functional morphology of lung parenchyma. In: Macklem P, Mead J (eds) Handbook of physiology: mechanics of breathing. American Physiological Society, Washington, pp 89–111
Weibel ER (1989) Lung morphometry and models in respiratory physiology. In: Chang HK, Paiva M (eds) Respiratory Physiological and analytical approach. Dekker, New York, pp 1–56
Weibel ER (1990) Morphometry: stereological theory and practical methods. In: Gil J (ed) Models of lung disease: microscopy and structural methods. Dekker, New York, pp 199–252
Weibel ER (1991) Fractal geometry: a design principle for living organisms. Am J Physiol 261: L361–L369
Weibel ER, Taylor CR, O'Neil JJ, Leith DE, Gehr P, Hoppeler H, Langman V, Baudinette RV (1983) Maximal oxygen consumption and pulmonary diffusing capacity: a direct comparison of physiologic and morphometric measurements in canids. Respir Physiol 54: 173–188
Weibel ER, Marques LB, Constantinopol M, Doffey F, Gehr P, Taylor CR (1987) Adaptive variation in the mammalian respiratory system in relation to energetic demand. V: the pulmonary gas exchanger. Respir Physiol 69: 81–100
Weibel ER, Taylor CR, Hoppeller H (1991) The concept of symmorphosis: A testable hypothesis of structure-function relationship. Proc Natl Acad Sci USA 88: 10357–10361
Weis-Fogh T (1961) Power in flapping flight. In: Ramasy L, Wigglesworth VB (eds) The cells and the organisms. Cambridge University Press, Cambridge, pp 89–112
Weis-Fogh T (1964a) Diffusion in insect wing muscle, the most active tissue known. J Exp Biol 41: 229–256
Weis-Fogh T (1964b) Transition regime diffusion and the structure of the insect tracheolar system. J Insect Physiol 20:947–956
Weis-Fogh T (1964c) Functional design of the tracheolar system of flying insects as compared with the avian lung. J Exp Biol 41: 207–227
Weis-Fogh T (1967) Respiration and tracheal ventilation in locusts and other flying insects. J Exp Biol 47: 561–587

Weis-Fogh T (1973) Quick estimates of flight fitness in hovering animals, including novel mechanisms for lift production. J Exp Biol 59: 169–230
Welsch U (1981) Fine structure and enzyme histochemical observations on the respiratory epithelium of the caecilian lungs and gills. A contribution to the understanding of the evolution of the vertebrate respiratory epithelium. Arch Histol Jpn 44: 117–133
Welsch U (1983) Phagocytosis in the amphibian lung. Anat Anz 154: 323–327
Welsch U, Aschauer B (1986) Ultrastructural observations on the lung of the emperor penguin (*Apternodytes forsteri*). Cell Tissue Res 243: 137–144
Welsch U, Storch V (1973) Die Feinstruktur verhornter und nichtverhornter ektodermaler Epithelien und der Hautdrüsen embryonaler und adulter Gymnophionen. Zool Jahrb Anat 90: 323–342
West JB (1974) Respiratory physiology: the essentials. Williams and Wilkins, Baltimore
West NH, Jones DR (1975) Breathing movements in the frog *Rana pipiens*. II. The power output and efficiency of breathing. Can J Zool 53: 345–353
West NH, Bamford OS, Jones DR (1977) A scanning electron microscope study of the microvasculature of the avian lung. Cell Tissue Res 176: 533–564
White FN (1978) Circulation: a comparison of reptiles, mammals, and birds. In: Piiper J (ed) Respiratory function in birds, adult and embryonic. Springer, Berlin Heidelberg New York, pp 51–60
Whitford WG (1973) The effects of temperature on respiration in the Amphibia. Am Zool 13: 305–312
Whitford WG, Hutchison VH (1965) Gas exchange in salamanders. Physiol Zool 38: 228–242
Whiting HP, Bone Q (1980) Ciliary cells in the epidermis of the larval Australian dipnoan, *Neoceratodus*. J Linn Soc Lond 68: 125–137
Whitten JM (1972) Comparative anatomy of the tracheal system. Annu Rev Entomol 17: 373–402
Wigglesworth VB (1935) The regulation of respiration in the flea, *Xenopsylla cheopsis*, Roths (Pulicidae). Proc R Soc Lond B 118: 397–419
Wigglesworth VB, Lee WM (1982) The supply of oxygen to the flight muscles of insects: a theory of tracheole physiology. Tissue Cell 14: 501–518
Williams DD, Rausch RL (1973) Seasonal carbon dioxide and oxygen concentrations in the dens of hibernating mammals (Sciuridae). Comp Biochem Physiol 44A:1227–1235
Willmer EN (1934) Some observations on the respiration of certain tropical freshwater fish. J Exp Biol 11: 283–306
Willmer EN (1970) Cytology and evolution, 2nd edn. Academic Press, London
Wit F (1932) Über den Einfluss der Luftfeuchtigkeit auf die Grösse der Atemöffnung bei Landpulmonaten. Z Vgl Physiol 18: 116–124
Wolcott TG (1988) Ecology. In: Burggren WW, McMahon BR (eds) Biology of the land crabs. Cambridge University Press, Cambridge, pp 55–56
Wolcott TG, Wolcott DL (1984) Extrarenal urine modification for salt conservation in the land crab *Cardisoma guanhami*. Am Zool 57: 318–324
Wolcott TG, Wolcott DL (1985) Extrarenal modifications of urine for ion conservation in ghost crabs, *Ocypode quadrata* Fabricius. J Exp Mar Biol Ecol 91: 93–107
Wood CM, Randall DJ (1981) Oxygen and carbon dioxide exchange during exercise in the land crab (*Cardisoma carnifex*). J Exp Zool 218: 7–22
Wood SC, Lenfant C (1976) Respiration, Mechanics, control and gas exchange. In: Gans C, Dawson WR (eds) Biology of Reptilia, vol 5. Academic Press, London, pp 225–274
Wood SC, Weber RE, Maloiy GMO, Johansen K (1975) Oxygen uptake and blood respiratory properties of the caecilian *Boulengelura taitanus*. Respir Physiol 24: 355–363
Yonge CM (1947) The pallial organs in the Apidobrach Gastropoda and their evolution throughout the Mollusca. Philos Trans R Soc Lond B 232: 443–518
Yonge CM (1952) The mantle cavity in *Siphonaria alternata* Say. Proc Malacol Soc Lond 29: 190–199
Yonge CM (1958) Observations on the pulmonate limpet *Trimusculus* (*Gadinia*) *reticularus* (Sowerby). Proc Malacol Soc Lond 33: 31–37

Youlson JH, Freeman PA (1976) Morphology of the gills of larval and parasitic adult sea lamprey, *Petromyzon marinus* L. J Morphol 149: 73–104
Ysseling MA (1930) Über die Atmung der Weinbergschnecke (*Helix pomatia*). Z Vgl Physiol 13: 1–60
Zaaijer JJP, Volvekamp HP (1958) Some experiments on the haemoglobin-oxygen equilibrium in the bold of the ramshorn (*Planorbis corneus*). Acta Physiol Pharmacol Neerl 7: 56–77
Zaccone G, Tagliafierro G, Goniakowska-Witalinska L, Fasulo S, Ainis L, Mauceri A (1989) Serotonin-like immunoreactive cells in the pulmonary epithelium of ancient fish species. Histochemistry 92: 6–36
Zar JH (1969) The use of allometric model for avian standard metabolism-body weight relationships. Comp Biochem Physiol 29: 227–234
Zeuthen E (1942) The ventilation of the respiratory tract in birds, Biol Medd Dan Vidensk Selsk 17: 1–15
Zeuthen E (1953) Oxygen uptake as related to body size in organisms. Q Rev Biol 28: 1–12
Zhao-Xian WS, Ning-Zhen S, Wen-Fang S (1989) Aquatic respiration in soft-shelled turtles, *Trionyx sinensis.* Comp Biochem Physiol 92A: 593–598
Zhao-Xian W, Zhen-Ning S, Wei-Pung M, Jie-Ping C, Gong-Qing H (1991) The breathing pattern and heart rates of *Alligator sinensis.* Comp Biochem Physiol 98A: 77–87
Zheng WB, Liu WS (1988) Morphology and histology of the swimbladder and ultrastructure of respiratory epithelium in the air-breathing catfish, *Pangasius sutchi* (Pangasiidae). J Fish Biol 33: 147–154
Zoond A (1931) Studies on the localization of respiratory exchange in invertebrates. III. the book lungs of the scorpion. J Exp Biol 8: 263–266

Chapter 5

Molecular Adaptations in Resistance to Penicillins and Other β-Lactam Antibiotics

J. Coyette, M. Nguyen-Distèche, J. Lamotte-Brasseur, B. Joris, E. Fonzé, and *J.-M. Frère*[1]

Contents

1 Introduction

The chance discovery by Fleming in 1928 of a metabolite produced by *Penicillium notatum* which exhibited bacteriolytic properties was followed by the heroic

[1] Centre d'Ingénierie des Protéines et Laboratoire d'Enzymologie, Université de Liège, Institut Chimie (B6) au Sart Tilman, B-4000 Liège 1, Belgium

Advances in Comparative and Environmental Physiology, Vol. 20

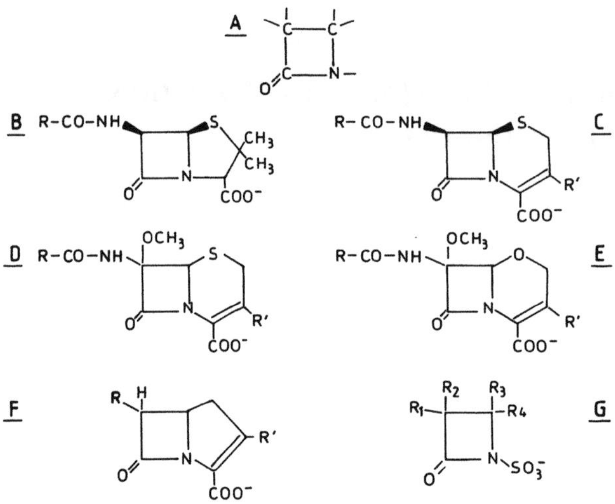

Fig. 1. Structure of the β-lactam ring (*A*) and of various antibiotics of the β-lactam family: penicillins (*B*), cephalosporins (*C*), cephamycins (*D*), oxocephamycins (*E*), carbapenems (*F*) and monobactams (*G*). Typical examples are: benzylpenicillin (*B*, $R = C_6H_5\text{–}CH_2\text{–}$); cephaloglycin [$C, R = C_6H_5\text{–}CH(NH_2)\text{–}$, $R' = CH_3\text{–}CO\text{–}O\text{–}CH_2\text{–}$]–, cefoxitin (*D*, $R =$ 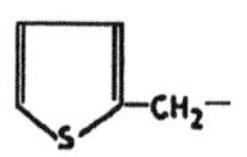

, $R' = H_2N\text{–}CO\text{–}O\text{–}CH_2\text{–}$); moxalactam (*E*, $R = HO\text{–}C_6H_4\text{–}CH(COOH)\text{–}$,
$R' =$ 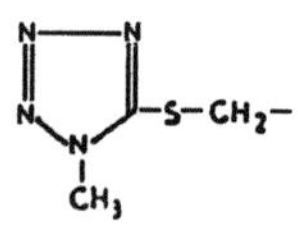

; imipenem (*F*, $R = CH_3\text{–}CHOH\text{–}$, $R' = HN = CH\text{–}NH\text{–}CH_2\text{–}CH_2\text{–}S$); and aztreonam (*G*, $R_1 =$ 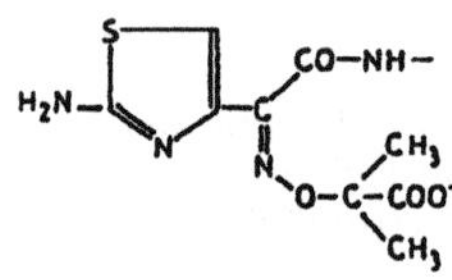

, $R_2 = H$, $R_3 = H$, $R_4 = CH_3$)

efforts of the Oxford group of Chain and Florey to isolate and identify the active molecule. This led to the introduction of benzylpenicillin in clinical trials about 50 years ago, probably one of the major breakthroughs in modern chemotherapy. Although some pathogenic bacteria were rapidly recognized as resistant to the new wonder drug, it was believed that a nearly ideal solution had been found to the problem of bacteria-mediated infectious diseases. Indeed, penicillin was extremely efficient and nearly completely innocuous to eukaryotic cells, which allowed the utilization of relatively high doses with little or no unwanted secondary effects. However, in the early 1950s, resistant strains started to appear in generally sensitive species such as *Staphylococcus aureus* and this phenomenon initiated a constant struggle between chemists, biochemists and microbiologists, on the one side, and bacteria, on the other. The former continuously isolated new molecules from natural sources and synthesized additional compounds, while the latter kept devising new strategies to escape the lethal action

A

B

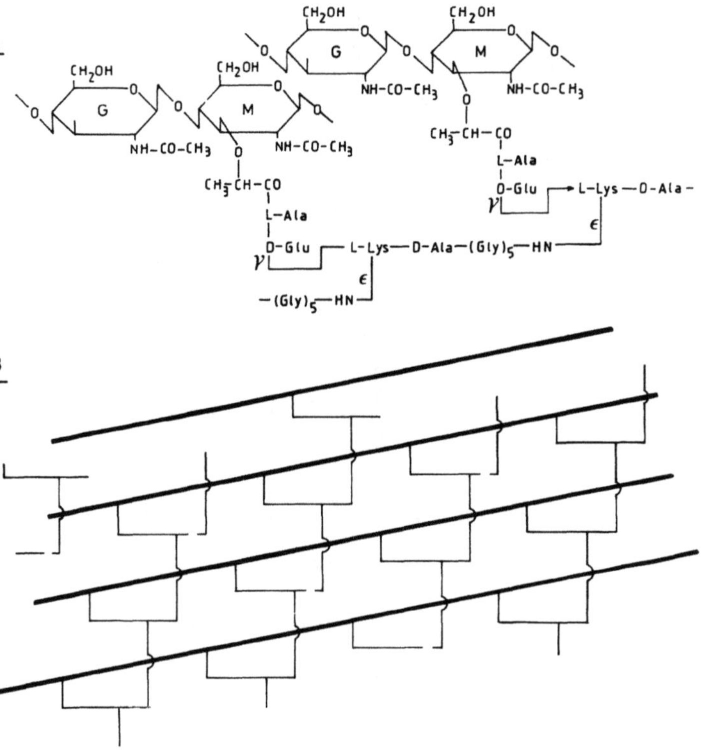

Fig. 2A, B. General structure of the bacterial cell wall peptidoglycan. **A** Detailed structure of a peptide cross-bridge in *Staphylococcus aureus.* Only one disaccharide unit is represented in each glycan strand; *G* = N–acetylglucosamine and *M* = N–acetylmuramic acid. **B** Scheme showing the general structure of a highly cross-linked peptidoglycan. The *heavy* and *light lines* represent the glycan strands and the peptide cross-bridges, respectively

of an ever expanding arsenal of drugs which exhibited one common chemical characteristic: the presence of the four-membered β-lactam ring shown in Fig. 1 (together with some members of the β-lactam family).

2 Peptidoglycan Structure and Biosynthesis

To understand the mechanism of action of these antibiotics and the various aspects of the resistance phenomena, it was necessary to unravel the structure and biosynthesis of the most important polymer of the bacterial cell wall: the peptidoglycan. The cell wall supplies the bacterium with mechanical protection against its own osmotic pressure, since most bacteria live in a hypotonic environment. The peptidoglycan, a net-like giant molecule which completely surrounds the cell, plays the major role in that protection. It is composed of

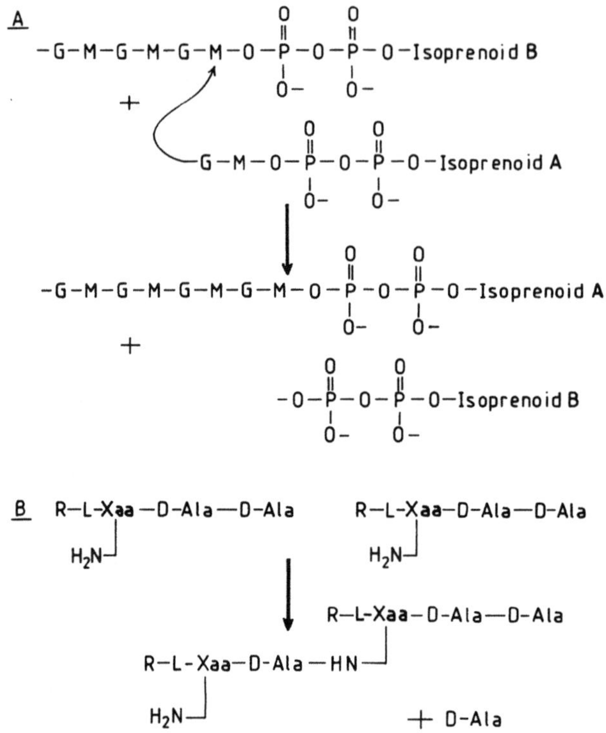

Fig. 3A, B. Final reactions in the biosynthesis of peptidoglycan. **A** Transglycosylation reaction; *G* = N–acetylglucosaminyl unit; *M* = N–acetylmuramyl peptide unit. **B** Transpeptidation reaction; *L-Xaa* represents the amino acid involved in forming the cross-bridge. In Gram-negative bacteria, it is *meso*-diaminopimelic acid. In *S. aureus*, it is an L-lysine residue carrying a pentaglycine chain on its ε-amino group

parallel glycan strands cross-linked by short peptides. The former consist of alternating units of N-acetylglucosamine and N-acetylmuramic acid whose lactyl side-chain is amidated by the N-terminal amino group of the peptide (Fig. 2). The latter have some unusual characteristics, such as the presence of D-amino acids and γ-glutamyl peptide bonds.

The biosynthesis of the polymer involves three major steps which occur in distinct cellular locations

1. The synthesis of activated monomers, UDP-N-acetylglucosamine and UDP-N-acetylmuramyl-peptide, takes place in the cytoplasm;

2. These precursors are successively transferred to a membrane-bound C55 isoprenoid alcohol, forming an isoprenoidyl-pyrophosphoryl-disaccharide-peptide which translocates the disaccharide-peptide across the cytoplasmic membrane;

3. The disaccharide-peptide units are assembled outside the bacterium (in the periplasmic space in Gram-negative organisms) with the help of membrane-bound transglycosylases and transpeptidases. The former enzymes utilize the

energy of the phosphoester bond to synthesize the new glycosidic bond (Fig. 3A), while the activity of the latter depends on the presence of an additional D-alanine at the C-terminus of the peptide: they utilize the energy of the D-alanyl-D-alanine peptide bond to create a new amide bond between the penultimate D-alanine and a free amino group of a second peptide, attached to another glycan strand (Fig. 3B).

Penicillins and other β-lactam antibiotics interfere with the transpeptidation reaction. At sublethal concentrations, they induce the formation of an abnormally high proportion of uncross-linked glycan chains. At higher concentrations, the mechanical resistance of the polymer is too low, which results in cell lysis, a phenomenon in which the bacterial autolysins also appear to be involved.

This short summary is understandably incomplete. More details on peptidoglycan synthesis and the mechanism of action of β-lactams can be found in specialized reviews (Ghuysen 1968; Mirelman 1979; Shockman and Barrett 1983; Frère et al. 1992).

3 The Targets: Transpeptidases and Penicillin-Binding Proteins

In the cytoplasmic membranes of all eubacteria studied so far, several proteins capable of covalently binding radioactive penicillin were found. These penicillin binding proteins (PBPs) were numbered in order of decreasing molecular weight as determined by SDS-polyacrylamide gel electrophoresis (Spratt 1975). In most cases, the high-molecular weight (HMW) PBPs play an essential physiological role and are composed of two domains, a penicillin-binding C-terminal domain responsible for the transpeptidase (or DD-peptidase) activity and an N-terminal region which in the cases of *Escherichia coli* PBPs 1a and 1b catalyzes the transglycosylation reaction (Suzuki et al. 1980; Nakagawa et al. 1984; Ghuysen 1991). In consequence, PBPs 1a and 1b synthesize cross-linked peptidoglycan with the isoprenoidyl-pyrophosphoryl disaccharide peptide as a substrate. The role of the N-terminal domain of other HMW-PBPs is still a matter of controversy, but it has been shown that *E. coli* PBPs 2 and 3 are involved in elongation of the cell wall and formation of the septum, respectively, with the collaboration of other membrane-bound proteins (Wientjes and Nanninga 1991; Frère et al. 1992). Conversely, low-molecular weight PBPs (LMW) are either soluble PBPs and secreted in the culture medium or membrane-bound. Their functions are still unclear, but many of them exhibit DD-peptidase activities (Frère and Joris 1985). They do not contain the N-terminal domain of the HMW-PBPs.

At this point, it is important to note that all PBPs and also most of the penicillin-destroying β-lactamases (see below) are members of a super-family of penicillin-recognizing proteins (PRPs) characterized by a series of structural features directly related to their interactions with β-lactam compounds (Joris et al. 1988, 1991; Ghuysen 1991).

Fig. 4. Structure of the TEM-1 β-lactamase, a class A enzyme. Only the α-carbon trace is shown together with the position of some conserved residues forming and surrounding the active site; ★ S70 and K73 of the SXXK tetrad, where *S* is the active serine residue; ■ S130, D131 and N132 of the SDN triad; ▲ K234, S235, G236 of the KS(T)G triad. Residues E166 and N170 (●) are also located. They play an important role in the catalytic mechanism of class A β-lactamases but the identity of the equivalent residues in the other active-site serine penicillin recognizing enzymes remains mysterious

In the known three-dimensional structures (several β-lactamases and one soluble LMW-PBP), the active site serine (S*) is situated at the N-terminus of an α-helix and is the first residue of a S*XXK tetrad, where the lysine residue is also strictly conserved in all PRPs (Fig. 4). One boundary of the cavity is formed by a loop, exhibiting a SXN or YXN conserved triad and facing, on the opposite side, a piece of β-strand generally containing a KTG sequence, sometimes replaced by the equivalent KSG, HTG or RSG triads (Joris et al. 1991).

4 Resistance Mechanisms

Logically, resistance can be increased by decreasing the sensitivity of the target enzymes or by keeping the antibiotic molecules from reaching them, either by impeding the diffusion of the aggressor or by destroying it. Bacteria have successfully utilized these three strategies.

4.1 Diffusion Barriers

In Gram-negative bacteria, the antibiotic molecule must cross the outer membrane and diffuse through the periplasm before reaching the target transpeptidases (Nikaido and Normark 1987). Depending upon the structure, and mainly upon the hydrophobicity of the R and R′ side chains (Fig. 1), the outer membrane can significantly slow down the penetration of the β-lactam. Moreover, the β-lactamases are mostly located in the periplasmic space and the combination of their activity with the permeability barrier can result in complete destruction of the antibiotic before it can reach the transpeptidases.

Gram-positive species do not enjoy the protection of an outer membrane and excrete the β-lactamases in the external medium. However, the mycolic acids of mycobacteria can seriously reduce the rates of out-diffusion of the β-lactamase and of in-diffusion of the β-lactam. Some cases have also been observed where the β-lactamase is bound to the outer face of the cytoplasmic membrane.

4.2 Sensitivity of the Targets

The affinity of the target PBPs for penicillins varies over a wide range, not only between species, but also between the different PBPs of the same bacterium. For instance, PBP 1 of *Bacillus licheniformis* is at least 100-fold more sensitive than PBPs 1a, 1b, 2 and 3 of *E. coli* and than its own PBP 3 (Frère and Joris 1985). Inactivation of PBP 1 being sufficient to kill *B. licheniformis* explains the high sensitivity of non β-lactamase producing strains of this species. But, if one assumes that one of the *E. coli* PBPs could replace PBP 1 in *B. licheniformis*, it becomes clear that the acquisition of the corresponding gene by the latter bacterium would significantly increase its resistance. The same result might be obtained if the *B. licheniformis* PBP 3 could be somehow modified and become capable of carrying out the function of PBP 1. The validity of these theoretical predictions has been verified in the recent past, as will be discussed in Sect. 6.

4.3 β-Lactamases

The first penicillin-destroying enzyme produced by a bacterium was discovered in 1940, well before the advent of the therapeutic utilization of β-lactam antibiotics (Abraham and Chain 1940). These enzymes, initially called "penicillinases" were later found to hydrolyze, often with high efficiency, the β-lactam amide bond yielding biologically inactive compounds (Fig. 5). In that respect, the first clinical problems due to β-lactamase-producing strains of *Staphylococcus*

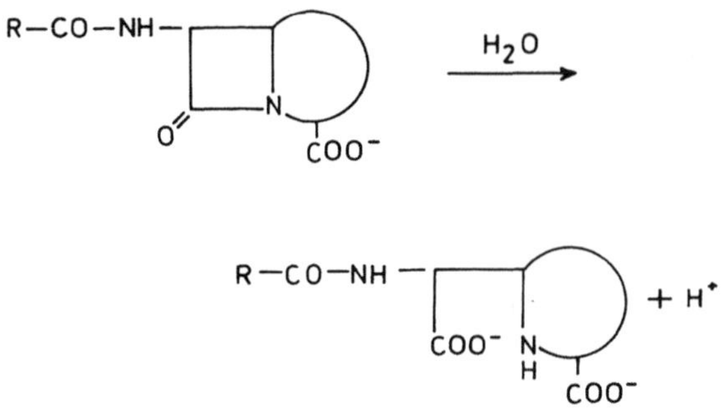

Fig. 5. Hydrolysis of a β-lactam antibiotic by a β-lactamase. The second ring fused to the β-lactam nucleus in structures *B-F* (Fig. 1) is only schematically represented

aureus, were solved by the introduction of methicillin, a semi-synthetic compound (Fig. 1B, R = 2,6 dimethoxyphenyl) which was poorly hydrolyzed by the staphylococcal β-lactamase. The widespread and sometimes abusive utilization of β-lactams favored the selection of β-lactamase-producing strains of many pathogenic species. Additional difficulties resulted from the fact that these enzymes were often plasmid or transposon-encoded, which facilitated the spreading of the genes throughout the bacterial world (for a review, see Bush 1989a).

In the following pages, we will present examples of each type of resistance mechanism and discuss them in the light of integrated models.

5 Permeability

In Gram-negative bacteria, only small (Mr < 1000) hydrophilic molecules, including β-lactam antibiotics can cross the outer membrane. They diffuse through water-filled channels of low specificity formed by porins (Nikaido 1989, 1992). Biochemical and X-ray crystallographic results indicate that these specialized proteins exist as stable trimers. Each subunit forms one channel and a trimer therefore contains three channels. Their structure and conformation explain their selectivity for the solutes on the basis of the size, the charge and hydrophilicity of the latter (Cowan et al. 1992).

In many Gram-negative organisms, (e.g., *E. coli*) this diffusion barrier does not *per se* significantly modify the susceptibility of the cell to most β-lactams and other types of small hydrophilic antibiotics but can acquire an increased importance when associated with the presence of a periplasmic β-lactamase. Some other species, such as *Pseudomonas aeruginosa*, however exhibit a higher natural resistance to the same compounds due to the general poor permeability of their outer membrane (Hancock et al. 1990). Resistance associated with a decreased permeability was also observed in various mutants of pathogenic strains (Nikaido 1989). Usually, this was related to the deficient expression of either two different major porins or of one porin presenting slightly larger pores than the other channel-forming proteins. The first and second types of mutations result in large (more than 100-fold) or low (2- to 10-fold) increases in the resistance and were observed both in laboratory mutants and in clinical isolates. Pleiotropic mutations, still poorly defined, can also decrease the outer membrane permeability. In Gram-positive bacteria which are devoid of outer membrane, secreted proteins are released in the external medium and, conversely, most β-lactam antibiotics freely reach their PBP targets exposed on the external face of the cytoplasmic membrane (Ghuysen 1991). However, mycobacteria and probably closely related species such as *Nocardia* spp. possess complex polymers layered outside their peptidoglycan. These are rich in lipids containing mainly long-chain fatty acids such as mycolic acids (Minnikin 1982) which act as a permeability barrier, just like an outer membrane (Jarlier and Nikaido 1990). Small hydrophilic molecules appear to diffuse through water-filled channels but the permeability coefficients are 3 orders of magnitude lower than those for the *E. coli* porins.

6 Target Sensitivity

Although the exact structural features which determine the target sensitivity remain undetermined, this field is understandably the subject of intense investigation and a lot of fundamental information has been accumulated, including the establishment of an adequate kinetic model and the measurement of the relevant constants, allowing the description of the interaction between some PBPs and various β-lactams in quantitative terms.

6.1 Kinetic model

The inactivation of the transpeptidases by β-lactams relies on the formation of a rather stable covalent intermediate as demonstrated by Fig. 6. On the basis of that model, the efficient immobilization of the enzyme would rely on good recognition of the antibiotic (low dissociation constant K), on a rapid formation of EC* and on a slow degradation of this covalent adduct (high and low k_2 and k_3 values, respectively) (Frère et al. 1975). If the second and third conditions are usually fulfilled, the dissociation constant is generally large (1 mM or higher) so that the factor which determines the rate of EC* formation is the k_2/K ratio,

$$E + C \overset{K}{\rightleftharpoons} EC \xrightarrow{k_2} EC^{x} \xrightarrow[H_2O]{k_3} E + P(s)$$

Fig. 6. Kinetic scheme representing the interaction between an active-site serine penicillin recognizing enzyme (*E*) and a β-lactam antibiotic (*C*). *EC* is a non-covalent Henri-Michaelis complex and EC^x the acylenzyme; *K* is the dissociation constant of *EC* and k_2 and k_3 first-order rate constants. In some interactions, more than one product (*Ps*) are formed upon degradation of the acylenzyme a second-order rate constant. In fact, the degradation rate (k_3) is often so slow that it becomes irrelevant, since the synthesis of new transpeptidase molecules by the cell is greater than the regeneration of immobilized targets, which leaves the k_2/K ratio as the sole important parameter.

In theory, the sensitivity of the transpeptidase could be decreased by either decreasing k_2/K or increasing k_3. In practice, only the first situation has been encountered, but the second remains a distinct (and frightening) possibility.

Table 1 illustrates the large dispersion of the k_2/K values which have been determined, and Table 2 shows that, for a given DD-peptidase, the same parameter also exhibits very different values depending upon the exact structure of the β-lactam.

Table 1. Properties of some penicillin-binding proteins

	Mr	Interaction with benzylpenicillin	
		k_2/K $(M^{-1}s^{-1})^a$	k_3 $(s^{-1})^a$
Streptomyces R61			
DD-peptidase	37500	13000 (25°)	1.4×10^{-4}
Actinomadura R39			
DD-peptidase	50000	300000	3×10^{-6}
Streptomyces K15			
DD-peptidase	26000	150	1.1×10^{-4}
Proteus mirabilis			
PBP4	46000	2–300000	4×10^{-5}
Staphylococcus aureus			
PBP4	46000	2000	8×10^{-3}
PBP1	115000	3800	1.3×10^{-4} (25°)
PBP2′	76000	10–20	ND
Bacillus licheniformis			
PBP4	46000	80	$< 6 \times 10^{-5}$
PBP1	123000	330000	2×10^{-3}
Streptococcus pneumoniae			
PBP2x	81000	58000	5×10^{-5}
Escherichia coli			
PBP6	40000	175	7×10^{-4}
PBP5	42000	175	2×10^{-3}
PBP4	49000	7000	4×10^{-4}
PBP3	60000	300–3000[b]	1.8×10^{-4}
PBP2	66000	130	1.4×10^{-4}
PBP 1a + 1b	90000[c]	200	3×10^{-5}

[a] Kinetic parameters were measured at 37 °C unless otherwise stated. The k_2/K ratio represents the second-order rate constant for PBP acylation and directly reflects its sensitivity to the antibiotic.
[b] Different results were obtained with different preparations.
[c] Most results were obtained with a mixture containing both PBPs.
For references, see Frère and Joris (1985).

Table 2. Influence of the antibiotic structure on the kinetic parameters for their interaction with the *S.* R61 DD-peptidase (at 37 °C unless otherwise stated)

	k_2/K ($M^{-1}s^{-1}$)	k_3 (s^{-1})
Benzylpenicillin	13000 (25 °C)	1.4×10^{-4}
6-Aminopenicillanate	0.25	$< 6 \times 10^{-5}$
Carbenicillin	830	1.4×10^{-4}
Ampicillin	110	1.4×10^{-4}
β-Iodopenicillanate	0.7	$< 3 \times 10^{-5}$
Cephalosporin C	1500	1×10^{-6}
Cephaloglycin	22	3×10^{-6}
Nitrocefin	460 (10 °C)	3×10^{-4}
Cefotaxime	16	$< 4 \times 10^{-6}$
Cefuroxime	350	4×10^{-6}
Cefoxitin	1500	5×10^{-5}
Imipenem	1000	7×10^{-6}
Clavulanate	21	5×10^{-6}
Aztreonam	$< 2 \times 10^{-3}$	ND

For references, see Frère and Joris (1985); ND, not determined

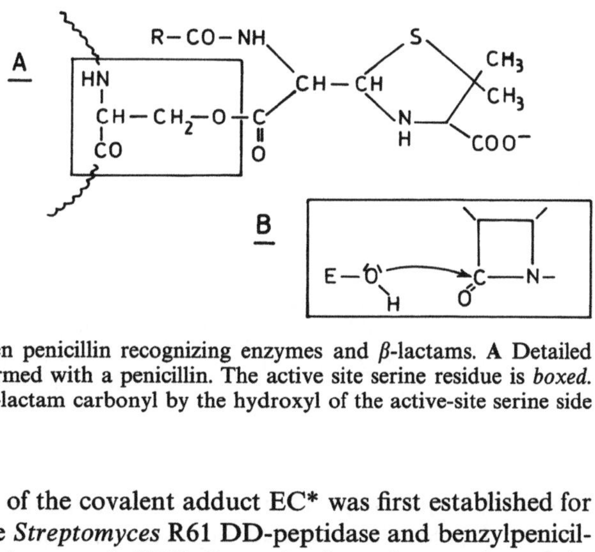

Fig. 7A, B. Interaction between penicillin recognizing enzymes and β-lactams. **A** Detailed structure of the acylenzyme formed with a penicillin. The active site serine residue is *boxed.* **B** Nucleophilic attack of the β-lactam carbonyl by the hydroxyl of the active-site serine side chain

The chemical structure of the covalent adduct EC* was first established for the interaction between the *Streptomyces* R61 DD-peptidase and benzylpenicillin (Frère et al. 1976; Degelaen et al. 1979). It results from the opening of the β-lactam ring by a nucleophilic attack of the β-lactam amide bond by the hydroxyl group of a serine side-chain of the enzyme (Fig. 7), which yields an acylenzyme intermediate, characterized by an ester bond between the enzyme and the "penicilloyl" moiety. The decay of the acylenzyme (k_3) involves the hydrolysis of this ester bond, but additional reactions can precede, accompany or follow this event. That the same serine residue was also involved in the interaction with the substrate was demonstrated in 1982 (Yocum et al.). In consequence, PBPs were recognized as "active site serine transpeptidases" and

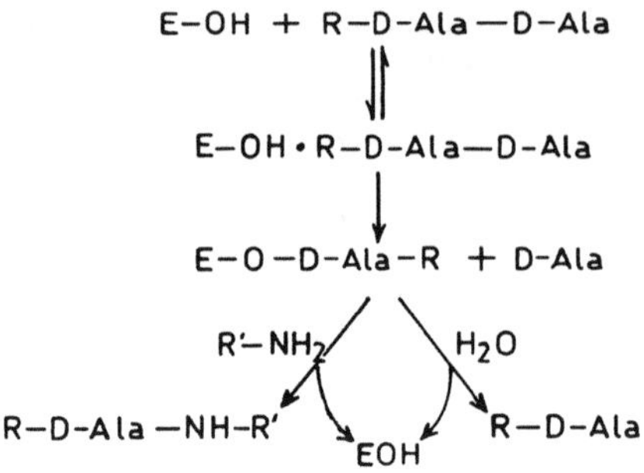

Fig. 8. Postulated mechanism for the transpeptidation (*left*) and carboxypeptidation (*right*) reactions catalyzed by DD-peptidases (*EOH*)

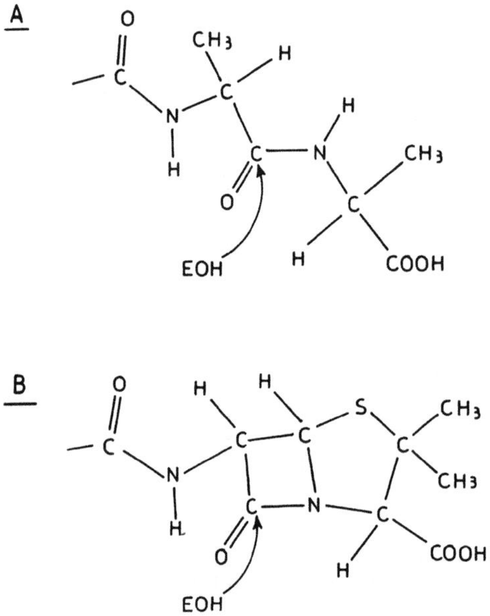

Fig. 9. Structural similarity between a D-alanyl-D-alanine-terminated peptide (**A**) and a penicillin (**B**). The nucleophilic attack by the enzyme hydroxyl group (*EOH*) is also shown

their catalytic pathway was hypothesized to also involve an acylenzyme intermediate (Fig. 8) which reacted in a second step with an aminated acceptor (transpeptidation) or with water (carboxypeptidation). The validity of this hypothesis was recently demonstrated, at least for the S.R61 DD-peptidase (Jamin et al. 1991).

The similarity of the interactions with penicillins and D-alanyl-D-alanine terminated substrates seems to corroborate the original substrate analogy hypothesis (Tipper and Strominger 1965) which assumed a structural similarity between penicillins and the D-alanyl-D-alanine peptide (Fig. 9). It is however important to note that the absolute validity of this simple and elegant concept has been challenged by recent results, obtained with modified proteins produced by site-directed mutagenesis. These proteins exhibited strongly decreased rates in their interactions with peptide substrates and no modification in their sensitivity to β-lactams (Malhotra and Nicholas 1992; Wilkin et al. 1993).

Interestingly, site-directed mutagenesis has, up to now, failed to yield an enzyme where the transpeptidase activity was less affected than the rate of penicillin binding, a situation which would increase the resistance of the bacteria producing the modified enzymes. Some mutations, which will be described below, do indeed increase that resistance, but their effects on the transpeptidase activity remain undetermined. The data presented in Table 1 show that the rates of acylation by penicillins can vary by several orders of magnitude, but the protein structures are so different that it remains impossible to attribute these variations to specific amino acid substitutions. Moreover, accurate data concerning the catalytic activities of these proteins on peptide substrates are only available in a very limited number of cases.

6.2 Multiple Targets

Rapid lysis of *E. coli* occurs when both PBPs 1a and 1b are inhibited. Inactivation of PBPs 2 or 3 causes the formation of spherical cells or filaments, respectively. Ampicillin kills *E. coli* by reacting with all 4 PBPs, but other antibiotics such as cephalexin, cefuroxime, mezlocillin, piperacillin and aztreonam exhibit a much higher affinity for PBP 3 than for the other essential targets and, therefore, kill the cells by inactivating the sole PBP 3. Conversely, inactivation of PBP 2 by mecillinam is sufficient to inhibit growth (Spratt 1983).

In rapidly dividing *Enterococcus hirae* (generation time: 31 min) and at the minimum inhibitory concentration (MIC), the apparent targets depend upon the antibiotic: PBP 3 for benzylpenicillin, PBPs 2 and 3 for ampicillin and cefotaxime and PBPs 1, 2 and 3 for methicillin. At the minimum bactericidal concentration (MBC), which is 10 to 300 times larger than the MIC, all the PBPs bind the antibiotics except the low affinity PBP 5 and the killing of the bacteria is faster. When *E. hirae* is growing more slowly (generation time: 50 min), the MIC is equal to the MBC for the four antibiotics and, at these concentrations, all the PBPs are inactivated, including the low affinity PBP 5 (del Mar Lleo et al. 1987).

In laboratory mutants and clinical isolates, resistance to β-lactam antibiotics can be associated with a decrease in PBP affinity or to the acquisition or overproduction of a resistant PBP (see the three following).

6.3 Single Point Mutations

Cephalexin-resistant *E. coli* mutants can be obtained by chemical mutagenesis followed by selection. Three different amino acid substitutions in the C-terminal transpeptidase domain of PBP 3 each result in a seven-fold increase in resistance.

The first mutation (T308P) is located in the active-site serine tetrad S307XXK, the second (N361S) in the S359XN triad and the third (V344G) between these two motifs (Hedge and Spratt 1985a).

High levels of resistance to cephalexin require a remodeling of the PBP 3 active site which does not impair its ability to process the peptide substrates. A highly cephalexin-resistant PBP 3 has been obtained by four amino acid substitutions resulting from a four-step mutagenesis process, steps which reduce the affinity of the enzyme for the antibiotic and decrease its thermostability alternating with steps which have little effect on the affinity but increase the thermostability of the protein. Two mutations (N361S and E349K) were located near the active site and two (V530I and Y541L) upstream of the KTG motif. The highly cephalexin resistant PBP 3 is also more resistant to the cephalosporins whose characteristic target is PBP 3 but not to the penicillins or monobactams, an observation which reflects a very specific effect of the acylating agent's structure.

A plasmid which carries the gene encoding the highly cephalexin-resistant PBP 3 can transform sensitive strains of *E. coli*, *Salmonella typhimurium* or *Klebsiella pneumoniae* to broad-spectrum cephalosporin resistance (Hedge and Spratt 1985b).

An amino acid insertion (D345A) in *Neisseria gonorrhoeae* PBP 2, a protein very similar to the *E. coli* PBP 3 decreases its affinity for benzylpenicillin. This position is very close to that of the modified residue in one of the cephalexin-resistant *E. coli* PBP 3 (V344G). Transformation by the modified gene can confer an increased level of resistance to a sensitive strain of *N. gonorrhoeae*. This insertion could represent a first step in the development of a highly resistant PBP 2 (Brannigan et al. 1990).

In *Streptococcus pneumoniae*, resistance is mediated by alterations in 4 PBPs (1a, 2x, 2a and 2b). Moderately resistant laboratory mutants were isolated by growing the bacteria in the presence of increasing β-lactam concentrations. Cefotaxime and piperacillin resistances were due to modifications of the sensitivities of PBP 2x and of PBPs 1a and 2b, respectively. These decreased affinities could be attributed to the probably rather localized effects of a small number of point mutations (Munoz et al. 1992). In one of the cefotaxime-resistant forms of PBP 2x, the substitution of three residues (T550A, S596L and G601E) induced a 300-fold decrease of the k_2/K parameter (from 162,000 to 510 $M^{-1} s^{-1}$), resulting in a 30-fold increase in the MIC (Jamin et al. 1993). The positions modified in this and other similar mutants were almost identical to those found in some *E. coli* cephalexin-resistant PBP 3 mutants.

6.4 Mosaic Genes

PBP-mediated resistance to penicillins has been described in several clinically important pathogens among which the best documented examples are *S. pneumoniae* (pneumococci) and *Neisseria* species (Spratt et al. 1991). Levels of resistance increased over the last two decades among pneumococci and gonococci (*N. gonorrhoeae*) and more recently among meningococci (*N. meningitidis*). The observed levels of resistance were much higher than those of the laboratory mutants discussed above. By contrast to the few mutations found in the latter, the clinical isolates were characterized by extensive modifications of the PBPs resting on large numbers of nucleotide substitutions in the corresponding genes. Detailed studies of the genes coding for the *S. pneumoniae* PBPs 1a, 2x and 2b indicated that the modifications were apparently due to recombinational events resulting in the replacement of whole portions of the original genes by the corresponding pieces of homologous genes (Dowson et al. 1989; Laible et al. 1991; Munoz et al. 1992). The recombined genes thus acquired a mosaic structure, where regions barely different from those found in the sensitive pneumococci alternated with markedly modified areas, exhibiting up to 23% of nucleotide substitutions. The origin of these DNA stretches is still unknown, but streptococci of the viridans group, which are closely related and naturally more resistant to penicillins, are good candidates as donor sources (Spratt et al. 1991). It is interesting to note that the structure of the peptidoglycan cross-bridges appears to be altered in highly resistant strains, probably reflecting different substrate specificities for the modified enzymes (Garcia-Bustos and Tomasz 1990).

Gonococci et meningococci have two physiologically important PBPs (1 and 2) whose affinity for penicillin is significantly reduced in highly resistant clinical isolates. This reduction also appears to be correlated to the emergence of mosaic genes formed by similar recombination phenomena (Spratt et al. 1991, 1992). Resistant commensal *N. flavescens* and *N. cinerea* species have been identified as potential gene donors. Comparisons between the sensitive and resistant PBP 2 genes again indicated 14 to 23% substitutions in the divergent regions of the mosaic genes (Spratt et al. 1989, 1992). PBP 1 genes remain to be sequenced, but it is likely that a similar mechanism is responsible for decreased acylation rates. The presence of low affinity PBPs is not sufficient to reach the highest levels of resistance observed in gonococci, but a simultaneous alteration of the outer membrane permeability can contribute to increasing the MIC by 3 orders of magnitude (Faruki and Sparling 1986).

Mosaic genes coding for more resistant PBPs have also been described in streptococci of the viridans group and *Haemophilus influenzae* strains. Not surprisingly, all the strains where mosaic genes have been found are naturally transformable (Spratt 1988; Dowson et al. 1990).

6.5 Overexpression

PBP-mediated resistance to β-lactams can have a different origin in other intrinsic or nosocomial Gram-positive pathogens. *Staphylococcus aureus* and nosocomial, coagulase-negative staphylococci (*S. epidermidis, S. haemolyticus, S. simulans*) are usually highly sensitive to β-lactams. As explained above, methicillin was introduced to counter the emergence of β-lactamase-producing strains of *S. aureus*, but this was rapidly followed by the appearance of highly methicillin-resistant strains, the famous MRSA (for methicillin resistant *Staphylococcus aureus*) which have become a major and worldwide problem (Hackbarth and Chambers 1989; Suzuki et al. 1992).

Resistance in MRSA strains is correlated with the production of a new, low affinity PBP (PBP 2′) which is absent in the normal *S. aureus* strains and which can apparently fulfill the functions of the sensitive essential PBPs when the latter are inactivated by methicillin.

The *mecA* gene which encodes PBP 2′, is always present in MRSA strains. Its genetic inactivation induces the loss of resistance and its introduction in a sensitive *S. aureus* strain transforms it into a MRSA (Matthews and Tomasz 1990). Its expression can be constitutive or inducible and is sometimes linked to the β-lactamase plasmid. Its origin remains mysterious (Ubukata et al. 1985; de Lencastre et al. 1991). No good correlation is found between the amount of PBP 2′ produced and the level of methicillin resistance which seems to involve products of additional genes (Berger-Bachi et al. 1989, 1992).

Enterococci, a major source of serious nosocomial infections, rarely synthesize β-lactamases (Moellering 1990; Boyce et al. 1992). They are usually more resistant than other Gram-positive pathogens and this seems to be correlated with the presence of a low affinity PBP (e.g., PBP5 in *Enterococcus hirae*) which like PBP 2′ in MRSA strains can take over the role of the sensitive PBPs. Here, the levels of penicillin resistance are however well correlated to the amount of PBP5 in the *E. hirae* membranes (Fontana et al. 1983; Al Obeid et al. 1990). Its direct implication in resistance is confirmed by the high penicillin sensitivity of mutants devoid of PBP5 (Fontana et al. 1985) and by the dramatic increase of resistance exhibited by these mutants after transformation by high copy number plasmids carrying the PBP5 gene (El Kharroubi and Coyette unpubl. results). A very significant degree of sequence similarity (30%) was found between the MRSA PBP 2′ and the *E. hirae* PBP5 (El Kharroubi et al. 1991). In contrast to the situation in pneumococci, the structure of the peptidoglycan does not appear to be modified in resistant staphylococci (de Jonge et al. 1992).

7 β-Lactamases

These penicillin-destroying enzymes were discovered even before the introduction of β-lactams as chemotherapeutic agents (Abraham and Chain 1940). A very large number of β-lactamases have been presently identified, the synthesis of which can be constitutive or inducible (Bush 1989a). They exhibit very diverse substrate specificities, and new enzymes are continuously discovered and described. The primary structures of these proteins can be very different and this precludes a recent divergence from a small number of "ancestral" enzymes. But this latter mechanism has not been negligible, so that β-lactamase-mediated resistance phenomena presently rest on two distinct processes:

1. The pressure exerted on the bacteria by the intensive utilization of β-lactams has led to the selection of strains producing a wide variety of β-lactamases (see below) and the acquisition of these enzymes by an increasing number of pathogenic strains has been facilitated by the fact that the genes coding for these enzymes are often plasmid-borne.

2. A similar selection procedure has resulted in the appearance and rapid spread of new proteins upon which a very limited number of point mutations conferred the capacity of efficiently hydrolyzing compounds which had the reputation of being "β-lactamase stable". The genes coding for those new enzymes are also plasmid-borne and, within a few years, they became an additional source of acute clinical problems.

It is thus interesting to note that several enzymes capable of hydrolyzing cefotaxime and ceftazidime, often considered as good examples of "β-lactamase stable" compounds, have been identified and on the basis of their primary structures it can be concluded that the same result can be obtained either by acquiring or manufacturing a very different enzyme or by slightly altering another one (see below).

7.1 The Four Classes of β-Lactamases

The chemistry of their catalytic mechanisms divides β-lactamases into two large groups. The first group contains metalloproteins whose activity relies on the presence of a Zn^{2+} ion in their active site. For historical reasons, they have been called "class B" enzymes. Only a limited number of these enzymes are known (Waley 1992). Until recently, they had only been encountered in innocuous strains of *Bacillus cereus*. But they have now been found in more dangerous genera, such as *Pseudomonas*, *Aeromonas* and *Bacteroides* (Thompson and Malamy 1990; Rasmussen et al. 1990). They will not be further discussed here, but they represent a potential threat and it should be underlined that a detailed study of their mode of action remains the best method to prepare for their expected emergence in the future.

All the other β-lactamases described so far are active-serine enzymes which according to their primary structures belong to three distinct classes A, C and D. But despite this diversity, they all appear to hydrolyze β-lactams according to the kinetic scheme represented by Fig. 6, involving an acylenzyme intermediate similar to that shown in Fig. 7 (Waley 1992). The main difference between β-lactamases and PBPs is that, with the latter, k_3 can reach values of several thousand s^{-1}, many orders of magnitude higher than those observed with the PBPs. This purely (but enormous) quantitative factor translates into a very sharp qualitative difference: β-lactams inactivate DD-peptidases while β-lactamases efficiently destroy β-lactams.

Moreover, X-ray diffraction studies have demonstrated clear similarities between the secondary and tertiary structures of the two types of enzymes (Kelly et al. 1986; Samraoui et al. 1986; Oefner et al. 1990; Herzberg 1991; Knox and Moews 1991). On this basis, a small number of conserved residues have been identified close to the active sites of all the members of the super family of active-serine penicillin recognizing enzymes (see Fig. 4). By constrast, the individual specificity profiles of β-lactamases can also exhibit spectacular variations. For instance, the k_3 values for oxacillin can be larger than 800 s^{-1} with some class D and class A, and as low as 10^{-2} s^{-1} with several class C enzymes.

Class A enzymes are synthesized by Gram-negative and positive bacteria. The genes which encode them can be found either on plasmids or on the chromosome. Their catalytic properties are extremely different, reflecting a wide diversity in their primary structures (Matagne et al. 1990; Ambler et al. 1991). Among those produced by pathogenic strains, the *Staphylococcus aureus*, TEM and SHV enzymes are the most studied. Some 3-D structures have been determined by X-ray crystallography but, at the present time, it has not been possible to establish reliable correlations between the sequences and the specificity profiles, so that the activity of one enzyme vs one set of substrates cannot be predicted on the sole basis of its primary structure.

Class C enzymes are generally chromosome-encoded although some plasmid-borne genes have recently been described (Lindberg et al. 1988; Payne and Amyes 1991) and are only produced by Gram-negative strains. They are generally known as cephalosporinases, but can exhibit very high k_{cat}/Km values with penicillins. Deacylation is usually slower than acylation, so that the acylenzyme is the most abundant intermediate at the steady-state. Oxacillin, cloxacillin, cefotaxime and cefuroxime are very poor substrates of these enzymes, and this is due to very low (10^{-2} s^{-1}) k_3 values (Galleni and Frère 1988; Galleni et al. 1988). Enzymes produced by *Enterobacter cloacae* P99 and 908R, *Citrobacter freundii, Pseudomonas aeruginosa* and several *Serratia* strains are typical representatives of this class. The 3-D structure of the *C. freundii* enzyme has been determined (Oefner et al. 1990).

Class D enzymes are all plasmid-encoded and produced by Gram-negative strains (Couture et al. 1992). The class D OXA-1, OXA-2 and PSE-2 enzymes are characterized by a high efficiency in the hydrolysis of isoxazolyl penicillins and a poorer activity vs cephalosporins (Ledent et al. 1993). The study of these enzymes has received less attention than that of their class A and class C counterparts.

7.2 Strategies in the Struggle Against β-Lactamases

Two strategies have been devised to fight β-lactamase-producing strains.

The first strategy tends to render them useless by preparing molecules which escape their hydrolytic action. As stated above, this approach was successful against β-lactamase-producing staphylococci.

Fig. 10. Structures of clavulanate (*A*), β-halogenopenicillanates (*B*, $X = I$ or Br) and sulbactam (*C*)

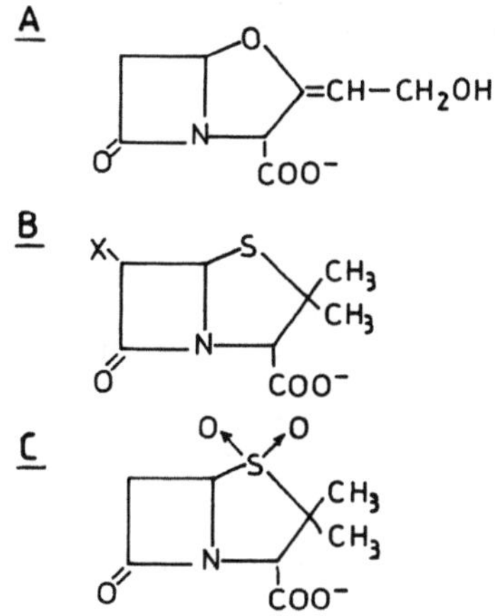

The second strategy is based on the discovery or synthesis of β-lactams such as clavulanate, β-halogenopenicillanates or sulbactam (Fig. 10) which inactivate the enzymes (Fisher et al. 1978; Brenner and Knowles 1984; De Meester et al. 1986; Pratt 1992). These compounds behave as mechanism based inactivators: the acylenzymes can rearrange into irreversibly inactivated adducts. Since they are rather poor inactivators of the PBPs, their clinical utilization rests on a concomitant administration of a classical, β-lactamase-sensititive compound such as ampicillin or amoxycillin. They are particularly efficient against class A enzymes and these mixtures have been widely and successfully utilized against the strains which produce the SHV and TEM-like enzymes. In Sects. 7.3 and 7.4 we show that a limited number of mutations have allowed bacteria to counter both strategies.

7.3 The "Extended Spectrum" β-Lactamases

The clinical utilization of compounds such as cefotaxime, ceftazidime and aztreonam became widespread around 1980, mainly in Europe, because these compounds appeared to be "β-lactamase stable". It should however be mentioned that such an appellation was somewhat misleading, since their rate of hydrolysis by several enzymes was slow, but not completely negligible (Matagne et al. 1990). They were efficient against most strains producing either the chromosome-encoded class C enzymes, or the most common plasmid-encoded β-lactamases such as TEM or SHV. Consequently, "buoyed up by false security, the infectious disease community relaxed and assumed that the problem of resistance due to β-lactamases was well under control " (Bush 1989b). But within a few years, the appearance of an increasing number of resistant strains was reported. In some cases, this resistance could be attributed to the combination of decreased permeability, due to the loss or modification of a specific porin, with

Table 3. Mutation in the SHV family

	Position		
	205	238	240
SHV1	Arg	Gly	Glu
SHV2		Ser	
SHV3	Leu	Ser	
SHV4	Leu	Ser	Lys
SHV5		Ser	Lys

For references, see Jacoby and Medeiros (1991).

Table 4. Mutations in the TEM family

	Position						
	39	104	164	237	238	240	265
TEM1	Gln	Glu	Arg	Ala	Gly	Glu	Thr
2	Lys						
3	Lys	Lys			Ser		
4		Lys			Ser		Met
5			Ser	Thr		Lys	
6		Lys	His				
7	Lys		Ser				
8	Lys	Lys	Ser		Ser		
9		Lys	Ser				Met
10			Ser			Lys	
11	Lys		His		ND		
12			Ser				
13	Lys						Met
14	Lys	Lys			Ser		Met
15		Lys			Ser		
16	Lys	Lys	His				
17		Lys			Ser		
18	Lys	Lys					
19					Ser		
23			Ser			Lys	
24	Lys	Lys	Ser	Thr		Lys	
26		Lys	Ser				

Silent mutations can also result in additional differences between the genes; ND, not determined.
For references, see Jacoby and Medeiros (1991).

the hyperproduction of a class C enzyme. In some of these "monsters", periplasmic concentrations of β-lactamase could reach 30–40 mg/ml (Hechler et al. 1989)! Other strains also produced two or more enzymes, for instance a chromosomal class C and a plasmid-encoded class A. But the discovery of transferable resistance to cefotaxime indicated that another phenomenon was involved and in 1988 Barthélémy et al. demonstrated that the cefotaxime-hydrolyzing SHV2 β-lactamase differed from the original SHV1 by the single substitution of Gly 238 by a Ser residue. Similar mutants in the SHV and TEM families were rapidly isolated, so that, at the present time, more than 5 SHV- and 23 TEM-like enzymes are known, all exhibiting hydrolytic properties vs one or more of the "β-lactamase-stable" compounds (Jacoby and Medeiros 1991; Courvalin pers. comm.). These new enzymes differ from the parent proteins by one to four residues (Tables 3 and 4) and have been isolated all over the planet, from Tunisia to Singapore, so that one cannot consider the problem to be "purely local".

The following discussion will be centered on the SHV and TEM enzymes, which are rather closely related, since the parent enzymes exhibit 68% of identical residues. It should be noted here that β-lactamases often exhibit "ragged N-termini", as shown for instance by Matagne et al. (1991), but this has little influence on the catalytic properties since this part of the protein is well away from the active site. This phenomenon results in the presence, even in highly purified preparations, of isozymes which differ only by their exact molecular sizes and isoelectric pH values.

7.3.1 The SHV Family

The SHV-2 β-lactamase was the first "extended spectrum" β-lactamase to be fully characterized. The sole Gly 238 Ser substitution was sufficient to greatly increase the activity of the new enzyme against cefotaxime, ceftazidime and aztreonam, compounds usually considered as "β-lactamase stable" (Table 5). Additional variants (SHV3, 4 and 5) were later discovered, exhibiting similar or increased activities against the same compounds (Labia et al. 1988). The same Gly 238 Ser mutation was present in all cases, together with two other variations: Arg 205 Leu and Glu 240 Lys (Table 3).

The 3-D structures are not available for the SHV enzymes, but in the known structures, residues 238 and 240 are on the β-3 strand, just after the highly conserved Lys 234-Thr(Ser) 235-Gly 236 triad (Fig. 11A).

7.3.2 The TEM Family

A similar phenomenon, although more complex, is observed in the TEM family. Despite the Gln to Lys mutation at position 39, the catalytic properties of the TEM1 and TEM2 enzymes appear to be identical and they are considered as the "original" enzymes. The mutation is responsible for the different isoelectric pH values of the two proteins. To various degrees, all the other variants exhibit increased activities against cefotaxime, ceftazidime or aztreonam (Sougakoff et al. 1988; Jacoby and Medeiros 1991).

The Gly 238 Ser mutation is present in some, but not all variants. Among the other positions, modifications of Glu 104, Arg 164 and Glu 240 appear as the most interesting. Unfortunately, detailed kinetic analyses are not available for

Table 5. Comparison of the kinetic parameters of some SHV and TEM β-lactamases

	SHV1			SHV2 (G238S)			TEM1 and TEM2			TEM3 (E104K/G238S)			E104K/R164S mutant of TEM1		
	kcat (s^{-1})	Km (μM)	kcat/Km ($mM^{-1} s^{-1}$)	kcat (s^{-1})	Km (μM)	kcat/Km ($mM^{-1} s^{-1}$)	kcat (s^{-1})	Km (μM)	kcat/Km ($mM^{-1} s^{-1}$)	kcat (s^{-1})	Km (μM)	kcat/Km ($mM^{-1} s^{-1}$)	kcat (s^{-1})	Km (μM)	kcat/Km ($mM^{-1} s^{-1}$)
Benzylpenicillin	1800	11	160000	600 (0.33)	4	150000 (0.9)	2000	22	90000	38	4	9500	32 (0.015)	6	5300 (0.06)
Cephalothin	1300	100	13000	640 (0.5)	15	40000 (3)	200	230	900	40	20	2000	–	–	–
Cephaloridin	–	–	–	–	–	–	750	660	1100	33	30	1100	30 (0.04)	90	330 (0.03)
Cefotaxime	7	> 500	< 15	420 (60)	40	10000 (> 600)	2	450	5	50 (25)	50	1000 (200)	2	84	24 (5)
Ceftazidime	< 5	> 500	< 10	40 (> 10)	24	1700 (> 200)	0.01	80	0.125	7 (700)	170	40 (320)	57 (600)	150	380 (3000)
Aztreonam	< 5	> 500	< 10	6	10	600 (> 100)	1.5	1500	1	0.2	1	200 (200)	10 (7)	64	160 (160)

The numbers in parentheses give the ratios between the measured values and those observed with the original (SHV1 or TEM1) enzymes.
For references, see Labia et al. (1988); Sirot et al. (1987); Sougakoff et al. (1988); and Sowek et al. (1991). Data for TEM3: Raquet and Frère (unpubl.); the values for cephalothin were computed on the basis of the data of Labia et al. (1988).

all the variants but Sowek et al. (1991) have produced and purified six proteins modified at these three positions by site-directed mutagenesis. Their careful study shows that the Arg 164 Ser substitution is the most efficient single mutation for increasing the activity vs cefotaxime (10-fold), aztreonam (15-fold) and mainly ceftazidime (more than 600-fold), but more spectacular effects are obtained by combining the Glu 104 Lys and Arg 164 Ser substitutions (Table 5). It is interesting to note that, in this case, the increased activity vs the "β-lactamase-stable compounds" is obtained at the expense of that against the classical, good substrates of the TEM1/2 enzymes, such as benzylpenicillin and cephaloridine. A similar cumulative effect is observed by combining the Arg 164 Ser and Glu 240 Lys mutations but the triple mutant does not exhibit an additional increase in activity, it only further loses efficiency against benzylpenicillin.

Figure 12 shows the position of the various "hot" residues in the 3-D structure of the TEM protein and Fig. 11C, D compares the structures of the modified and original side-chains.

Various attempts have been made to explain the new properties of the variants by "docking" the substrate molecules into the active sites of the proteins, but these interpretations should be considered with caution since they do not rely on an accurate knowledge of the 3-D structures and the individual rate constants have only been determined for a very limited number of interactions between the TEM2 enzyme and a few substrates.

Finally, it is interesting to note that a high level of catalytic efficiency against aztreonam has been found with other class A β-lactamases which do not exhibit any of the substitutions described here, but whose sequences differ greatly from those of the TEM or SHV enzymes (Matagne et al. 1993; Franceschini et al. 1993), an observation which shows that the same result can be obtained by apparently unrelated variations on the same original theme.

7.3.3 Decreased Sensitivity to Inactivators

The SHV and TEM variants described above remain sensitive to the usual inactivators of class A β-lactamases, clavulanate and sulbactam, so that the second strategy described in Sect. 7.2 retains its efficiency against the strains that produce them. Recently, however, new TEM variants were found which conferred resistance to the amoxycillin/clavulanate combination (Belaaouaj et al. 1992; Thomson and Amyes 1992).

Producing strains were isolated nearly at the same time in Paris and in Scotland. Sequencing of the genes revealed that one single residue of the TEM1 enzyme had been substituted: Arg 244. Further studies of enzymes modified in the same position by site-directed mutagenesis indeed showed drastic decreases in the rates of inactivation by clavulanate and sulbactam (Delaire et al. 1992). The α-carbon of Arg 244 is on the β4 strand, rather far away from the active site, but the orientation of its long side-chain is such that it could interact with the

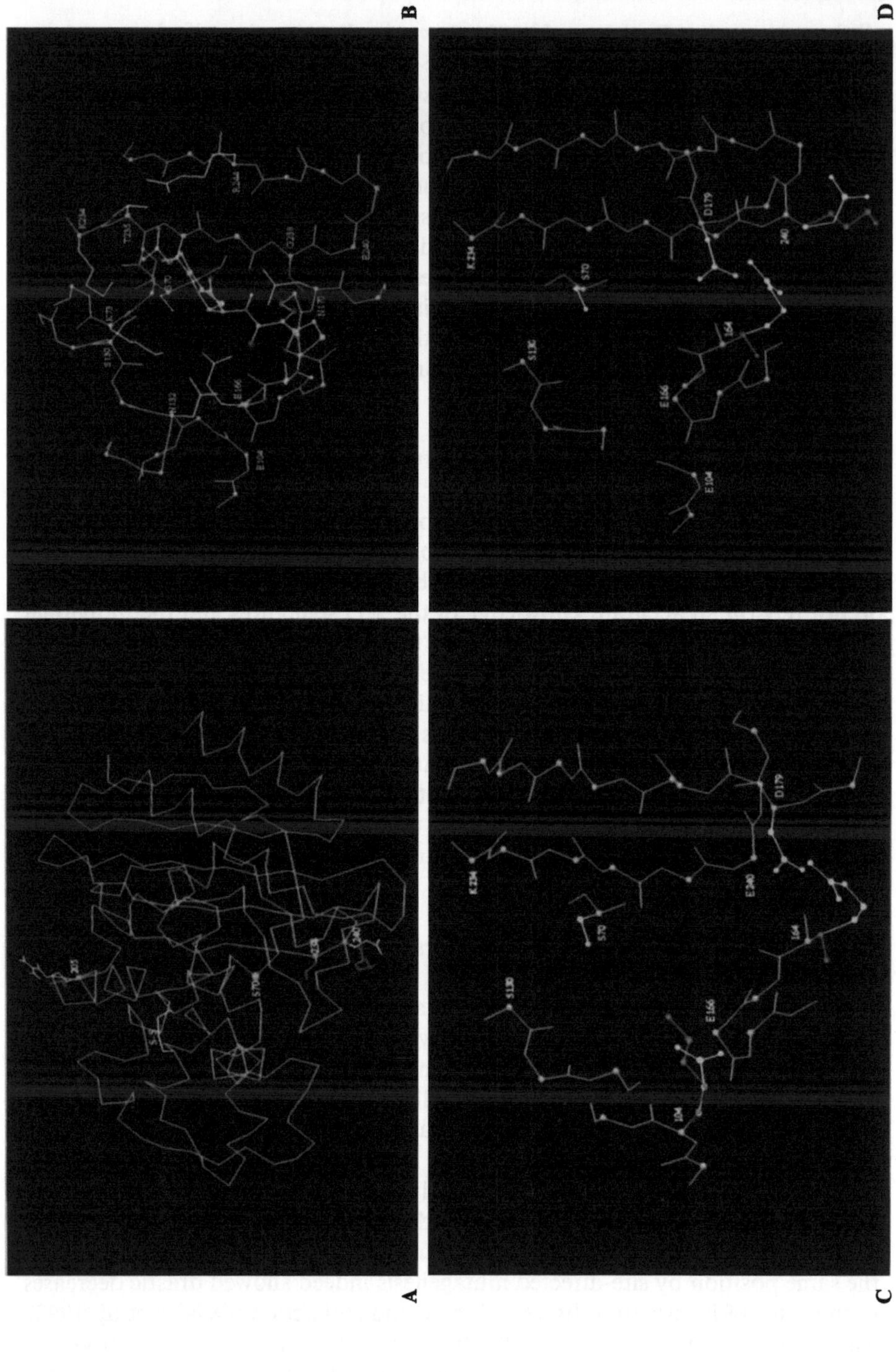
A
B
C
D
K234
S70
S130
E166
E104
D179
E240
164
104
240

Fig. 11. A Positions of the modified residues in the SHV variants. The side-chains of the SHV1 enzyme are shown in *yellow*, those of the variants in *red*. Only the α-carbon trace is represented. **B** Docking of cefotaxime (Fig. 1C, *R* =

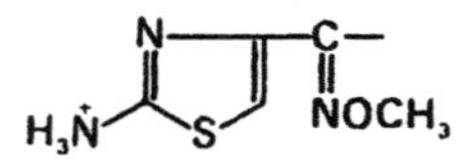

R' = CH_3–CO–O–CH_2–) in the active site of the SHV2 enzyme. The polypeptide backbone is shown in *blue*, the important conserved side-chains in *white*, and the G238→S mutation in *red*. The skeleton of the cefotaxime molecule is in *yellow* with the following color code for the individual atoms: carbon *pale green*; nitrogen *blue*; oxygen *red*; sulfur *yellow*. The proximity of the = N–O–CH_3 group of the antibiotic and of the S238 side-chain is evident. **C** Enlarged view of the active site of the double (E104K, R164S) mutant of the TEM β-lactamase. The side-chains of the original TEM-1 enzyme are shown in *white*, those of the mutants in *red*. In the former protein, a salt bridge is formed between R164 and D179, which is usually conserved in class A β-lactamases. Its disappearance in the mutant increases the flexibility of the loop extending between R164 and N170, which possibly contributes to the accommodation of the large side-chains of ceftazidime, cefotaxime and aztreonam. It can be seen on Fig. 11B that the R group of these antibiotics extends into a part of the enzyme comprising residues 104, 164 and 240. **D** Similar view of the active site of the double (R164S, E240K) mutant of the TEM β-lactamase. For details, see **C**

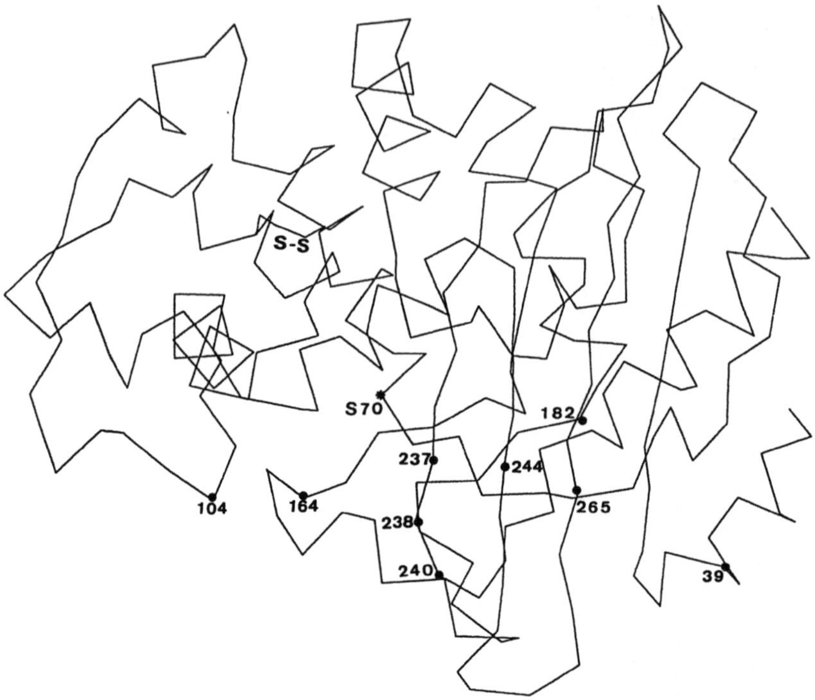

Fig. 12. Location, on the α-carbon trace, of the residues which have been found to be modified in the TEM family. The active serine (*) is also indicated. Differences at residue *39* have probably little influence on the activity, and the importance of those at residues *182* and *265* is not established with certainty. All the other residues (*104, 164, 237, 238, 240* and *244*) are seen to be very close to the active site (see also Fig. 13 for R244)

substrate or with other side-chains which are in direct contact with it. Accordingly, the modified enzymes also exhibit decreased activity against the usual good substrates of the parent enzyme but, surprisingly, hydrolyze imipenem with a somewhat increased efficiency (Zafaralla and Mobashery 1992; Zafaralla et al. 1992).

Interestingly, residue Arg 244 is not conserved in all class A β-lactamases but in several cases it seems that the guanidinium group of another Arg residue, Arg 220, occupies the same position, near the β-3 strand, between residues Thr 235 and Gly 236 (Fig. 13; Jacob-Dubuisson et al. 1991).

8 Integrated Models

The interplay between the various factors which influence the sensitivity of the bacterial cells to β-lactam antibiotics has been studied with the help of several

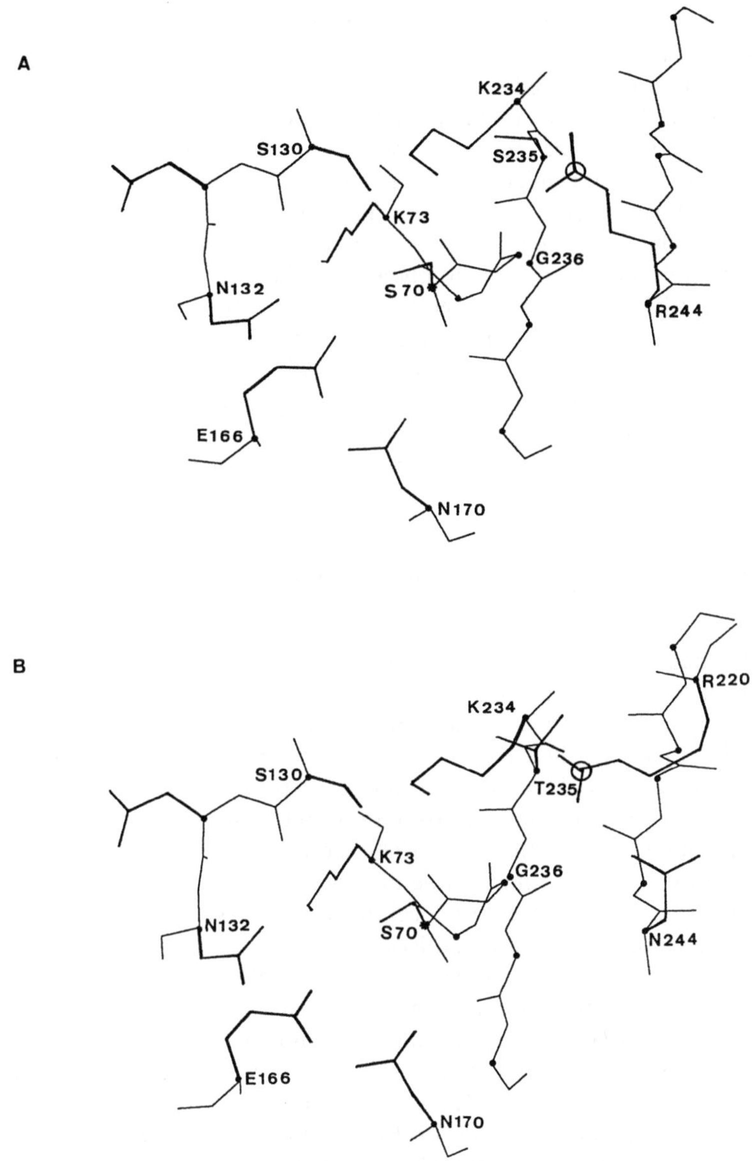

Fig. 13. Equivalent positions of the guanidinium group (○) of the R244 and R220 residues in the TEM-1 (*A*) and *Streptomyces albus* G (*B*) β-lactamases, respectively, as determined from the X-ray crystallography data. This group is seen to be in both cases very close to the side chain of residue *235*, *S* in the TEM-1 and T in the *S. albus* enzymes. The side-chains of the major conserved residues bordering the active sites of both enzymes are also shown

mathematical models based on a simple equation derived in 1977 by Zimmermann and Rosselet. Good correlations were found between the MIC values and the protection represented by the outer membrane permeability barrier and the presence of a β-lactamase (Nikaido and Normark 1987; Waley 1987; Frère 1989). In *E. coli* it appears that the former only becomes significant if a β-lactamase is also produced, but this might not be true when the outer membrane constitutes a more efficient barrier as in *Pseudomonas.* The contribution of the sensitivity of the PBPs is more difficult to estimate. Indeed, the exact proportion of active PBP necessary for cell survival or division is not known and the mechanisms by which one or several other PBPs might take over the role of an inactivated transpeptidase remain mysterious. Although some progress has been made in understanding how *E. coli* regulates the balance between cell elongation and division (Begg et al. 1990), the picture is still not clear and the situation in Gram-positive bacteria is even more obscure. The building of more accurate models will require a more profound knowledge of the transpeptidation reactions as they occur in the growing cells.

9 Glycopeptide Antibiotic Resistance

Although somewhat beside the general scope of this review, recent discoveries in the field of resistance to vancomycin and related glycopeptides provide another example of the surprising adaptability of bacteria. Whereas penicillins covalently interact with the enzyme, vancomycin binds non-covalently to the substrate (i.e., to D-alanyl-D-alanine-terminated peptides) with a very high affinity, thereby preventing the transglycosylation and transpeptidation reactions (Willams and Butcher 1981; Reynolds 1989; Courvalin 1990). Due to its size, it cannot penetrate the outer membrane of Gram-negative bacteria and is mainly useful against streptococci, enterococci (in cases of resistance or allergy to β-lactams) and methicillin-resistant staphylococci.

Plasmid-mediated resistance to these glycopeptides was first detected in 1986. The phenomenon is complex and involves a cluster of at least five genes. However, the underlying mechanism is now well understood: the substrate has been modified. The enzymatic activities of two of the genes products have been characterized. One, VanH, is a D-specific α-ketoacid reductase, the second, VanA, a ligase (Bugg et al. 1991a, b). In contrast to the "normal" ligase which can only synthesize D-alanyl-D-alanine dipeptides, VanA can form an ester bond between the D-alanine carboxylate and the α-hydroxyacid alcohol oxygen (Fig. 14). The adduct thus formed is introduced in the peptidoglycan precursor (Messer and Reynolds 1992) and numerous studies have shown that DD-peptidases efficiently utilize substrates of general structure

$$\text{R}'\text{–D alanyl–X–}\underset{\displaystyle \text{R}}{\underset{|}{\text{CH}}}\text{–COO}^-,$$

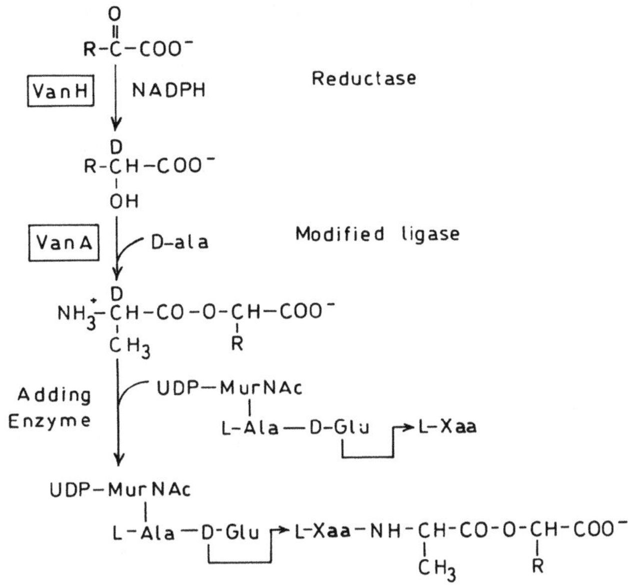

Fig. 14. Synthesis of the UDP-N-acetylmuramylpentapeptide precursor in vancomycin-resistant bacteria; *MurNAc* N-acetylmuramyl residue

where X is O or S in addition to their usual substrates where X = NH (Pratt and Govardhan 1984; Jamin et al. 1991). However, the replacement of NH by O drastically decreases (more than 1000-fold) the affinity of the molecules for vancomycin by eliminating one of the hydrogen-bonds crucial for the stability of the complex. The fact that the structure of the ligase appears to be related to those of similar enzymes found in bacteria possessing intrinsic resistance to glycopeptides (*Leuconostoc*, *Lactobacillus*) suggests a possible origin for some of the genes carried by the resistance plasmid.

10 Conclusions

The results presented in this contribution demonstrate the exquisite adaptation potential of bacteria. Thanks to their number and short generation times, they appear to be capable of solving most problems rather rapidly. In the field of chemotherapy, these observations should constitute an important warning. It is always dangerous to assume that the resistance problem is under control. Indeed, it seems that the only safe prediction which can be made is that the introduction of a new drug will, sooner or later, elicit the appearance of a new resistance mechanism which might spread to a large variety of bacteria. In turn, microbiologists and biochemists can look forward to future interesting challenges. It might be appropriate to conclude this review by quoting Nathan Citri "Many bacteria owe their survival to β-lactamases. So do some of us".

Acknowledgments. The authors wish to thank Patrice Courvalin and all their coworkers for access to unpublished material and Peter Reynolds, Henry Grosjean and Keith Tipton for kindly reading the manuscript.

References

The authors wish to underline the fact that a reference list including all original articles would have been exceedingly long. In consequence, the cited works are quite often reviews, where more references can be found, or research papers considered as particularly representative

Abraham EP, Chain E (1940) An enzyme from bacteria able to destroy penicillin. Nature 146: 837

Al Obeid S, Gutmann L, Williamson R (1990) Correlation of penicillin-induced lysis of *Enterococcusfaecium* with saturation of essential penicillin-binding proteins and release of lipoteichoic acid. Antimicrob Agents Chemother 34: 1901–1907

Ambler RP, Coulson AF, Frère JM, Ghuysen JM, Jaurin B, Joris B, Levesque R, Tiraby G, Waley SG (1991) A standard numbering scheme for the class A beta-lactamases. Biochem J 276: 269–272

Barthélémy M, Peduzzi J, Yaghlane HB, Labia R (1988) Single amino acid substituion between SHV-1 β-lactamase and cefotaxime-hydrolysing SHV-2 enzyme. FEBS Lett 231: 217–220

Begg KJ, Takasuga A, Edwards DH, Dewar SJ, Spratt BG, Adachi H, Ohta T, Matsuzawa H, Donachie WD (1990) The balance between different peptidoglycan precursors determines whether *Escherichia coli* will elongate or divide. J Bacteriol 172: 6697–6703

Berger-Bächi B, Barbis-Maino L, Strassle A, Kayser FH (1989) *FemA*, a host-mediated factor essential for methicillin resistance in *Staphylococcus aureus*; molecular cloning and characterization. Mol Gen Genet 219: 263–269

Berger-Bächi B, Strassle A, Gustason JE, Kayser FH (1992) Mapping and characterization of multiple chromosomal factors involved in methicillin-resistance in *Staphylococcus aureus*. Antimicrob Agents Chemother 36: 1367–1373

Boyce JM, Opal SM, Potter-Bynoe G, La Forge RG, Zorvos MJ, Furtado G, Victor G, Medeiros AA (1992) Emergence and transmission of ampicillin-resistant enterococci. Antimicrob Agents Chemother 36: 1032–1039

Brannigan JA, Tirodimos IA, Zhang QY, Dowson CG, Spratt BG (1990) Insertion of an extra amino acid is the main cause of the low affinity of penicillin-binding protein 2 in penicillin-resistant strains of *Neisseria gonorrhoeae*. Mol Microbiol 4: 913–919

Brenner DG, Knowles JR (1984) Penicillanic acid sulfone: nature of irreversible inactivation of RTEM β-lactamase from *E. coli* Biochemistry 23: 5833–5839

Bugg TDH, Dutka-Malen S, Arthur M, Courvalin P, Walsh CT (1991a) Identification of vancomycin resistance protein VanA as D-alanine ligase of altered substrate specificity. Biochemistry 30: 2017–2021

Bugg TDH, Wright GD, Dutka-Malen S, Arthur M, Courvalin P, Walsh CT (1991b) Molecular basis for vancomycin resistance in *Enterococcus faecium* BM4147: biosyntheis of depsipeptide peptidoglycan precursor by vancomycin resistance proteins VanH and VanA. Biochemistry 30: 10408–10415

Bush K (1989a) Characterization of β-lactamases. Antimicrob Agents Chemother 33: 259–263

Bush K (1989b) Excitement in the beta-lactamase area. J Antimicrob 24: 831–836

Courvalin P (1990) Resistance of enterococci to glycopeptides. Antimicorb Agents Chemother 34: 2291–2296

Couture F, Lachapelle J, Levesque RC (1992) Phylogeny of LCR-1 and OXA-5 with class A and class D β-lactamases. Mol Microbiol 6: 1693–1705

Cowan SW, Schirmer T, Rummel G, Ghosh R, Pauptit RA, Jansonius JN, Rosenbusch JP (1992) Crystal structures explain functional properties of two *E. coli* porins. Nature 358: 727–733

Degelaen J, Feeney J, Roberts GC, Burgen AS, Frère JM, Ghuysen JM (1979) NMR evidence for the structure of the complex between penicillin and the DD-carboxypeptidase of *Streptomyces* R61. FEBS Lett 98: 53–56

de Jonge BL, Chang YS, Gage D, Tomasz A (1992) Peptidoglycan comparison of a highly methicillin-resistant *Staphylococcus aureus* strain. The role of penicillin-binding protein 2A. J Biol Chem 267: 11248–11254

Delaire M, Labia R, Samama JP, Masson JM (1992) Site-directed mutagenesis at the active site of *Escherichia coli* TEM-1 β-lactamase. Suicide inhibitor-resistant mutants reveal the role of arginine 244 and methionine 69 in catalysis. J Biol Chem 267: 20600–20606

de Lencastre H, Sa Figueiredo AM, Urban C, Rahal J, Tomasz A (1991) Multiple mechanism of methicillin resistance and improved methods for detection in clinical isolates of *Staphylococcus aureus.* Antimicrob Agents Chemother 35: 632–639

del Mar Lleo M, Canepari P, Cornaglia G, Fontana R, Satta G (1987) Bacteriostatic and bactericidal activities of β-lactams against *Streptococcus* (*Enterococcus*) *faecium* are associated with saturation of different penicillin-binding proteins. Antimicrob Agents Chemother 31: 1618–1626

De Meester F, Frère JM, Waley SG, Cartwright SJ, Virden R, Lindberg F (1986) 6-β-Iodopenicillanate as a probe for the classification of β-lactamases. Biochem J 239: 575–580

Dowson CG, Hutchison A, Brannigan JA, George RC, Hansman D, Linares J, Tomasz A, Smith JM, Spratt BG (1989) Horizontal transfer of penicillin-binding protein genes in penicillin-resistant clinical isolates of *Streptococcus pneumoniae.* Proc Natl Acad Sci USA 86: 8842–8846

Dowson CG, Hutchison A, Woodford N, Johnson AP, George RC, Spratt BG (1990) Penicillin-resistant viridans streptococci have obtained altered penicillin-binding protein genes from penicillin-resistant strains of *Streptococcus pneumoniae.* Proc Natl Acad Sci USA 87: 5858–5862

El Kharroubi A, Jacques P, Piras G, Van Beeumen J, Coyette J, Ghuysen JM (1991) The *Enterococcus hirae* R40 penicillin-binding protein 5 and the methicillin-resistant *Staphylococcus aureus* penicillin-binding protein 2′ are similar. Biochem J 280: 463–469

Faruki H, Sparling RF (1986) Genetics of resistance in non-β-lactamase producing gonococcus with relatively high-level penicillin resistance. Antimicrob Agents Chemother 30: 856–860

Fisher J, Charnas RL, Knowles JR (1978) Kinetic studies on the inactivation of *Escherichia coli* RTEM beta-lactamase by clavulanic acid. Biochemistry 17: 2180–2184

Fontana R, Cerini R, Longoni P, Grossato A, Canepari P (1983) Identification of a streptococcal penicillin-binding protein that reacts very slowly with penicillin. J Bacteriol 155: 1343–1350

Fontana R, Grossato A, Rossi L, Cheng YR, Satta G (1985) Transition from resistance to hypersusceptibility to β-lactam antibiotics associated with loss of a low-affinity penicillin-binding protein in a *S. faecium* mutant highly resistant to penicillin. Antimicrob Agents Chemother 28: 678–683

Franceschini N, Galleni M, Frère JM, Oratore A, Amicosante G (1993) A class A β-lactamase from *Pseudomonas stutzeri* highly active against monobactams and cefotaxime. Biochem J 292: 697–700

Frère JM (1989) Quantitative relationship between sensitivity to beta-lactam antibiotics and beta-lactamase production in Gram-negative bacteria. I. Steady-state treatment. Biochem Pharmacol 38: 1415–1426

Frère JM, Joris B (1985) Penicillin-sensitive enzymes in peptidoglycan biosynthesis. CRC Crit Rev Microbiol 11: 299–396

Frère JM, Ghuysen JM, Iwatsubo M (1975) Kinetics of interaction between the exocellular DD-carboxypeptidase-transpeptidase from *Streptomyces* R61 and β-lactam antibiotics. A choice of models. Eur J Biochem 57: 343–351

Frère JM, Duez C, Ghuysen JM, Vandekerckhove J (1976) Occurrence of a serine residue in the penicillin-binding site of the exocellular DD-carboxypeptidase-transpeptidase from *Streptomyces* R61. FEBS Lett 70: 257–260

Frère JM, Nguyen-Distèche M, Coyette J, Joris B (1992) Mode of action: interaction with the penicillin-binding proteins. In: Page MI (ed) The chemistry of beta-lactams. Chapman and Hall, Glasgow, pp 148–195

Galleni M, Frère JM (1988) A survey of the kinetic parameters of class C β-lactamases. I. Penicillins. Biochem J 255: 119–122

Galleni M, Amicosante G, Frère JM (1988) A survey of the kinetic parameters of class C β-lactamases. II. Cephalosporins and other β-lactam compounds. Biochem J 255: 123–129

Garcia-Bustos J, Tomasz A (1990) A biological price of antibiotic resistance: major changes in the peptidoglycan structure of penicillin-resistant pneumococci. Proc Natl Acad Sci USA 87: 5415–5419

Ghuysen JM (1968) Use of bacteriolytic enzymes in determination of wall structure and their role in cell metabolism. Bacteriol Rev 32: 425–464

Ghuysen JM (1991) Serine beta-lactamases and penicillin-binding proteins. Annu Rev Microbiol 45: 37–67

Hackbarth CJ, Chambers HF (1989) Methicillin resistant staphylococci: detection methods and treatment of infections. Antimicrob Agents Chemother 33: 995–999

Hancock REW, Siehnel R, Martin N (1990) Outer membrane proteins of *Pseudomonas*. Mol Microbiol 4: 1069–1075

Hechler U, Van Den Weghe M, Martin HH, Frère JM (1989) Overproduced β-lactamase and the outer membrane barrier as resistance factors in *Serratia marcescens* highly resistant to β-lactamase stable β-lactam antibiotics. J Gen Microbiol 135: 1275–1290

Hedge PJ, Spratt B (1985a) Amino acid substitutions that reduce the affinity of penicillin-binding protein 3 of *E. coli* for cephalexin. Eur J Biochem 151: 111–121

Hedge PJ, Spratt BG (1985b) Resistance to beta-lactam antibiotics by re-modelling the active site of an *E. coli* penicillin-binding protein. Nature 318: 478–480

Herzberg O (1991) Refined crystal structure of β-lactamase from *Staphylococcus aureus* PC1 at 2.0 Å resolution. J Mol Biol 217: 701–719

Jacob-Dubuisson F, Lamotte-Brasseur J, Dideberg O, Joris B, Frère JM (1991) Arginine 220 is a critical residue for the catalytic mechanism of the *Streptomyces albus* G β-lactamase. Protein Eng 4: 811–819

Jacoby GA, Medeiros AA (1991) More extended-spectrum β-lactamases. Antimicrob Agents Chemother 35: 1697–1704

Jamin M, Adam M, Damblon C, Christiaens L, Frère JM (1991) Accumulation of acyl-enzyme in DD-peptidase-catalysed reactions with analogues of peptide substrates. Biochem J 280: 499–506

Jamin M, Hakenbeck R, Frère JM (1993) Penicillin Binding Protein 2x as a major contributor to intrinsic β-lactam resistance in Streptococcus pneumoniae. FEBS Letters 331: 101–104

Jarlier V, Nikaido H (1990) Permeability barrier to hydrophilic solutes in *Mycobacterium chelonei*. J Bacteriol 172: 1418–1423

Joris B, Ghuysen JM, Dive G, Renard A, Dideberg O, Charlier P, Frère JM, Kelly J, Boyington J, Moews P, Knox J (1988) The active-site-serine penicillin-recognizing enzymes as members of the *Streptomyces* R61 DD-peptidase family. Biochem J 250: 313–324

Joris B, Ledent P, Dideberg O, Fonzé E, Lamotte-Brasseur J, Kelly JA, Ghuysen JM, Frère JM (1991) Comparison of the sequences of class-A beta-lactamases and of the secondary structure elements of penicillin-recognizing proteins. Antimicrob Agents Chemother 35: 2294–2301

Kelly JA, Dideberg O, Charlier P, Wery J, Libert M, Moews P, Knox J, Duez C, Fraipont C, Joris B, Dusart J, Frère JM, Ghuysen JM (1986) On the origin of bacterial resistance to penicillin: comparison of a β-lactamase and a penicillin target. Science 231: 1429–1431

Knox JR, Moews PC (1991) beta-Lactamase of *Bacilluslicheniformis* 749/C refinement at 2 Å resolution and analysis of hydration. J Mol Biol 220: 435–455
Labia R, Morand A, Tiwari K, Sirot J, Sirot D, Petit A (1988) Interactions of new plasmid-mediated β-lactamase with third-generation cephalosporins. Rev Infect Dis 10: 885–891
Laible G, Spratt BG, Hakenbeck R (1991) Interspecies recombinational events during the evolution of altered PBP2x genes in penicillin-resistant clinical isolates of *Streptococcus pneumoniae*. Mol Microbiol 5: 1993–2002
Ledent P, Raquet X, Joris B, Van Beeumen J, Frère JM (1993) A comparative study of class D β-lactamases. Biochem J 292: 555–562
Lindberg F, Lindquist S, Normark S (1988) Genetic basis of induction and overproduction of chromosomal class I β-lactamase in nonfastidious Gram-negative bacilli. Rev Infect Dis 10: 782–785
Malhotra KT, Nicholas RA (1992) Substitution of lysine 213 with arginine in penicillin-binding protein 5 of *Escherichia coli* abolishes D-alanine-carboxypeptidase activity without affecting penicillin binding. J Biol Chem 267: 11386–11391
Matagne A, Misselyn-Bauduin AM, Joris B, Erpicum T, Granier B, Frère JM (1990) The diversity of the catalytic properties of class A β-lactamases. Biochem J 265: 131–146
Matagne A, Joris B, Van Beeumen J, Frère JM (1991) Ragged N-termini and other variants of class A β-lactamases analysed by chromatofocusing. Biochem J 273: 503–510
Matagne A, Lamotte-Brasseur J, Frère JM (1993) Interactions between active-site serine β-lactamases and so-called β-lactamase-stable antibiotics. Eur J Biochem 217: 61–67
Matthews P, Tomasz A (1990) Insertional inactivation of the *mec* gene in a transposon mutant of a methicillin-resistant clinical isolate of *Staphylococcus aureus*. Antimicrob Agents Chemother 34: 1777–1779
Messer J, Reynolds PE (1992) Modified peptidoglycan precursors produced by glycopeptide-resistant enterococci. FEMS Microbiol. Lett 94: 195–200
Minnikin DE (1982) Lipids: complex lipids, their chemistry, biosynthesis and roles. In: Ratledge C, Standford J (eds) The biology of mycobacteria. Academic Press, London, pp 95–184
Mirelman D (1979) The biosynthesis and assembly of cell wall peptidoglycan. In: Inouye M (ed) Bacterial outer membranes. Wiley, New York, pp 115–166
Moellering J (1990) The enterococci: an enigma and a continuing therapeutic challenge. Eur J Clin Microbiol Infect Dis 9: 73–74
Munoz R, Dowson CG, Daniels M, Coffey TJ, Martin C, Hakenbeck R, Spratt BG (1992) Genetics of resistance to third generation cephalosporins in clinical isolates of *Streptococcus pneumoniae*. Mol Microbiol 6: 2461–2465
Nakagawa J, Tamaki S, Tomioka S, Matsuhashi M (1984) Functional biosynthesis of cell wall peptidoglycan by polymorphic bifunctional polypeptides. J Biol Chem 259: 13937–13946
Nikaido H (1989) Outer membrane barrier as a mechanism of antimicrobial resistance. Antimicrob Agents Chemother 33: 1831–1836
Nikaido H (1992) Porins and specific channels of bacterial outer membranes. Mol Microbiol 6: 435–442
Nikaido H, Normark S (1987) Sensitivity of *Escherichia coli* to various β-lactams is determined by the interplay of outer membrane permeability and degradation by periplasmic β-lactamase: a quantitative predictive treatment. Mol Microbiol 1: 29–36
Oefner C, Darcy A, Daly JJ, Gubernator K, Charnas RL, Heinze I, Hubschwerlen C, Winkler FK (1990) Refined crystal structure of beta-lactamase from *Citrobacter freundii* indicates a mechanism for beta-lactam hydrolysis. Nature 343: 284–288
Payne DJ, Amyes SGB (1991) Transferable resistance to extended-spectrum β-lactams: a major threat or a minor inconvenience? J Antimicrob Chemother 27: 255–261
Pratt RF (1992) β-Lactamase: inhibition. In: Page MI (ed) The chemistry of β-lactams. Chapman and Hall, Glasgow, pp 229–271
Pratt RF, Govardhan CP (1984) β-Lactamase-catalyzed hydrolysis of acyclic depsipeptides and acyl transfer to specific amino aicd acceptors. Proc Natl Acad Sci USA 81: 1302–1306

Rasmussen BA, Gluzman Y, Tally FP (1990) Cloning and sequencing of class A β-lactamase gene (CCRA) from *Bacteroides fragilis* TAL3636. Antimicrob Agents Chemother 34: 1590–1592

Reynolds PE (1989) Structure, biochemistry and mechanism of action of glycopeptide antibiotics. Eur J Clin Microbiol Infect Dis 8: 943–950

Samraoui B, Sutton B, Todd R, Artymiuk P, Waley SG, Phillips D (1986) Tertiary structure similarity between a class A β-lactamase and a penicillin-sensitive D-alanyl-carboxypeptidase-transpeptidase. Nature 320: 378–380

Schockman GD, Barrett JF (1983) Structure, function and assembly of cell walls of Gram-positive bacteria. Annu Rev Microbiol 37: 501–527

Sirot D, Sirot J, Labia R, Morand A, Courvalin P, Darfeuille-Michaud A, Perroux R, Cluzel R (1987) Transferable resistance to third-generation cephalosporins in clinical isolates of *Klebsiella pneumoniae*: identification of CTX-1, a novel β-lactamase. J Antimicrob Chemother 20: 323–334

Sougakoff W, Goussard S, Gerbaud G, Courvalin P (1988) Plasmid-mediated resistance to third-generation cephalosporins caused by TEM-type penicillinase genes. Rev Infect Dis 10: 879–884

Sowek JA, Singer SB, Ohringer S, Malley MF, Dougherty TJ, Gougoutas JZ, Bush K (1991) Substitution of lysine at position 104 or 240 of TEM-1 pTZ18R β-lactamase enhances the effect of serine-164 substitution on hydrolysis or affinity for cephalosporins and the monobactam aztreonam. Biochemistry 30: 3179–3188

Spratt BG (1975) Distinct penicillin-binding proteins involved in the division, elongation and shape of *Escherichia coli* K12. Proc Natl Acad Sci USA 72: 2999–3003

Spratt BG (1983) Penicillin-binding proteins and the future of beta-lactam antibiotics. J Gen Microbiol 129: 1247–1260

Spratt BG (1988) Hybrid penicillin-binding proteins in penicillin-resistant strains of *Neisseria gonorrhoeae*. Nature (Lond) 332: 173–176

Spratt BG, Zhang QY, Jones DM, Hutchison A, Brannigan JA, Dowson CG (1989) Recruitment of a penicillin-binding protein gene from *Neisseria flavescens* during the emergence of penicillin resistance in *Neisseria meningitidis*. Proc Natl Acad Sci USA 86: 8988–8992

Spratt BG, Dowson CG, Zhang QY, Bowler LD, Brannigan JA, Hutchison A (1991) Mosaic genes, hybrid pencillin-binding proteins, and the origins of penicillin resistance in *Neisseria meningitidis* and *Streptococcus pneumoniae*. In: Perspectives on cellular regulations: from bacteria to cancer. Wiley-Liss, New York, pp 73–83

Spratt BG, Bowler LD, Zhang QY, Zhou J, Maynard-Smith J (1992) Role of interspecies transfer of chromosomal genes in the evolution of penicillin resistance in pathogenic and commensal *Neisseria* species. J Mol Evol 34: 115–125

Suzuki E, Hiramatsu K, Yokota T (1992) Survey of methicillin-resistant clinical strains of coagulase-negative staphylococci for *mecA* gene distribution. Antimicrob Agents Chemother 36: 429–434

Suzuki H, van Heijenoort Y, Tamura T, Mizoguchi J, Hirota Y, van Heijenoort J (1980) In vitro peptidoglycan polymerization catalysed by penicillin-binding protein 1b of *Escherichia coli*. FEBS Lett 110: 245–249

Thompson JS, Malamy MH (1990) Sequencing the gene for imipenem-cefoxitin-hydrolysing enzyme (CfiA) from *Bacteroides fragilis* TAL2480 reveals strong similarity between CfiA and *Bacillus cereus* β-lactamase II. J Bacteriol 172: 2584–2593

Thomson CJ, Amyes SGB (1992) TRC-1: emergence of a clavulanic acid-resistant TEM β-lactamase in a clinical strain. FEMS Microbiol Lett 91: 113–118

Tipper DJ, Strominger JL (1965) Mechanism of action of penicillins: a proposal based on the structural similarity to acyl-D-alanyl-D-alanine. Proc Natl Acad Sci USA 54: 1131–1141

Ubukata K, Yamashita N, Konno M (1985) Occurrence of β-lactamase inducible penicillin-binding protein in methicillin-resistant staphylococci. Antimicrob Agents Chemother 27: 851–857

Vedel G, Belaaouaj A, Gilly L, Labia R, Philippon A, Nevot P, Paul G (1992) Clinical isolates of *Escherichia coli* producing TRI β-lactamases: novel TEM-enzymes conferring resistance to β-lactamase inhibitors. J Antimicrob Chemother 30: 449–462

Waley SG (1987) An explicit model for bacterial resistance: application to beta-lactam antibiotics. Microbiol Sci 4: 143–146
Waley SG (1992) β-Lactamase: mechanism of action. In: Page MI (ed) The chemistry of β-lactams. Chapman and Hall, Glasgow, pp 198–228
Wientjes FB, Nanninga N (1991) On the role of the high molecular weight penicillin-binding proteins in the cell cycle of *Escherichia coli*. Res Microbiol. 142: 333–344
Wilkin JM, Jamin M, Joris B, Frère JM (1993) The mechanism of action of DD-peptidases. The role of asparagine 161 in the *Streptomyces* R61 DD-peptidase. Biochem J 293: 195–201
Williams DH, Butcher DW (1981) Binding site of the antibiotic vancomycin for a cell-wall peptide analogue. J Am Chem Soc 103: 5697–5700
Yocum RR, Amanuma H, O'Brien TA, Waxman DJ, Strominger JL (1982) Penicillin is an active site inhibitor for four genera of bacteria. J Bacteriol 149: 1150–1153
Zafaralla G, Mobashery S (1992) Facilitation of the Δ2–Δ1 pyrroline tautomerization of carbapenem antibiotics by the highly conserved arginine-244 of class A β-lactamases during the course of turnover. J Am Chem Soc 114: 1505–1506
Zafaralla G, Manavathu EK, Lerner SA, Mobashery S (1992) Elucidation of the role of arginine-244 in the turnover processes of class A β-lactamases. Biochemistry 31: 3847–3852
Zimmermann W, Rosselet A (1977) Function of the outer membrane of *Escherichia coli* as a permeability barrier to beta-lactam antibiotics. Antimicrob Agents Chemother 12: 368–372

Waley SG (1987) An explicit model for bacterial resistance: application to β-lactam antibiotics. Microbiol Sci 4:143–146
Waley SG (1992) β-Lactamase: mechanism of action. In: Page MI (ed) The chemistry of β-lactams. Chapman and Hall, Glasgow, pp 198–228
[illegible]

Chapter 6

Molecular Adaptations of Enzymes from Thermophilic and Psychrophilic Organisms

J.L. Arpigny, G. Feller, S. Davail, S. Génicot, E. Narinx, Z. Zekhnini, and *Ch. Gerday*[1]

Contents

[1] Laboratoire de Biochimie-Inst. Chimie (B6) Université de Liège au Sart Tilman, B-4000 Liège 1, Belgium

Advances in Comparative and Environmental Physiology, Vol. 20

1 Introduction

Although the average temperature on earth is probably of the order of 15 °C, the environments with extreme temperature in which life can be considered possible are very common on the surface of the planet.

Environments of high temperature are always associated with volcanic activities; hot springs, geysers, undersea hydrothermal vents are such habitats and cover temperatures ranging from 50–350 °C. The lower temperature of 50 °C constitutes, with few exceptions, the upper limit for survival of multicellular organisms, either animals or plants. Above this limit only microorganisms survive (see Brock 1985). The upper limit corresponds to temperature of hot water jets recorded in submarine hydrothermal activity in the Eastern Pacific (Spiess et al. 1980). As far as life is concerned this upper limit is not very realistic since peptide and phosphodiester bonds involved in proteins and nucleic acids have a half-life not exceeding 1 s at 250 °C. The latter temperature is the highest recorded for growing "black smoker" bacteria (Baross and Deming 1983) but is probably an artefact. It is therefore more reasonable to admit that the upper limit for life is of the order of 110 °C, temperature at which methanogenic archaebacteria and others were found to grow with reasonably low doubling time (Huber et al. 1989).

Environments with low temperature are certainly the most abundant if one takes into account not only the polar regions but also the oceans which, below a depth of 1000 m, currently display temperatures lower than 5 °C.

All the habitats of extreme temperature have been successfully colonized by numerous ectothermic organisms showing remarkable temperature adaptations at the ultrastructural and molecular levels which have been brought to the attention of the scientific community by the pioneering works of Peter W. Hochachka and Georges N. Somero (reviews: Somero 1977; Hochachka and Somero 1984; Hochachka 1991).

The temperature adaptations concern in particular enzymes displaying, in most cases, activities strongly dependent on temperature. One can therefore first ask the general question: How do ectothermic organisms living in extreme temperature environments succeed in maintaining adequate metabolic fluxes despite the temperature sensitivity of their enzymes? Clearly, their enzymes have molecular structures adapted to the temperature of the environment. A further question follows: What are the molecular bases of these adaptations?

For a non-catalyzed chemical reaction $S \rightleftharpoons P$ the rate of reaction is given by the equation:

$$V = \frac{kKT}{h}[S]\,e^{-\Delta G*/RT},$$

in which k is the transition factor often considered to be equal to one, K is the Boltzmann's constant (1.38×10^{-23} J/K), h the Planck's constant (6.63×10^{-34} J/Hz), T the temperature (in degrees Kelvin), S the concentration of the initial substance or substrate, ΔG^* the activation energy and R the gas constant (8.31 J/K.mol). This indicates that the rate constant depends exponentially on the temperature but in a way strongly dependent on the value of the activation energy: the greater the activation energy ΔG^*, the stronger the temperature dependence of the reaction rate. Reactions with low values of activation energies (10 kJ/mol) have rates that increase only slightly with temperature contrarily to reactions with large activation energies (50 kJ/mol) which display rates depending strongly on the temperature. The Q_{10} value (V_{T+10}/V_T), representing the ratio between reaction rates measured at an interval temperature of 10 °C is commonly used to describe the effect of temperature upon reaction rates: for most single biochemical reactions or sequences of reactions (i.e., metabolic pathways) the Q_{10} values are generally between 2 and 3.

For enzyme-catalyzed reactions, the situation is much more complicated since the temperature dependence of the reaction rate will depend not only on the enzyme concerned but also on the enzyme/substrate ratios. Indeed, the function of enzymes is modulated by conformational changes. Whereas covalent linkages of amino acids provide the backbone of the polypeptide chain, weak bonds are responsible for almost all biochemical properties of the protein because they stabilize the folded (i.e., active) conformation. These weak interactions are involved in the formation of enzyme – substrate complexes, they direct the catalytic mechanism and its regulation by cofactor binding for example, but also determine the folding of secondary structure, domains packing, intramolecular interactions and ion binding. Covalent bonds are virtually unaffected within the considered temperature range. For enzymes showing elevated k_{cat} value and low activation energy, the activity at saturated substrate concentration is almost independent of the temperature and Q_{10} values are close to 1. By contrast, less efficient enzymes characterized by elevated activation energy have Q_{10} values which can be higher than 3. The situation can be completely different at low substrate concentration, a common situation, especially in the intracellular space. In this case, the rate of reaction becomes strictly dependent on the affinity of the substrate for the enzyme. If one considers that in most cases the affinity can be correctly expressed by the K_M, the thermodependency of the K_M becomes the primary determinant of the rate of reaction. In this context, the interaction between substrate and enzyme essentially occurs through few weak bonds which can be affected in a differential way by lowering or increasing the temperature of the system. Indeed, electrostatic and hydrogen bonds which can be implicated in the enzyme – substrate interaction are formed exothermically; heat is released and these bonds are stabilized by a decrease in temperature. So if this type of bonds is essential for the enzyme – substrate interaction, low temperatures will favor the affinity of the substrate for the enzyme, tending to lower Q_{10} value. On the other hand, at elevated temperatures, the affinity of the substrate for the enzyme will decrease contributing to increase Q_{10} value. Contrarily, hydrophobic bonds are formed endothermically (i.e., with absorption of heat); these bonds will therefore be stabilized by an increase of the temperature of the medium. If the interaction between enzyme and substrate is mainly driven by this type of bonds, any decrease in temperature will also decrease the affinity of the substrate for the enzyme. This will contribute to a high Q_{10} value resulting from the superposition of two factors: low kinetic energy and disrupting effects of low temperatures on the interaction.

In conclusion, generalizations concerning the effect of low and high temperatures on enzyme-catalyzed reactions occurring in living organisms are unrealistic.

It is also clear that the problem of the adaptation is completely different when one considers enzymes from psychrophilic or thermophilic organisms. Both have to maintain high or at least adequate catalytic efficiency at the temperature of the environment. For enzymes from psychrophilic organisms the problem is probably, as discussed below, to ensure adequate flexibility of the molecule at low temperatures, whereas for enzymes from thermophilic organisms the problem is to maintain the 3D structure of the enzyme at high temperatures as well as the integrity of the enzyme – substrate complex. In a 3D structure of a protein, slight thermal rotations or vibrations of interconnecting bonds confer a significant flexibility to the whole molecule. This dynamic is essential for catalysis since the interaction between enzyme and substrate can be considered as induced: the protein adopts the shape of the substrate and this implies structural constraints fragilizing the bonds which have to be broken. For any given protein from organisms adapted to different temperatures the flexibility of the protein structure has more or less the same value at the temperature of the environment and depends on weak bonds that stabilize both secondary and tertiary structure. The number, type and configuration of these bonds ought to be in accordance with disruptive forces experienced by the enzyme. In the present context, the temperature of the environment will be considered as the major influence.

2 Temperature and Protein Stability

The thermodependency of protein stability has been discussed in numerous excellent reviews (e.g., Privalov 1979; Baldwin and Eisenberg 1987; Privalov and Gill 1988; Creighton 1991). Therefore we limit the discussion of this problem to the essential features pointed out by the above-mentioned authors.

As stated by Privalov and Gill (1988), the unfolding of globular proteins such as enzymes is a two state phenomenon implicating the folded (N) and unfolded polypeptide chains (D), while intermediate states only represent a very small fraction of the two fundamental states. The unfolding process appears as a cooperative process occurring generally after slight modifications of the environmental conditions and leads to the absorption of a definite energy. On the contrary, large multidomain proteins can unfold according to discrete states corresponding to the disruption of individual structural modules which compose the whole protein.

The free energies of the folded and unfolded states are very similar and are close to $5\ 10^7$ kJ mol^{-1} for a protein of 100 amino acids, including the covalent bonds (Baldwin and Eisenberg 1987). The Gibbs energy difference between the two states is therefore small and has been tentatively evaluated to 20–60 kJ mol^{-1} (Privalov 1979); it corresponds to the energy required to achieve the transition from the folded to the unfolded state. It is a measure of the stability of the protein structure or of the cooperative unit if the protein is a large multidomain protein.

Stabilizing forces are intramolecular hydrogen bonding, electrostatic, Van der Waals and hydrophobic interactions. Other factors, such as disulfide bonds, tend to decrease the conformational entropy of the unfolded chains (Flory 1956; Matsumura et al. 1989). Their stabilizing contribution is a function of the size of the loop formed and, of course, of the compatibility of the cross-link with the folded structure (Goldenberg 1985).

Other amino acid residues, as discussed later on, affect the conformational entropy: Pro, with only one dihedral angle, decreases the entropy of the unfolded state; on the contrary. Gly, which has no side chain, increases the degrees of freedom of an unfolded polypeptide backbone. The replacement of a Gly residue reduces the backbone flexibility and can destabilize the unfolded state by as much as 3.3 kJ mol^{-1} at 65 °C (Nemethy et al. 1966).

Destabilizing forces consist in favorable interactions of the unfolded state with the solvent as well as in the suppression of stereochemically unfavorable interactions in the unfolded molecule. However, the main factor favoring the unfolded state is the gain in conformational entropy resulting from the transition from the folded to the unfolded state.

The discrete balance mentioned above, between stabilizing and destabilizing factors also confers on the protein, at the physiological temperature (i.e., the temperature usually experienced by the organisms producing the protein) an adequate flexibility essential for functions.

As illustrated by Privalov (1979) and Privalov and Gill (1988), the temperature-dependent profile of the free energy of stabilization of globular proteins takes the overall shape of a parabole (Fig. 1). Both heating and cooling from the temperature of maximum stability lead to a decrease in stability eventually giving rise to the transition between the native (N) to the denatured state (D). Low temperatures weaken hydrophobic bonds so that cold denaturation is accompanied by a release of heat. On the contrary, at elevated temperature the denaturation is accompanied by absorption of heat and

$$\Delta_N^D H(T) = H^D - H^N$$

is positive. The difference in enthalpy varies as a function of temperature as does the difference in entropy. This is a main consequence of the significant increase

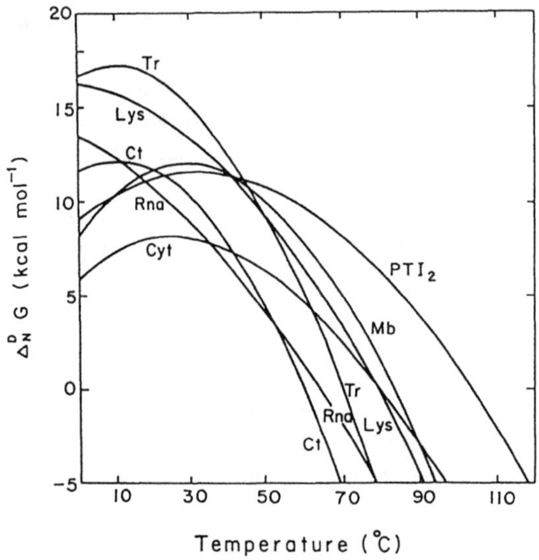

Fig. 1. $\Delta_N^D G$ values as a function of temperature for lysozyme (*Lys*), ribonuclease a (*Rna*), α-chymotrypsin (*Ct*), cytochrome C (*Cyt*), *met-myo-globin* (*Mb*), dimer of pancreatic trypsin inhibitor (*PTI*$_2$). (Privalov 1979)

of the partial heat capacity accompanying the transition of a globular protein from the native to the denatured state (Privalov and Gill 1988). Indeed,

$$\Delta_N^D H(T_2) = \Delta_N^D H(T_1) + \Delta_N^D C_p(T_2 - T_1),$$

with $\Delta_N^D C_p$ being roughly constant between 0 and 80 °C.

As a consequence, the change in free energy Δ_N^D G (Gibbs' energy) occurring upon unfolding is a complex function of temperature which can be expressed as follows (Privalov and Gill 1988):

$$\Delta_N^D G(T) = \Delta_N^D H(T_G^{den}) \frac{(T_G^{den} - T)}{T_G^{den}} - \Delta_N^D C_p (T_G^{den} - T)$$
$$+ \Delta_N^D C_p T . \ln (T_G^{den}/T),$$

in which T_G^{den} is the transition temperature at which the populations of molecules in the native and the denatured states are equal.

3 Enzymes from Thermophilic Organisms

The interest in enzymes produced by thermophilic organisms started with the growing importance of biotechnology and the need for more thermostable proteins which could improve the efficiency of many industrial processes (Klibanov 1979; Fontana 1988; Steel and Walker 1991). From the point of view of fundamental research, thermophilic enzymes provide an excellent model for investigating the different factors governing the stability and the flexibility of protein structure, which are essential for maintaining the proper biological activity of the molecule in its natural environment.

3.1 General Stability of Thermophilic Enzymes

In order to investigate the remarkable stability of thermophilic enzymes, a number of studies were carried out on the resistance of these proteins to denaturing agents other than heat. Homologous enzymes from mesophilic and thermophilic organisms were compared in experiments performed at moderate temperatures (25–40 °C). All these works led to the conclusion that thermophilic enzymes resist much higher concentrations of denaturants such as urea (Chaffotte et al. 1992), guanidinium chloride (Sanbongi et al. 1989), detergents (Fujita et al. 1976), and organic solvents (Veronese et al. 1984).

In addition, other studies have shown the positive correlation between the thermostability of proteins and their resistance to enzymatic proteolysis (Daniel et al. 1982; Suzuki and Imai 1982; Veronese et al. 1984).

Fontana (1988) has demonstrated that enzymatic cleavages of the extremely thermostable thermolysin occur preferentially in loops or random-coiled regions of the three-dimensional structure of the protein, which strongly suggests that the substrate recognition by proteases takes place more easily in the most flexible regions of globular proteins. This is consistent with the observation that protein denaturation, and thus the flexibility of the polypeptide chain, favors proteolysis (Privalov 1979).

All these facts indicate that thermophilic enzymes have a tight and cohesive structure with very few mobile elements sensitive to adverse factors coming from the neighboring solution.

The relation between the thermostability of proteins and their flexibility was studied by Vihinen (1987). The author selected proteins (of six different families) for which refined crystallographic structures were known. A flexibility index was calculated for each protein using atomic temperature factors which indicate thermal fluctuations in the molecule. The results confirmed the hypothesis that a higher thermostability is associated with a reduced flexibility of the polypeptide chain.

In their study on glyceraldehyde-3-phosphate dehydrogenase (GAPDH) from the extreme thermophile *Thermotoga maritima*, Wrba et al. (1990) recorded the hydrogen-deuterium exchange of peptide hydrogens in comparison with mesophilic GAPDHs. This criterion gives an estimate of the lability of the numerous hydrogen bonds which stabilize secondary structures. In this case, it was also concluded that the thermophilic enzyme is characterized by a more rigid structure.

It is important to stress the fact that, as a general rule, the overall topology of thermophilic enzymes is highly conserved as compared to their mesophilic homologs (Perutz and Raidt 1975; Grütter et al. 1979; Mozhaev and Martinek 1984) as well as their catalytic mechanism, since the amino acid residues involved in substrate (or cofactor) binding, and in reaction catalysis, remained unchanged (Wrba et al. 1990; Dekker et al. 1991; Imada et al. 1991).

The remarkable stability which characterizes thermophilic enzymes is commonly attributed to subtle, and sometimes conservative, changes in their primary structure which improve the intramolecular binding and increase the free energy difference between the folded and the unfolded states of the protein. As stated above, this difference is of the order of 20–60 kJ/mol. In the case of thermophilic enzymes, it is raised by 20 to 40 kj/mol (Privalov 1979; Mozhaev and Martinek 1984). These values indicate that the extra energy of stabilization of thermostable proteins is small and can be provided by very few additional weak bonds. Indeed a single hydrogen bond or a salt bridge for example typically contributes 4 to 12 kJ/mol to the conformation stability (Privalov 1979).

In the following, we shall focus our interest on the different molecular features which are capable of enhancing the cohesion and the stability of protein structures. Natural stabilizing factors as well as some improvements in protein stability obtained by genetic engineering will be discussed (see also Fontana 1991; B.W. Matthews 1991). No extrinsic influences, such as immobilization or chemical modifications, will be examined here (for reviews on these topics see Mozhaev et al. 1988 and Gupta and Mattiasson 1992).

3.2 The Diversity of Internal Stabilizing Factors

The conformation of a protein is submitted to two strong and opposing influences: on the one hand, the tendency to a higher entropy state leads to an unfolded structure; on the other hand, a large number of intramolecular links stabilizes the native form.

Upon unfolding, the protein molecule undergoes a free-energy change which can be attributed to several factors described by the following equation (Tanford 1970):

$$\Delta G = \Delta G_{conf} + \sum_i \Delta g_{i,int} + \sum_i \Delta g_{i,s} + \Delta W_{el}, \tag{3.1}$$

where ΔG_{conf} is the change in conformational free energy related to the rotations about peptide and side-chain bonds, $\Delta g_{i,int}$ and $\Delta g_{i,s}$ concern respectively short-range interactions inside the molecule (hydrogen bonds and van der Waals forces) and with the solvent (hydrophobic effects, hydration), and ΔW_{el} accounts for the long-range electrostatic interactions (Coulomb interactions between charged groups).

These four categories of forces are involved in the stabilization of thermophilic enzymes either by decreasing the free energy of the native state (mainly through electrostatic bonds and hydrophobic interactions) or by increasing the free energy of the unfolded state (essentially through entropic effects obtained by decreasing the degrees of freedom of the unfolded chain).

3.3 Factors Affecting the Entropy of the Unfolded Chain

The process of protein denaturation is usually represented by the following equation in which the native form (N) is reversibly unfolded into a random coil (U) before being submitted to an irreversible denaturation (D state) by proteolysis, deamidation, aggregation, etc.

$$N \rightleftharpoons U \rightarrow D\,. \qquad (3.2)$$

The main force that favours the first denaturation step is the considerable increase of entropy arising from the unfolding of the polypeptide chain (Dill et al. 1989).

Any structural characteristic of the protein which reduces this gain of entropy will drive the equilibrium backwards and will stabilize the native form. Such structural features occur naturally in some thermostable enzymes and appear to be important stabilizing factors.

3.3.1 Disulfide Bonds

The fact that intramolecular cross-linking reduces the conformational freedom of a random coil is easily conceivable and the subsequent decrease in entropy of the molecule was theoretically demonstrated by Kuhn and Majer (1956).

As a confirmation of this result, disulfide links have been proved experimentally to reduce the unfolding entropy of proteins especially in the well-documented cases of trypsin and of pancreatic trypsin inhibitor (Privalov 1979).

More recently, the thermostability of aqualysin I produced by the thermophilic strain *Thermus aquaticus* appeared to rely partly on the presence of two disulfide bonds which cross-link the polypeptide chain (Kwon et al. 1988; Takagi et al. 1990). However, this strategy seems to rarely contribute to thermophily probably because of the difficulty to obtain adequate configurations of the disulfide bonds. Indeed, the favorable orientation of the cysteine side-chains involved is restricted by severe spatial constraints (Wetzel 1987). All other orientations induce high-energy states which counterbalance the effect on the unfolding entropy. A disulfide bond is therefore less prone to stabilize a protein than non-covalent links (electrostatic and hydrophobic interactions) which are more tolerant with respect to the geometry of the molecule. Consequently, the introduction of disulfide bridges in proteins by site-directed mutagenesis rarely provides a higher resistance to thermal inactivation. Seven different disulfides were introduced separately in subtilisn BPN' (Mitchinson and Wells 1989) and one in dihydrofolate reductase (Villafranca et al. 1987) at sites carefully chosen by computer modelling, but none of them actually stabilized the proteins.

On the contrary, an engineered disulfide bond could be introduced in T4 lysozyme (Wetzel et al. 1988) which increases the melting temperature of the enzyme from 65 to 68 °C at pH 6.5. Similarly, the thermal stability of subtilisin E was enhanced by a disulfide link (the melting temperature was shifted from

58.5 to 63 °C at pH 6.2) which was positioned on the basis of structural homology with the thermophilic aqualysin I (Takagi et al. 1990).

An argument concerning the mechanism of stabilization should be mentioned here: Wetzel et al. (1988) proposed from their experiments on T4 lysozyme that the disulfide could act kinetically by slowing down the irreversible inactivation process [see Eq. (3.2)] rather than by shifting the thermodynamic equilibrium towards the native state.

All these results show that the effect of a disulfide bond on protein stability is not unique and that predicting or even interpreting its influence is not an easy task.

3.3.2 Proline Residues

The cyclic structure of proline severely impairs the rotations about its $N{-}C^{\alpha}$ bond. Consequently, the presence of this residue in a protein greatly reduces the number of possible local conformations of the molecule backbone. This implies that proline confers more rigidity to the native protein and reduces the conformational entropy of its unfolded state.

Watanabe et al. (1991) compared the oligo-1,6-glucosidases from the thermophile *Bacillus thermoglucosidasius* and from the mesophile *Bacillus cereus.* They found that the higher thermostability of the thermophilic enzyme could be explained by the presence of 14 additional prolines (besides 18 conserved prolines) as compared to the mesophilic enzyme. Interestingly, most of these extra prolines (11 out of 14) are situated in loops or β-turns joining adjacent secondary structures. They stabilize the relative orientations of β-strands and α-helices by rigidifying the connecting segments and therefore give more cohesion to the overall structure.

In addition, the entropic effect of these prolines is not negligible. It was evaluated by Matthews et al. (1987) who increased the thermostability of T4 lysozyme by inserting a proline in position 82 (i.e., in a β-turn). The decrease of the unfolding entropy due to this substitution (≈ -12 J/mol/K) could raise the unfolding free energy by approximately 4 kJ/mol at the melting temperature of the wild-type enzyme (65 °C).

More information on recent studies supporting the effect of proline on the thermostability of enzymes is available from the paper by Watanabe et al. (1991).

3.3.3 Other Residues

Owing to the relative bulkiness of their side-chain some amino acid residues have a backbone conformational freedom which is restricted with respect to other residues when they are involved in a polypeptide chain. It has been proposed by Matthews et al. (1987) that they could therefore reduce the backbone entropy of the unfolded protein and thermodynamically stabilize its native form.

For example, the judicious replacement of a glycine (no side-chain) by an alanine stabilized T4 lysozyme by 1.6 kJ/mol at pH 6.5 as a result of a decrease of its unfolding entropy (Matthews et al. 1987).

This substitution, as well as the introduction of residues with branched β-carbons (threonine, valine) proposed by these authors, could constitute a strategy exploited by naturally thermostable enzymes since these preferences also appear from the statistical study of thermophilic enzymes by Menendez-Arias and Argos (1989; see below).

3.4 Electrostatic Interactions

Electrostatic effects involve permanent dipoles or interactions between charged groups. As pointed out by Perutz (1978), they often play an important role in protein folding, in enzyme catalysis and in conformational stability. Numerous examples show that the temperature adaptation of thermophilic enzymes frequently relies on a strengthening of electrostatic bonds or on an increase of their number. Since these bonds form with a concomitant release of energy, they specifically stabilize the native conformation of the protein and increase the activation free energy of the unfolding process. Their effect on protein stability is thus of kinetic nature.

3.4.1 Hydrogen Bonds

After submitting subtilisin BPN' to random mutagenesis, Bryan et al. (1986) obtained a subtilisin variant with an increased thermostability and differing from the wild-type enzyme by a single mutation: an asparagine at position 218 (hydrogen-bonded to Ser-204) is replaced by a serine. From the refined crystal structures of both proteins, the authors could ascribe the gain in stability firstly to a better geometry of the hydrogen bond between the mutated Ser-218 and Ser-204, and secondly to the smaller side-chain of serine which allows a closer packing of the local secondary structure. It is worth mentioning that three thermostable proteases (thermitase, proteinase K and thermomycolase, which are structurally and enzymatically related to subtilisin) also have a serine at position 218, whereas five subtilisins from mesophilic *Bacillus* species conserve the asparagine at this position (Bryan et al. 1986). This experiment shows that a subtle change in the strength of a single hydrogen bond can be crucial for the stabilization of subtilisin and that the same structural feature probably contributes to the thermostability of thermophilic subtilisin-like proteases.

In the case of rubredoxin from the extreme thermophile *Pyrococcus furiosus* (optimal growth at 100 °C), the main structural difference with its mesophilic homologs lies in the three-stranded β-sheet of the molecule (Blake et al. 1991). This domain includes one more residue (Ala-1) and is further stabilized by an additional H-bond between the carbonyl of Ala-1 and the amide hydrogen of

Glu-14. Moreover, the possible confinement of the N-terminal amine between the carboxyl groups of Glu-14 and Glu-52 probably contributes to the hyperthermostability of this enzyme.

The large number of H-bonds usually found in proteins induces a cooperative behavior between these bonds and makes them almost equally important for the overall stability of the molecule. However, as shown by the above mentioned examples, single H-bonds are able to compensate for structural weaknesses. Furthermore, the countless possibilities to form hydrogen bonds between amino acids confer on these bonds a high potential to play such a stabilizing role.

3.4.2 Salt Bridges and Arginine Residues

In their early crystallographic study on ferredoxins from mesophilic and thermophilic bacteria, Perutz and Raidt (1975) showed that the greater heat stability of the thermophilic enzyme mainly arises from two salt bridges linking residues situated at opposite ends of the polypeptide chain.

More recently, Tomazic and Klibanov (1988) explored the mechanism of stabilization of the thermostable α-amylase from *Bacillus licheniformis* and compared it to the α-amylases from mesophilic *Bacillus* strains. It appeared that at least one extra salt bridge is responsible for the increased thermal stability and, more precisely, that this bond slows down the reversible unfolding of the thermophilic enzyme at elevated temperatures.

Besides this, it seems that arginine is generally better qualified than lysine for the formation of salt bridges (Mozhaev and Martinek 1984; Mrabet et al. 1992). In addition, Merkler et al. (1981) and Menendez-Arias and Argos (1989) show in their comparative studies that thermophilic enzymes contain a higher proportion of arginines with respect to lysines when compared to their mesophilic counterparts.

There are at least three possible reasons for this. Firstly, the side-chain of arginine is somewhat longer than that of lysine and can therefore stretch out further to bind to a distant Asp or Glu. Secondly, the charge resonance which is possible in the guanidinium group of arginine distributes the charge between three atoms; this has a stabilizing effect and gives arginine the possibility to make more than one electrostatic bond. Thirdly, the larger hydrophilic head of the arginine side-chain covers a shorter hydrophobic arm and both these characteristics render the interaction with water molecules more favorable. This last argument is particularly important since the great majority of Lys and Arg are situated at the surface of globular proteins, where a contact with water occurs (Schultz and Schirmer 1979), and of course it also applies when the basic residue is not involved in a salt bridge.

A remarkable demonstration of these properties based on crystallographic structures and on computer modeling was described in a recent report by Mrabet et al. (1992): Lys → Arg substitutions improve the thermal stability of three structurally unrelated proteins (xylose isomerase, CuZn-superoxide dismutase and glyceraldehyde-3-phosphate dehydrogenase) thanks to additional H-bonds due to the arginine guanidinium group. Moreover, the interaction between the subunits of the tetrameric xylose isomerase is reinforced by an arginine which can bind directly to an aspartate in the facing subunit, whereas the shorter lysine side-chain in the wild-type enzyme binds less firmly via a bridging water molecule.

3.4.3 Calcium Ions

The binding of calcium ions (one to four) to subtilisin-like proteases and to thermolysin proved to be determining for the kinetic thermal stability of these enzymes (Voordouw et al. 1976). The role of calcium in these molecules is to maintain flexible elements, such as surface loops, tightly bound to the globule surface. Furthermore, the high resistance of thermitase and thermolysin towards denaturation and autolysis (Fontana 1988; Siezen et al. 1991) is at least partly due to the presence of respectively three and four calcium ions in these thermophilic enzymes, whereas mesophilic subtilisins have only two calcium-binding sites.

3.4.4 Electrostatic Stabilization of α-Helices

It has been suggested that the parallel alignment of the dipolar peptide bonds in α-helices induces the formation of a macroscopic dipole which would be equivalent to the presence of a positive 0.5 elemental charge at the N-terminus of the helix and a corresponding negative charge at the C-terminus (Hol et al. 1978). It is therefore supposed that a charged residue situated at either end of the helix would interact favorably or not with the helix dipole depending on the relative signs of the charges.

Sanbongi et al. (1989) compared two cytochrome molecules, one from the thermophile *Hydrogenobacter thermophilus* and the other one from the mesophile *Pseudomonas aeruginosa.* The higher stability of the thermophilic enzyme was partly attributed to the replacement of the unfavorable charged lysine at the N-terminus of helix 2 by a non-charged Ala and to the introduction of a lysine at the C-terminus of helix 3. Several charged residues were mutated at both ends of α-helices in another cytochrome (Caffrey and Cusanovich, 1991), in barnase (Sancho et al. 1992) and in a neutral protease (Eijsink et al. 1992). All these studies tend to demonstrate that a stabilization of proteins can be achieved by a favorable interaction between α-helix dipoles and charged amino acids.

3.4.5 Aromatic–Aromatic Interactions

Burley and Petsko (1985, 1988) pointed out a special kind of interaction which arises between phenyl rings: the polarization of C–H bonds in these rings can induce an electrostatic attraction between the δ^- π-electron cloud of one ring and the δ^+ hydrogen atoms of a second ring. The interaction is maximized when both rings form a dihedral angle close to 90° and when their centers are approximately 5 Å apart. Burley and Petsko therefore proposed that pairs of aromatic residues (Phe, Tyr, Trp) with appropriate configurations can have a stabilizing effect on protein structures. Their study showed that 60% of aromatic side-chains in proteins are involved in such associations and are highly conserved in different protein families. The free-energy contribution of an aromatic interaction to the stability of protein structures was estimated at about 6 kJ/mol, which is non-negligible in comparison with the contribution from other weak bonds.

As an example, the extra thermal stability of a λ-repressor variant (melting temperature increased by 6 °C; Hecht et al. 1984) can be explained by a Gln → Tyr mutation allowing an additional aromatic interaction (Burley and Petsko 1985).

In addition, thermitase (a thermophilic subtilase) has been shown to contain at least 10 aromatic pairs (Teplyakov et al. 1990), which is twice as many as observed in the mesophilic subtilisins BPN' and Carlsberg (Siezen et al. 1991). This comparison suggests that aromatic interactions could also contribute to the high stability of thermophilic enzymes.

3.5 Hydrophobic Interactions

These interactions appear in the three-dimensional structure of proteins from the clustering of hydrophobic amino acid side-chains and are maintained by van der Waals contacts between non-polar atoms. This behavior reduces the extent of unfavorable contacts with polar water molecules and may be a major driving force intervening in the process of protein folding (C.R. Matthews 1991).

Each non-polar group exposed to an aqueous solvent is surrounded by a shell of ordered water molecules. Consequently, the gathering of several hydrophobic groups will release a number of water molecules and will cause a favorable increase of the solvent entropy. However, the rupture of the tight arrangement of water molecules surroundings each hydrophobic group requires thermal energy. An increase of the temperature will therefore favor the formation of hydrophobic bonds. This suggests that hydrophobic interactions are particularly well suited for stabilizing thermophilic enzymes.

Indeed, Suzuki et al. (1987) noted a linear correlation between the thermostability of five oligo-1,6-glucosidases and their content in amino acid residues which contribute to hydrophobic interactions. This tendency was confirmed by extensive statistical studies (see below) and is further supported by some recent works: the α-amylase of *Bacillus caldovelox* (Bealin-Kelly et al. 1991) and the xylose isomerase of *Thermus thermophilus* (Dekker et al. 1991) owe their high resistance against thermal unfolding to an increased content in hydrophobic residues. A major factor for the remarkable thermostability of the dimeric

3-isopropylmalate dehydrogenases from *Thermus thermophilus* and from *Thermus aquaticus*, as compared to seven mesophilic homologues, is the Glu → Ile substitution at the subunit interface which reinforces the hydrophobic interaction between the subunits in the thermophilic enzymes (Imada et al. 1991).

As shown by crystallographic structures, cavities exist in the hydrophobic core of proteins. They usually contain water molecules which cause destabilizing interactions. The stability of proteins might thus be increased by improving the compaction of their structures. This could be achieved by the judicious replacement of some residues by bulkier ones which would exclude undesirable water molecules from the interior of the protein and which would improve van der Waals contacts within the molecule (Mozhaev and Martinek 1984; Creighton 1991; Zülli et al 1991). However, there should be no major change in the chemical properties of the residues concerned and the configuration of the polypeptide chain should be respected. This mechanism of stabilization seems to be in agreement with some amino acid exchanges observed by Menendez-Arias and Argos (1989) between mesophilic and thermophilic enzymes (e.g., Gly → Ala; Ser → Thr).

3.6 Statistical Approaches

Several attempts have been made to find some general rules leading to an increase in the thermostability of proteins through amino acid exchanges. Ikai (1980) related the thermostability of proteins to the relative volume occupied by their aliphatic amino acid residues. Merkler et al. (1981) showed that proteins from thermophilic organisms are characterized by a greater hydrophobicity, an increased Arg/(Arg + Lys) ratio and a higher content in residues which stabilize α-helices and β-sheets. As more protein sequences and three-dimensional structures were known, Menendez-Arias and Argos (1989) could define preferred amino acid exchanges in the transition from mesophilic to thermophilic proteins and localize these substitutions in the tertiary structures. The general conclusion of this study is that the replacements greatly favor the arginine and alanine residues, increase the internal hydrophobicity of the molecule, decrease its backbone flexibility and occur mainly in helices and at domain interfaces.

Hence, rigidity and hydrophobic contacts between structural elements seem to play a major role in protein stability as well as helical structures since alanine, an excellent helix-forming residue, frequently replaces serine and glycine, i.e., two very poor helix-formers (Chou and Fasman 1978). Menendez-Arias and Argos (1989) also highlighted the importance of arginine. The properties of this particular residue have been discussed above.

The results of the statistical studies converge, despite some small differences, and they take into account the properties of a large number of proteins. However, as stated by Menendez-Arias and Argos themselves, the rules defined in their study do not apply systematically and entirely to all thermostable proteins. In particular, they do not account for the increased proline content of some thermostable enzymes (see Sect. 3.3.2). Moreover, Querol and Parrilla (1987) reported results which are in contradiction with these statistics: their studies on tyrosinases, lactate dehydrogenases and ferredoxins point to stabilizing amino acid replacements which occur mainly at the surface of proteins and do not lead to an increased internal hydrophobicity.

The main purpose of statistical studies is to provide some general hints for engineering more stable proteins by site-directed mutagenesis. It is clear however that the knowledge of refined three-dimensional structures remains a major asset for determining a rational strategy in this field and that each particular enzyme requires its own appropriate solution.

3.7 Summary

We have tried to show throughout this section the wide diversity of interactions which may contribute to the thermal stability of thermophilic enzymes. None of these factors is typical of thermophily. On the contrary, they belong to the range of structural features usually found in proteins. Actually, the accession to thermophily seems to derive from the accumulation of slight mutational changes which favor the rigidity and the compactness of proteins.

4 Enzymes from Psychrophilic Organisms

Cold adaptations of ectothermic organisms cover a wide range of aspects related to their metabolism: seasonal hibernation, freeze avoidance or tolerance, through the production of antifreeze proteins or synthesis of cryoprotectants, or regulation of membrane fluidity properties (see Hazel and Prosser 1974; Franks 1985; Storey and Storey 1988). These studies underline the diversity and complexity of the biochemical responses to environments with low temperature, e.g., during short acclimation periods or over evolutionary time. We will focus here on the molecular adaptations of enzymes to low temperatures with special reference to structural features of proteins from ectothermic psychrophilic Antarctic species.

Ectothermic species (or poikilotherms) from cold environments such as psychrophilic bacteria from polar regions, are mostly aquatic and permanently face water temperatures ranging from -2.2 to $+2\,°C$ as in the case of Antarctic sea water.

We have seen that reaction velocities of numerous enzymes from homeotherms and mesophilic organisms are reduced by a factor of about 10 at temperatures close to 0 °C. Considering this trivial observation, life would be hardly conceivable at low temperatures without specific adaptations. However, metabolic rates of Antarctic fishes are only slightly lower than those of temperate water species (Clarke 1983) and the generation time of psychrophilic bacteria

Table 1. Kinetic parameters of the hydrolysis of L-BAPA by bovine and fish trypsins and of E_t-G_7-PNP by bacterial α-amylases

Tryhsin[a]

	t(°C)	K_{cat} (min^{-1})	K_m (μM)	K_{cat}/K_m ($min^{-1}\,nM^{-1}$)
Homeotherm: bovine	35	177.0	2.700	65
	5	5.4	170	32
Ectotherm: trout	10	27.8	137	155
	5	16.6	107	105
Ectotherm: antarctic cod	5	29.0	121	243

α-Amylase[b]

	t(°C)	K_{cat} (s^{-1})	K_m (mM)	K_{cat}/K_m ($s^{-1}\,mM^{-1}$)
Mesophile *B. amyloliquefaciens*	30	122	2.0	61
	5	7	2.4	3
Psychrophile *A. haloplanctis*	5	67	0.2	335

[a] Data from Génicot et al. (1988).
[b] Data from Feller et al. (1992).

near 0 °C are of the same order as those of mesophilic microorganisms at 37 °C (Morita 1975; Mohr and Krawiec 1980). This clearly indicates that mechanisms of temperature compensation are involved.

4.1 General Properties of Cold-Adapted Enzymes

The basic observations and hypotheses regarding thermal adaptations of enzymes over evolutionary time have been described in detail by Somero (1977) and Hochachka and Somero (1984). Therefore, they will be only briefly outlined here. The dominating character of cold-adapted enzymes is probably their high specific activity or turnover number (k_{cat}). A comparison of substrate turnover number indicated that enzymes from ectothermic organisms are considerably more active than the homologous enzymes produced by homeotherms (Low et al. 1973). This observation is also made when k_{cat} values from psychrophilic and mesophilic bacteria are compared (Table 1). The finding that k_{cat} is inversely correlated with the temperature of the environment led to the suggestion that organisms living in systems where only limited thermal energy is available possess enzymes that are highly effective in lowering the activation energy barrier of their reaction (Somero 1969). Indeed, Low et al. (1973) have shown that enzymes from cold-adapted species are more efficient in reducing ΔG^*, the free energy of activation, which determines the reaction velocity of a chemical reaction, by adjusting either the enthalpic (ΔH^*) or the entropic (ΔS^*) contribution to ΔG^*.

However, k_{cat} reflects the catalytic potential of enzyme–substrate complexes at saturating substrate concentrations. This is not as stated in the introduction, certainly a physiological situation because most intracellular regulatory enzymes catalyze their reactions at low substrate levels ($\leqslant K_M$). This restriction is probably also valid for the case of enzymes excreted into the extracellular medium. Therefore the physiological efficiency (k_{cat}/K_M) is generally a better indication of catalytic evolution than k_{cat} alone. Such measurements are currently lacking for the vast majority of cold-adapted enzymes, but some available data (Table 1) support the view that enhanced k_{cat}/K_M values contribute to the compensation of reaction rate reduction near 0 °C in order to maintain adequate metabolic fluxes. In addition, it has been shown that organisms living in different thermal environments display very similar K_M values at their respective physiological temperatures in spite of wide variations in K_M below and above these temperatures (Somero 1977). Such an adaptation of the Michaëlis constant tends to optimize k_{cat}/K_M values at the temperature of the habitat but this mechanism of compensation is by no means universal (see Hall 1985).

Another common structural trait of proteins and enzymes isolated from cold-blooded species and psychrophilic microorganisms is their low thermal stability (Stokes 1967; Morita 1975). Heat lability has been correlated with environmental conditions (Ushakov 1964) because organisms evolving in the absence of the challenge of elevated temperature would not require thermostable proteins. However, Hochachka and Somero (1984) have convincingly argued that, in order to reduce the activation energy and to perform catalysis at low temperatures, these enzymes must possess an appropriate folding flexibility. They suggest that enzymes with enhanced abilities to undergo conformational changes during catalysis are capable to achieve their function with a reduced energy demand. Accordingly, thermal instability should be regarded as the consequence of the highly flexible structure of cold-active enzymes. However, one can not exclude that, besides the structural flexibility imposed by catalysis, the lack of thermal selective pressure is involved in heat lability.

A typical investigation of thermal stability characterization is illustrated in Fig. 2 for α-amylases and subtilisins excreted by psychrophilic Antarctic bacteria and their mesophilic counterparts. The half-time for denaturation ($t_{1/2}$) at 50 °C is approximately ten times lower for psychrophilic enzymes than for enzymes from mesophilic Bacillus. The thermally induced activity decay allows a rough estimate of the difference in free energy of activation ($\Delta(\Delta G) = RT \;\; 2.3 \log \Delta t_{1/2}$; Perutz and Raidt 1975) between heat-labile and mesophilic proteins. This difference amounts to about 6 kJ mol^{-1} and is rather small, but can be accounted for by the loss of a quite limited number of stabilizing weak bonds.

The improved structural flexibility of cold-adapted enzymes is further demonstrated by the drastic shift of the apparent optimal temperature of activity towards low temperatures, by the low resistance to protein denaturants and by the susceptibility of secondary structures to unfolding at moderate temperatures (Kobori et al. 1984; Feller et al. 1992).

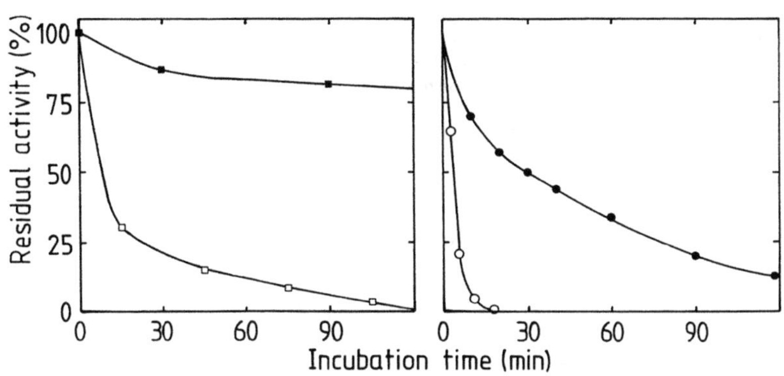

Fig. 2. Thermal stability of α-amylases (*left panel*) and subtilisins (*right panel*) from psychrophilic (*open symbols*) and mesophilic (*closed symbols*) bacteria. α-Amylases are from the Antarctic strain *Alteromonas haloplanctis* and from *Bacillus amyloliquefaciens*. Subtilisins are from the Antarctic strain *Bacillus* TA41 and from *Bacillus licheniformis*

4.2 Molecular Modelling of Cold-Adapted Enzymes

A detailed analysis of molecular adaptations to low temperatures remains a rather difficult task because the forces driving and maintaining the folded protein conformation are not properly understood and their relative contribution is still under debate (Creighton 1991). Nonetheless, if clues to cold adaptation are to be found at the molecular level, a critical step will be the availability of 3-D structures of enzymes produced by ectothermic or psychrophilic organisms living at low temperatures. Unfortunately, no refined crystallographic structure of such enzymes has been established so far, but we will describe here some recent results gained from molecular modelling. The data presented below concern proteins from Antarctic species showing 55–65% sequence identity with their mesophilic counterparts. The description will be limited to the factors potentially implicated in structure flexibility.

4.2.1 Subtilisin

Subtilisin is a proteolytic enzyme excreted by various *Bacillus* species. The enzyme belongs to the subtilase family which differs from the chymotrypsin family by the architecture of the active site (Siezen et al. 1991). The subtilisin molecule displays a globular shape and is generally composed of eight α-helices and nine β-sheets. The primary structure of subtilisin from the Antarctic psychrophile *Bacillus* TA41 reveals several long amino acid sequence insertions, located in surface loops, and responsible for a 4 kDa-higher molecular weight (Davail 1992; Davail et al. 1992). Moreover, a surprisingly large number of polar residues, mainly acidic, is found on the external shell of the protein providing

Table 2. Conserved ionic and aromatic interactions in subtilisins from a psychrophile (subtilisin S41), two mesophiles (subtilisins BPN′ and Carlsberg) and a thermophile (thermitase). (After Davail 1992; Siezen et al. 1991)

		S41	BPN′	CAR	THER
Ionic	10–184	–	–	–	K–D
	87–22	–	–	–	K–E
	94–49	K–D	–	–	R–D
	94–52	–	–	–	R–D
	136–140	–	K–D	K–D	–
	141–112	–	K–E	–	–
	145–116	–	–	–	K–D
	170–195	–	K–E	K–E	–
	185–181	–	–	–	–
	247–197	R–E	R–D	R–E	R–D
	247–251	–	R–E	–	R–E
	267–184	–	–	–	R–D
	267–255	–	–	–	R–D
	272–255	–	–	–	K–D
Aromatic	4–206	–	–	–	Y–Y
	4–214	–	–	–	Y–Y
	4–17	–	–	–	F–W
	48–50	–	–	–	W–F
	48–113	–	–	–	W–Y
	50–113	–	–	F–W	F–Y
	91–113	–	Y–W	Y–W	–
	167–170	–	–	–	Y–Y
	167–171	–	Y–Y	Y–Y	Y–Y
	171–195	–	–	–	Y–W
	192–262	–	–	–	Y–Y
	261–262	–	F–Y	F–Y	–
	262–263	–	Y–Y	Y–Y	Y–W

Residue numbering according to the reference sequence of subtilisin BPN′.

a very hydrophilic surface. These two features, namely the increased size of surface loops and the improved solvent interactions with the hydrophilic surface, can potentially destabilize the enzyme molecule. Whereas the overall hydrophobicity of the protein core seems unaltered, electrostatic and aromatic interactions are significantly reduced in the psychrophilic subtilisin. As shown in Table 2, aromatic interactions between side-chains of aromatic residues are totally lacking and only two ion pairs still persist within the *Bacillus* TA41 enzyme. Because the number of interactions decreases in the order thermophile > mesophile > psychrophile, there is little doubt that the removal of electrostatic and weakly polar interactions contribute to the conformational flexibility of the cold-adapted enzyme.

All amino acids pointing their side-chain towards the catalytic cavity are strictly conserved in the psychrophilic subtilisin but two substitutions in the substrate binding sites are worth mentioning. A proline residue is lacking on an interactive loop; this position corresponds to Pro 131 in Savinase, which provides substantial rigidity to the loop in the thermostable enzyme (Betzel et al. 1992). The second substitution corresponds to the replacement of a tyrosine residue in mesophilic subtilisins by an alanine residue. This substitution reduces the bulkiness of the side-chains surrounding a substrate binding sub-site and possibly also reduces the energy required for conformational changes during the catalytic process.

The binding of Ca^{2+} ions, generally occurring in external loops, is known to stabilize many proteases by restriction of the molecule flexibility (Siezen et al. 1991). The dissociation constant of the high-affinity Ca^{2+} binding site of subtilisins is in the range of 10^{-8} to 10^{-10} M but rises to 10^{-6} M in the case of the psychrophilic enzyme. The looser coordination of the Ca^{2+} ion is also indicative of a less compact conformation of the psychrophilic enzyme.

4.2.2 α-Amylase

The α-amylase excreted by the Antarctic bacterium *Alteromonas haloplanctis* A23 (Feller et al. 1992) displays 52% sequence identity with the porcine pancreatic enzyme. The latter consists of a central (α β) 8 domain, a C-terminal globular domain and a domain composed of antiparallel β-sheets (Buisson et al. 1987). The active site is located in a cleft between the central domain and the β-sheet domain and is stabilized by a Ca^{2+} ion coordinated by ligands from both domains. This psychrophilic α-amylase displays several deletions, essentially localized in connecting loops, and accounting for a 6 kDa lower molecular weight. Analysis of the hydrophobic clusters that form the core of the psychrophilic enzyme reveals that, out of 25 substitutions noted, 19 positions are occupied by amino acids having a lower hydrophobicity. Most replacements also increase the main chain flexibility as defined by Karplus and Schulz (1985). Lowering the core hydrophobicity (i.e., one of the main driving forces of protein folding) could possibly destabilize the psychrophilic enzyme.

The α-amylase from *A. haloplanctis* also differs by a very low proline content arising from deletions or replacements, mainly by alanine, in loops at the corresponding positions in the porcine sequence. Because the pyrrolidine ring of proline severely restricts the available dihedral angles of the preceding residue, one can expect that a low proline content would favor the loop flexibility and would increase the backbone entropy of unfolding.

Additional evidence for structural flexibility is given by the disappearance of two-thirds of the salt bridges located at the equivalent positions in the porcine sequence. Moreover, the low arginine content of the α-amylase from *A. haloplanctis* could contribute to the lability of this enzyme, since arginine residues have been shown to be stabilizing elements in proteins (Mrabet et al. 1992).

α-Amylases possess a particularly strongly bound Ca^{2+} ion ($Kd = 10^{-11}$ M) essential for the tertiary structure of the enzyme and its catalytic activity. In the case of the psychrophilic enzyme, the dissociation constant rises up to 10^{-8} M and certainly reflects looser interactions between the two domains forming the catalytic cleft, which in turn should allow easier conformational changes during catalysis.

4.2.3 Triosephosphate Isomerase

Rentier-Delrue et al. (1993) have recently described a structural analysis of psychrophilic and thermophilic triosephosphate isomerases. This enzyme is a dimer of identical subunits composed of typical $(\beta/\alpha)_8$ barrel. The authors observe two small insertions in loops involved in dimerization of the psychrophilic enzyme which could lead to dimer destabilization, and they also note a low proline and arginine content.

Interestingly, the charge–dipole interactions in α-helices of the psychrophilic triosephosphate isomerase are less favorable than those in the homologous thermophilic enzyme. Stabilization of α-helices through charge–dipole interactions has been proposed for various proteins (Shoemaker et al. 1987). Therefore, alteration of either the charge or the nature of amino acids located in the first and last turns of helices can weaken the charge–dipole interaction and is probably another factor that promotes enzyme flexibility.

4.2.4 *β*-Lactamase

Bacterial β-lactamases hydrolyze the amide bond of the β-lactam ring in penicillin-derived antibiotics, which leads to their inactivation. The gene of a class C β-lactamase excreted by the Antarctic bacterium *Psychrobacter immobilis* A5 has been cloned and sequenced (Zekhnini 1993). Its analysis was limited to informations derived from sequence alignment with six other mesophilic homologous enzymes including the β-lactamase of *Citrobacter freundi*, the 3-D structure of which has been solved (Oefner et al. 1990). The β-lactamase of *P. immobilis* has no specific insertions or deletions and the structural motives distinguishing the psychrophilic enzyme from its mesophilic counterparts are in agreement with some of the features listed above. These include a low proline and arginine content, a tendency to substitute arginine lysine for residues, and a reduced overall hydrophobicity. The latter arises from amino acid replacements occurring mainly in α-helices. Moreover, it is interesting to note that these replacements cause an increase of the main chain flexibility index.

4.2.5 Trypsin

Trypsin from the Antarctic fish *Paranotothenia magellanica* offers another example of cold-adapted enzyme showing high catalytic efficiency and pronounced heat lability with respect to bovine pancreatic trypsin (Génicot et al. 1988; Génicot 1993). The fish and bovine enzymes are globular proteases essentially formed by *β*-sheets and bridged by six disulfide bonds. Sequence alignment of both enzymes reveals 65% identity with only one amino acid deleted in the fish trypsin. Its proline content is unchanged and nine ion pairs have been located within the fish structural model compared to ten in the bovine crystal structure. This Antarctic fish trypsin also contains twice as many aromatic interactions. By contrast, the most relevant element leading to an increase in the flexibility of the fish trypsin corresponds to an increased hydrophilicity of the protein surface as already noted in cold-adapted subtilisin. Whereas buried residues are highly conserved in the fish trypsin, numerous surface amino acids are replaced by charged residues. Finally, the Antarctic fish trypsin shows an 18% lower potential for intramolecular hydrogen bonding when compared with the bovine protease. This property could be crucial for enzyme flexibility, in accordance with the increasing importance attached to these weak interactions in the stabilization of the folded conformation of proteins (Creighton 1991). However, modeled proteins of protein structures do not generally allow the evaluation of the exact number of hydrogen bonds.

5 Conclusions

The molecular changes observed in cold-adapted enzymes and the factors involved in the stabilization of thermophilic enzymes concern a multitude of altered intramolecular interactions at the level of protein 3D structure following various substitutions at the level of the primary structure. In the case of psychrophilic enzymes, molecular modifications seem to have induced looser surface loops, more polar surfaces exposed to the solvent or a lower hydrophobicity of the protein core, lower affinity of Ca^{2+} binding sites as well as reduced numbers of salt bridges, less stabilized α-helices and eventually weaker hydrogen bonding. Interestingly, these changes represent roughly the opposite of strategies adopted by proteins from thermophilic organisms.

Probably all possible structural motives or interactions have been modified during the course of evolution of these enzymes in order to harmonize enzyme flexibility and stability with the ambient thermal conditions. The molecular adaptations to either cold or hot environments indicate a general molecular mechanism, namely a change of the overall 3-D compaction of the enzymes. Thermophilic enzymes are stabilized by increased intramolecular interactions, and psychrophilic enzymes display conformations with an increased flexibility.

References

Baldwin RL, Eisenberg D (1987) Protein stability. In: Oxender DL, Fox CF (eds) Protein engineering. Liss, New York, pp 127–148

Baross JA, Deming JW (1983) Growth of black smoker bacteria at temperature of at least 250°C. Nature 303: 423–426

Bealin-Kelly F, Kelly CT, Fogarty WM (1991) Studies on the thermostability of the α-amylase of *Bacillus caldovelox*. Appl Microbiol Biotechnol 36: 332–336

Betzel C, Klupsch S, Papendorf G, Hastrup S, Branner S, Wilson KS (1992) Crystal structure of the alkaline proteinase Savinase from *Bacillus lentus* at 1.4 Å resolution. J Mol Biol 223: 427–445

Blake PR, Park JB, Bryant FO, Aono S, Magnuson JK, Eccleston E, Howard JB, Summers MF, Adams MW (1991) Determinants of protein hyperthermostability: purification and amino acid sequence of rubredoxin from the hyperthermophilic archaebacterium *Pyrococcus furiosus* and secondary structure of the zinc adduct by NMR. Biochemistry 30: 10885–10895

Brock TD (1985) Life at high temperature. Science 230: 132–138

Bryan PN, Rollence ML, Pantoliano MW, Wood J, Finzel BC, Gilliland GL, Howard AJ, Poulos TL (1986) Proteases of enchanced stability: characterization of a thermostable variant of subtilisin. Proteins Struct Funct Gen 1: 326–334

Buisson G, Duee E, Haser R, Payan F (1987) Three dimensional structure of porcine pancreatic α-amylase at 2.9 Å resolution. Role of calcium in structure and activity. EMBO J 6: 3909–3916

Burley SK, Petsko GA (1985) Aromatic-Aromatic Interaction: a mechanism of protein strucutre stabilization. Science 229: 23–28

Burley SK, Petsko GA (1988) Weakly polar interactions in proteins. Adv Protein Chem 39: 125–189

Caffrey MS, Cusanovich MA (1991) Lysins in the amino-terminal α-helix are important to the stability of *Rhodobacter capsulatus* cytochrome c2. Biochemistry 30: 9238–9241

Chaffotte AF, Guillou Y, Goldberg ME (1992) Inclusion bodies of the thermophilic endoglucanase D from *Clostridium thermocillium* are made of native enzyme that resist 8M urea. Eur J Biochem 205: 369–373

Chou PY, Fasman GD (1978) Empirical predictions of protein conformation. Annu Rev Biochem 47: 251–276

Clarke A (1983) Life in cold water: the physiological ecology of polar marine ectotherms. Oceanogr Mar Biol Annu Rev 21: 341–453

Creighton TE (1991) Stability of folded conformations. Curr Opin Struct Biol 1: 5–16

Daniel RM, Corvan DA, Morgan HW, Curran MP (1982) A correlation between protein thermostability and resistance to proteolysis. Biochem J 207: 641–644

Davail S (1992) Adaptations moléculaires d'une protéase isolée d'une bactérie du milieu marin antarctique. PhD Thesis, University of Liége, Belgium

Davail S, Feller G, Narinx E, Gerday C (1992) Sequence of the subtilisin-encoding gene from an antarctic psychrotroph *Bacillus* TA41. Gene 119: 143–144

Dekker K, Yamagata H, Sakagushi K, Udaka S (1991) Xylose (Glucose) isomerase gene from the thermophile *Thermus thermophilus*: cloning, sequencing, and comparison with other thermostable xylose isomerases. J Bacteriol 173: 3078–3083

Dill KA, Alonso DO, Hutchinson K (1989) Thermal stabilities of globular proteins. Biochemistry 28: 5439–5449

Eijsink VG, Vriend G, van den Burg B, van der Zee JR, Venema G (1992) Increasing the thermostability of a neutral protease by replacing positively charged amino acids in the N-terminal turn of α-helices. Protein Eng 5: 165–170

Feller G, Thiry M, Arpigny JL, Gerday C (1991) Cloning and expression in *Escherichia coli* of three lipase-encoding genes from the psychrotropic antarctic strain *Moraxella* TA144. Gene 102: 111–115

Feller G, Lonhienne T, Deroanne C, Libioulle C, van Beeumen J, Gerday C (1992) Purification, characterization and nucleotide sequence of the thermo-labile α-amylase from the antarctic psychrotroph *Alteromonas haloplanctis* A23. J Biol Chem 267: 5217–5221

Flory JG (1956) Theory of elastic mechanisms in fibrous protein. J Am Chem Soc 78: 5222–5235

Fontana A (1988) Structure and stability of thermophilic enzymes. Studies on thermolysin. Biophys Chem 29: 181–193

Fontana A (1991) Analysis and modulation of protein stability. Curr Opin Biotech 2: 551–560

Franks F (1985) Biophysics and biochemistry at low temperatures. Cambridge University Press, Cambridge

Fujita SC, Oshima T, Imahori K (1976) Purification and properties of D-glyceraldehyde-3-phosphate dehydrogenase from an exteme thermophile, *Thermus thermophilus* strain HB8. Eur J Biochem 64: 57–68

Génicot S (1993) Proposed 3D structure of trypsin from trout (*Salmo gairdneri*) and antarctic fish (*Notothenia neglecta*). PhD Thesis, University of Liége, Belgium

Génicot S, Feller G, Gerday C (1988) Trypsin from antarctic fish (*Paranotothenia magellanica*) as compared with trout (*Salmo gairdneri*) trypsin. Comp Biochem Physiol 90B: 601–609

Goldenberg DP (1985) Dissecting the roles of individual interactions in protein stability: lessons from a cricularized protein. J Cell Biochem 29: 321–335

Grütter MG, Hawkes RB, Matthews BW (1979) Molecular basis of thermostability in the lysozyme from bacteriophage T4. Nature 277: 667–669

Gupta MN, Mattiasson B (1992) Unique applications of immobilized proteins in bioanalytical systems. Methods Biochem Anal 36: 1–34

Hall JG (1985) The adaptation of enzymes to temperature: catalytic characterization of glucose-phosphate isomerase homologues isolated from *Mytilus edulis* and *Isognomon alatus*, bivalve molluscs inhabiting different thermal environments. Mol Biol Evol 2: 251–269

Hazel JR, Prosser CL (1974) Molecular mechanisms of temperature compensation in poikilotherms. Biol Rev 54: 620–677

Hecht MH, Sturtevant JM, Sauer RT (1984) Effect of single amino acid replacements on the thermal stability of the NH_2-terminal domain of phage repressor. Proc Natl Acad Sci USA 81: 5685–5689

Hochachka P (1991) Temperature: the ectothermy option. In: Hochachka P, Mommsen T (eds) Biochemistry and molecular biology of fishes, vol 1. Elsevier, New York, pp 313–322

Hochachka P, Somero GN (1984) Biochemical Adaptation. Princeton University Press, New Jersey, pp 355–449

Hol WG, van Duinen PT, Berendsen HJ (1978) The α-helix dipole and the properties of proteins. Nature 273: 443–446

Huber R, Kurr M, Jannasch HW, Stetter KO (1989) A novel group of abyssal methanogenic archaebacteria (*Methanopyrus*) growing at 110 °C. Nature 342: 833–834

Ikai A (1980) Thermostability and aliphatic index of globular proteins. J Biochem (Tokyo) 88: 1895–1898

Imada K, Sato M, Tanaka N, Katsube Y, Matsuura Y, Oshima T (1991) Three-dimensional structure of a highly thermostable enzyme, 3-iso-propylmalate dehydrogenase of *Thermus thermophilus* at 2.2 Å resolution. J Mol Biol 222: 725–738

Karplus PA, Schulz GE (1985) Prediction of chain flexibility in proteins. Naturwissenschaften 72: 212–213

Klibanov AM (1979) Enzyme stabilization by immobilization. Anal Biochem 93: 1–25

Kobori H, Sullivan CW, Shizuya H (1984) Heat-labile alkaline phosphatase from antarctic bacteria: rapid 5' end-labelling of nucleic acids. Proc Nat Acad Sci USA 81: 6691–6695

Kuhn W, Majer H (1956) The self cross-linking of macromolecules. Makromol Chem 16–19: 239–245

Kwon ST, Matsuzawa H, Ohta T (1988) Determination of the positions of the disulfide bonds in aqualysin I (a thermophilic alkaline serine protease) of *Thermus aquaticus* YT-1. J Biochem (Tokyo) 104: 557–559

Low PS, Bada JL, Somero GN (1973) Temperature adaptation of enzymes: roles of the free energy, the enthalpy and the entropy of activation. Proc Natl Acad Sci USA 70: 430–432

Matsumura M, Becktel WJ, Levitt M, Matthews BW (1989) Stabilization of phage T4 lysozyme by engineered disulfide bonds. Proc Natl Acad Sci USA 86: 6562–6566

Matthews BW (1987) Genetic and structural analysis of the protein stability problem. Biochemistry 26: 6885–6888

Matthews BW (1991) Mutational analysis of protein stability. Curr Opin Struct Biol 1: 17–21

Matthews BW, Nicholson H, Becktel WJ (1987) Enhanced protein thermostability from site-directed mutations that decrease the entropy of unfolding. Proc Natl Acad Sci USA 84: 6663–6667

Matthews CR (1991) The mechanism of protein folding. Curr Opin Struct Biol 1: 28–35

Menendez-Arias L, Argos P (1989) Engineering protein thermal stability. Sequence statistics point to residue substitutions in α-helices. J Mol Biol 206: 397–406

Merkler DJ, Farrington GK, Wedler FC (1981) Protein thermostability. Int J Pept Protein Res 18: 430–442

Mitchinson C, Wells JA (1989) Protein engineering of disulfide bonds in subtilisin BPN′. Biochemistry 28: 4807–4815

Mohr PW, Krawiec S (1980) Temperature characteristics and Arrhenius plots for nominal psychrophiles, mesophiles and thermophiles. J Gen Microbiol 121: 311–317

Morita RY (1975) Psychrophilic bacteria. Bacteriol Rev 39: 144–167

Mozhaev VV, Martinek K (1984) Structure-stability relationships in proteins: new approaches to stabilizing enzymes. Enzyme Microb Technol 6: 50–59

Mozhaev VV, Berezin IV, Martinek K (1988) Structure-stability relationships in proteins: fundamental tasks and strategy for the development of stabilized enzyme catalysts for biotechnology. CRC Crit Rev Biochem 23: 235–281

Mrabet NT, van den Broeck A, van den Brande I, Stanssens P, Laroche Y, Lambeir A-M, Matthijssens G, Jenkins J, Chiadmi M, van Tilbeurgh H, Rey F, Janin J, Quax WJ, Lasters I, de Maeyer M, Wodak SJ (1992) Arginine residues as stabilizing elements in proteins. Biochemistry 31: 2239–2253

Nemethy G, Leach SJ, Scheraga HA (1966) The influence of amino acid side chains on the free energy of helix-coil transition. J Phys Chem 70: 998–1004

Oefner C, Darcy A, Daly JJ, Gubernator K, Charnas RL, Heinze I, Hubschwerlen C, Winkler FK (1990) Refined crystal structure of β-lactamase from *Citrobacter freundii* indicates a mechanism for β-lactam hydrolysis. Nature 343: 284–288

Perutz MF (1978) Electrostatic effects in proteins. Science 201: 1187–1191

Perutz MF, Raidt H (1975) Stereochemical basis of heat stability in bacterial ferredoxins and in haemoglobin A2. Nature 255: 256–259

Privalov PL (1979) Stability of proteins. Adv Protein Chem 33: 167–241

Privalov PL, Gill SJ (1988) Stability of protein structure and hydrophobic interaction. Adv Protein Chem 39: 191–234

Querol E, Parrilla A (1987) Tentative rules for increasing the thermostability of enzymes by protein engineering. Enzyme Microb Technol 9: 238–244

Rentier-Delrue F, Mande SC, Moyens S, Terpstra P, Mainfroid V, Goraj K, Lion M, Hol WGJ, Martial J (1993) Cloning and overexpression of the triosephosphate isomerase genes from psychrophilic and thermophilic bacteriae. Structural comparison of the predicted protein sequences. J Mol Biol 229: 85–93

Sanbongi Y, Igarashi Y, Kodama T (1989) Thermostability of cytochrome c-552 from the thermophilic hydrogen-oxidizing bacterium *Hydrogenobacter thermophilus* Biochemistry 28: 9574–9578

Sancho J, Serrano L, Fersht AR (1992) Histidine residues at the N- and C-termini of α-helices: perturbed pK_as and protein stability. Biochemistry 31: 2253–2258

Schultz GE, Schirmer RH (1979) Principles of Protein Structure. Springer, Berlin Heidelberg New York

Shoemaker KR, Kim PS, York EJ, Stewart JM, Baldwin RL (1987) Tests of the helix dipole model for stabilisation of α-helices. Nature 326: 563–567

Siezen RJ, de Vos WM, Leunissen JA, Dijkstra BW (1991) Homology modelling and protein enginnering strategy of subtilases, the family of subtilisin-like serine proteinases. Protein Eng 4: 719–737

Somero GN (1969) Enzymatic mechanisms of temperature compensation: immediate and evolutionary effects of temperature on enzymes of aquatic poïkilotherms. Am Nat 103: 517–530

Somero GN (1977) Temperature as a selective factor in protein evolution: the adaptational strategy of "compromise". J Exp Zool 194: 175–188

Spiess FN, MacDonald KL, Atwater T (1980) East pacific rise: hot springs and geophysical experiments. Science 207: 1421–1432

Steel DM, Walker JM (1991) Thermostable proteins. Life Chem Rep 8: 49–96

Stokes JL (1967) Heat sensitive enzymes and enzyme synthesis in psychrophilic microorganisms. Publ Am Assoc Adv Sci 84: 311–323

Storey KB, Storey JM (1988) Freeze tolerance in animals. Physiol Rev 68: 27–84

Suzuki Y, Imai T (1982) Abnormally high tolerance against proteolysis of an exo-oligo-1,6-glucosidase from a thermophile *Bacillus thermoglucosidius* KP1006, compared with its mesophilic counterpart from *Bacillus cereus* ATCC7064. Biochim Biophys Acta 705: 124–126

Suzuki Y, Oishi K, Nakano H, Nagayama T (1987) A strong correlation between the increase in number of proline residues and the rise in thermostability of five *Bacillus* oligo-1,6-glucosidases. Appl Microbiol Biotechnol 26: 546–551

Takagi H, Takahashi T, Momose H, Inouye M, Maeda Y, Matsuzawa H, Ohta T (1990) Enhancement of the thermostability of subtilisin E by introduction of a disulfide bond engineered on the basis of structural comparison with a thermophilic serine protease. J. Biol Chem 265: 6874–6878

Tanford C (1970) Protein denaturation. Adv Protein Chem 24: 1–95

Teplyakov AV, Kuranova IP, Harutyhunyan EH, Vainshtein BK, Frömmel C, Höhne WE, Wilson KS (1990) Crystal structure of thermitase at 1.4 Å resolution. J Mol Biol 214: 261–279

Tomazic SJ, Klibanov AM (1988) Why is one *Bacillus* α-amylase more resistant against irreversible thermoinactivation than another? J Biol Chem 263: 3092–3096

Ushakov B (1964) Thermostability of cells and proteins in poikilotherms. Physiol Rev 44: 518–560

Veronese FM, Boccu E, Schiavon O, Grandi C, Fontana A (1984) General stability of thermophilic enzymes: studies on 6-phosphogluconate dehydrogenase from *Bacillus stearothermophilus* and yeast. J Appl Biochem 6: 39–47

Vihinen M (1987) Relationship of protein flexibility to thermostability. Protein Eng 1: 477–480

Villafranca JE, Howell EE, Oatley SJ, Xuong N, Kraut J (1987) An engineered disulfide bond in dihyrofolate reductase. Biochemistry 26: 2182–2189

Voordouw G, Milo C, Roche RS (1976) Role of bound calcium ions in thermostable, proteolytic enzymes. Separation of intrinsic and calcium ion contributions to the kinetic thermal stability. Biochemistry 15: 3716–3723

Watanabe K, Chishiro K, Kitamura K, Suzuki Y (1991) Proline residues responsible for thermostability occur with high frequency in the loop regions of an extremely thermostable oligo-1,6-glucosidase from *Bacillus thermoglucosidasius* KP 1006. J Biol Chem 266: 24287–24294

Wetzel R (1987) Harnessing disulfide bonds using protein engineering. Trends Biochem Sci 12: 478–482

Wetzel R, Perry LJ, Baase WA, Becktel WJ (1988) Disulfide bonds and thermal stability in T4 lysozyme. Proc Natl Acad Sci USA 85: 401–405

Wrba A, Schweiger A, Schultes V, Jaenicke R (1990) Extremely thermostable D-glyceraldehyde-3-phosphate dehydrogenase from the eubacterium *Thermotoga maritima*. Biochemistry 29: 7584–7592

Zekhnini Z (1993) Adaptations moléculaires d'une β-lactamase produite par une bactérie antarctique. PhD Thesis, University of Liège, Belgium

Zülli F, Schneiter R, Urfer R, Zuber H (1991) Engineering thermostability and activity of lactate dehydrogenases from bacilli. Biol Chem Hoppe-Seyler 372: 363–372

Subject Index

MIX
Papier aus verantwortungsvollen Quellen
Paper from responsible sources
FSC® C105338

If you have any concerns about our products,
you can contact us on
ProductSafety@springernature.com

In case Publisher is established outside the EU,
the EU authorized representative is:
Springer Nature Customer Service Center GmbH
Europaplatz 3, 69115 Heidelberg, Germany

Printed by Libri Plureos GmbH
in Hamburg, Germany